Environment and Geochemistry of Sediments

Environment and Geochemistry of Sediments

Editors

Marianna Kulkova
Dmitry Subetto

MDPI • Basel • Beijing • Wuhan • Barcelcna • Belgrade • Manchester • Tokyo • Cluj • Tianjin

Editors
Marianna Kulkova
Department of Geography
Herzen State Pedagogical
University of Russia
Saint-Petersburg
Russia

Dmitry Subetto
Department of Geography
Herzen State Pedagogical
University of Russia
Saint-Petersburg
Russia

Editorial Office
MDPI
St. Alban-Anlage 66
4052 Basel, Switzerland

This is a reprint of articles from the Special Issue published online in the open access journal *Minerals* (ISSN 2075-163X) (available at: https://www.mdpi.com/journal/minerals/special_issues/environment_geochemistry_sediments).

For citation purposes, cite each article independently as indicated on the article page online and as indicated below:

LastName, A.A.; LastName, B.B.; LastName, C.C. Article Title. *Journal Name* **Year**, *Volume Number*, Page Range.

ISBN 978-3-0365-8108-8 (Hbk)
ISBN 978-3-0365-8109-5 (PDF)

Cover image courtesy of Marianna Kulkova

Contents

Preface to "Environment and Geochemistry of Sediments"

This reprint considers a vast range of questions devoted to environmental geochemistry. The Special Issue "Environment and Geochemistry of Sediments" includes 10 papers. The themes of research cover different problems of the geochemistry of sediments, such as the reconstruction of the paleoenvironment in the lake basins in the high latitude zones on the base of geochemical indicators, the processes of the sedimentary environment and climate evolution during the Late Carboniferous–Early Permian period in the central Junggar Basin (China), and the study of the paleoenvironment and bio-sequence stratigraphy in the Cretaceous pelagic carbonate succession part of the eastern Tethys. Several articles discuss the features of the geochemistry of sedimentation in places of prehistoric anthropogenic activity, for example, based on lithological and geochronological analyses, magnetic susceptibility, and microcharcoal studies, the anthropogenic sources of metals in south-eastern Baltic lake sediments from the Neolithic to the Medieval Age were determined. A new approach using the geochemical indication was developed for the determination of the functional zones of prehistoric archaeological sites in Eastern Europe. The application of geochemical multi-element and geochronological analyses of the Iron Age and early Roman cultural layers was considered at the archaeological site in the Netherlands. Some papers explore to the pollution hazard problem concerning for example the accumulation and distribution of a hazardous contaminant, mercury (Hg), in the basin of Onega Lake, Russia, the second largest lake in Europe, and the distribution of arsenic in the soils of the Verkhnekamskoe potassium salt deposit, Perm Krai, Russia. The ecological state of basins as a result of technogenic processes using a geochemical approach is illuminated in the papers devoted to the mineralogical and geochemical contents of the bottom sediments of Al-Kharrar and Al-Shuaiba Lagoons, Red Sea, Saudi Arabia. The Guest Editors of this Special Issue would like to thank all of the authors for their valuable contributions.

Marianna Kulkova and Dmitry Subetto
Editors

minerals

MDPI

Editorial

Editorial for the Special Issue "Environment and Geochemistry of Sediments"

Marianna Kulkova * and Dmitry Subetto

Department of Geography, Herzen State Pedagogical University, 191186 St. Petersburg, Russia; subetto@mail.ru
* Correspondence: kulkova@mail.ru

The geochemical characteristics of environments can be determined via geochemical studies of sedimentary rocks. Different geochemical indicators can be used for the paleoenvironmental reconstruction of sedimentation processes. Trace and major elements in sedimentary rocks are extremely sensitive to paleoenvironmental changes, making them informative in studying the paleoclimate, paleoenvironment, as well as ancient and modern anthropogenic activity. The distribution of pollutants in sediments is important for modern geoecological processes. Isotopic research of sediments is essential for paleoreconstructions, geochronology, and ecology. The application of different analytic methods to study the geochemistry and mineralogy of sediment processes is also important in the context of these investigations.

The Special Issue "Environment and Geochemistry of Sediments" includes ten papers devoted to these problems. The topic of the first paper, by Bezrukova et al. [1], is the geochemistry of lacustrine deposits from high-latitude regions, for example, Lake Kaskadnoe-1 in the East Sayan Mountains, South Siberia, Russia (2080 m above sea level). The lake deposits contain a unique record of geochemical cycles pertaining to the last 13,200 years. They reflect sedimentation depending on environmental changes during the Late Glacial and Early Holocene. The Late Glacial sediments (13,200–12,800 cal yr BP) are characterized by high values of CIA, Mg/Al, K/Al, and Mn/Fe, and are depleted in Si/Al, Fe/Al, and Ca/Al. During the cold episode of the Younger Dryas, LOI enrichment was probably caused the presence of by less oxic conditions, as seen in the lower Mn/Fe values, due to a long period of lake ice cover. The Early Holocene (12,000–7500 cal yr BP) is associated with a decreasing trend of mineral matter with fluvial transport to Lake Kaskadnoe-1 (low K/Al, Mg/Al) and increasing chemical weathering in the lake basin. The increase in Ti/Al, K/Al, and CIA values over the last 7500 years suggests increasing terrigenous input into the lake. These data allow a reconstruction of the paleoenvironment in the lake basin during its development.

The second paper, by Malov et al. [2], is devoted to the problem of accumulation and distribution of a hazardous contaminant, mercury (Hg), in the basin of Lake Onega, Russia, the second largest lake in Europe. In this paper, the first investigations of the lateral distribution of total mercury in the water-suspended matter-bottom sediments system are presented. The total mercury content in the water of Lake Onega averages 0.32 ± 0.07 μg/L. The most common form of Hg in water is the dissolved and colloid form, except for water samples from the Kondopoga and Povenetsky Bays. In the material of the sedimentation traps, the Hg content is 0.5 ± 0.3 μg/g, and in the upper and lower parts of the bottom sediment section, this value is 0.067 ± 0.003 and 0.041 ± 0.001 μg/g, respectively. The main factors responsible for the increased Hg include anthropogenic pollution, the migration of Hg, and its redeposition at the geochemical barrier together with Fe and Mn.

In the third paper, by Shuncun Zhang et al. [3], the processes of the sedimentary environment and climate evolution of the Mosuowan area in the central Junggar Basin (China) during the Late Carboniferous–Early Permian are reconstructed. The investigative methods used in this paper include petrothermal methods, the use of a lipid biomarker,

Citation: Kulkova, M.; Subetto, D. Editorial for the Special Issue "Environment and Geochemistry of Sediments". *Minerals* **2023**, *13*, 709. https://doi.org/10.3390/min13050709

Received: 16 May 2023
Revised: 18 May 2023
Accepted: 19 May 2023
Published: 22 May 2023

and a consideration of the isotopic composition of mud shale core samples, thus enabling the exploration of the tectonic–climatic events and Central Asian orogenic belt evolution that contributed to the sedimentary environment. The authors conclude that the Late Carboniferous–Early Permian period was influenced by global changes, Paleo-Asian Ocean subduction, and continental splicing, which resulted in a continuous increase in levels of terrestrial organic matter, water desalination, and oxidation-rich sediments in the Mosuowan region, but a P–T biological mass extinction event was not recorded.

The fourth paper, by Druzhinina et al. [4], considers the enrichment of Pb, Ni, Zn, As, Co, and Cu in sediments from Kamyshovoe Lake, Kaliningrad Oblast, Russian Federation, in the process of prehistoric anthropogenic activity. The enrichment factors (EF) combined with the results of the lithological, geochronological, magnetic susceptibility, and microcharcoal studies reveal the possible anthropogenic sources of metals in south-eastern Baltic Lake sediments from the Neolithic to the Medieval period. Increasing Co EF values and peaks of the Pb EF in Kamyshovoe Lake sediments, starting from ~6000 cal yr BP in the Neolithic period are considered evidence of the usage of metals as dyes and fixatives during that period. Since ~3100 cal yr BP, at the end of the Bronze Age, a simultaneous increase in the content of indicators of metallurgical production, Pb, Ni, Zn, and As, coinciding with the growth of the microcharcoal curve, indicates the expanding demand for metal objects in the southeastern Baltic region and the input of the local or regional ancient metallurgy into the lake sediments.

In the fifth article, by Ghandour et al. [5], a study of the mineralogical and geochemical contents of the bottom sediments of Al-Kharrar Lagoon, Rabigh, Saudi Arabia, is conducted in order to assess the ecological state of the lagoon. Three main elemental groups were determined via statistical analysis. The first group includes the positively correlated SiO_2, TiO_2, Al_2O_3, Fe_2O_3, MnO, MgO, K_2O, Na_2O, V, Cr, Ni, Zn, Rb, and Ba, which are concentrated in the southern sector of the lagoon and associated with siliciclastic-related minerals (quartz, clay minerals, k-feldspars, and plagioclase along with traces of amphiboles). The second group consists of the carbonate-related elements (CaO and Sr) that dominate the northern sector. The distribution patterns of P_2O_5 and Cu vary significantly across the lagoon. Enrichment factors reveal moderate levels of Cu in some sites, supporting its anthropogenic source. The results indicate that the ecological status of the bottom of the lagoon is hospitable despite the presence of local anthropogenic stressors such as the influx of flood water that contains a mixture of lithogenic and dissolved Cu from local agriculture.

The sixth paper, by Kulkova [6], represents a new approach with which to determine the functional zones of prehistoric archaeological sites in Eastern Europe via geochemical indication. The application of mathematical statistics for processing the geochemical data of cultural deposits at archaeological sites allowed the identification of groups of interrelated chemical elements that reflect the processes of natural sedimentation and anthropogenic activity. Abnormal concentrations of P_2O_{5antr}, CaO_{antr}, and Sr_{antr}, which are associated to each other, in sediments are attributed to the zones of accumulation of bone remains. Anomalous concentrations of a group of elements (K_2O_{antr} and Rb_{antr}) in deposits are associated with wood ash and fireplaces, ash residues from ritual activities, and fires. The group of elements Ba, MnO, and C_{org} reflects the accumulation of humus and organic remains, and may characterize areas with food residues, skins, and rotten wood. Aided by the distribution of the main lithological elements (SiO_2 and Al_2O_3) in sediments, it is possible to consider the paleorelief at the sites. These functional zones were reconstructed at the sites of the Neolithic period, Early Metal Age, and Bronze–Early Iron Age in Eastern Europe.

In the seventh article, by Karavaeva et al. [7], the distribution of arsenic in the soils of the Verkhnekamskoe potassium salt deposit, Perm Krai, Russia, is discussed. The danger of arsenic (As) pollution is determined by its high toxicity and carcinogenic hazard. The content of arsenic in soils was determined via inductively coupled plasma mass spectrometry (ICP-MS). Statistical methods were used to analyze the features of As distribution in the soils of background areas and potash mining areas near production facilities. Three types of landscapes were studied within each territory, each of which was distinguished by the leading processes of substance migration. Arsenic concentrations in both the background areas and the potash mining territories vary considerably, ranging from n × 10^{-1} to n × 10. The arsenic concentrations in the soils of saline areas were found to be higher than those in the rest of the territories. Outside of saline areas, the identified patterns of As distribution in the soils of the Verkhnekamskoe potassium salt deposit indicate that potash operations are not a determinant of the technophilic accumulation of As.

The eighth paper, by Ullah et al. [8], focuses on studying the Goru Formation in the Chutair Section, Sulaiman Range, Pakistan, representing part of the eastern Tethys. The paleoenvironment and bio-sequence stratigraphy were studied in the Cretaceous pelagic carbonate succession. Based on the facies variations and planktonic foraminiferal biozones, the sea level curve in the data from the Goru Formation could be reconstructed. Fluctuations between the outer ramp and deep basin, showing the overall transgression in the second-order cycle in the study area, coincide with the global sea-level curve. At the same time, the third-order cycle represents the local tectonic process during the deposition of the strata.

The authors of the ninth article [9] present the results of geochemical multi-element, LOI, MS, and geochronological analyses of a cultural layer at the Voorthuizen archaeological site, The Netherlands, which was occupied in the Iron Age and early Roman periods. The study has provided a deeper insight into the agricultural techniques applied on the site and on the so-called Celtic fields, known as ancient field systems dating from the same period that were widespread throughout Northwestern Europe. It seems that household waste was not used as a fertilizer in Voorthuizen, while the application of manure is characteristic of Celtic fields. However, the phosphorous values in the Voorthuizen agricultural horizon are comparable to those in the Celtic fields, suggesting similar sources of P in both cases. Elevated Si and "mobile" Fe, Mn, V, Pb, As, and Sn, along with higher MS measurements, are indicative of the use of extra mineral matter for the fertilization of the ancient arable field in Voorthuizen.

The final paper [10] presents the results from a study of the ecological state of the northern and southern Al-Shuaiba lagoons, Red Sea, Saudi Arabia. The authors consider the impact of technogenic processes and building on the geochemical pollution of the lagoons of the Red Sea. The lower unit of the core is enriched in elements (Mo, As, U, and Re) suggesting the occurrence of deposition under anoxic conditions, this being possibly related to the Medieval Climate Anomaly. The middle unit is enriched in carbonate-related constituents ($CaCO_3$, Ca, and Sr). The upper unit is enriched in elements that co-vary significantly with Al, thus implying the presence of an increased terrigenous supply associated with the construction of a road between the two lagoons. The enrichment of elements in the lower and middle units was naturally driven, whereas the enrichment of lithogenic elements in the upper unit, though they are of geogenic origin, was induced after the road's construction.

The guest editors of this Special Issue would like to thank all of the authors and reviewers for their valuable contributions.

Funding: This research was funded by Russian Science Foundation (RSF), project no. 22-18-00065, "Cultural and historical processes and paleoenvironment in the Late Bronze-Early Iron Age of the North-Western Black Sea region: interdisciplinary approach", and project 2023-2025 (VRFY-2023-0010 of the Ministry of Education of the Russian Federation).

Conflicts of Interest: The author declares no conflict of interest.

References

1. Bezrukova, E.V.; Amosova, A.A.; Chubarov, V.M. Geochemical Records of the Late Glacial and Holocene Paleoenvironmental Changes from the Lake Kaskadnoe-1 Sediments (East Sayan Mountains, South Siberia). *Minerals* **2023**, *13*, 449. [CrossRef]
2. Malov, V.; Subetto, D.; Strakhovenko, V.; Ovdina, E.; Belkina, N. Distribution of Mercury in the Water-Suspended Matter-Bottom Sediments System of the Lake Onego Water Area. *Minerals* **2022**, *12*, 1410. [CrossRef]

3. Zhang, S.; Wang, T.; Guo, H.; Zhang, S.; Chen, B. Lipid Biomarker and Stable Isotopic Profiles through Late Carboniferous–Early Triassic of the Deepest Well MS-1 in the Junggar Basin, Northwest China. *Minerals* **2022**, *12*, 1299. [CrossRef]
4. Druzhinina, O.; Gedminienė, L.; van den Berghe, K. Metals in Lake Sediments as Indicators of Human Activities in Prehistory: Case Study of the Southeastern Baltic, Kamyshovoe Lake. *Minerals* **2022**, *12*, 1216. [CrossRef]
5. Ghandour, I.M.; Mannaa, A.; Alharbi, O.; Masoud, A.A.; Aljahdali, M.H. Major Factors Controlling the Elemental Composition of Al-Kharrar Lagoon Bottom Sediments, Rabigh, Saudi Arabia. *Minerals* **2022**, *12*, 1096. [CrossRef]
6. Kulkova, M. Geochemical Indication of Functional Zones at the Archaeological Sites of Eastern Europe. *Minerals* **2022**, *12*, 1075. [CrossRef]
7. Karavaeva, T.; Menshikova, E.; Belkin, P.; Zhdakaev, V. Features of Arsenic Distribution in the Soils of Potash Mines. *Minerals* **2022**, *12*, 1029. [CrossRef]
8. Ullah, S.; Jan, I.U.; Hanif, M.; Latif, K.; Mohibullah, M.; Sabba, M.; Anees, A.; Ashraf, U.; Vo Thanh, H. Paleoenvironmental and Bio-Sequence Stratigraphic Analysis of the Cretaceous Pelagic Carbonates of Eastern Tethys, Sulaiman Range, Pakistan. *Minerals* **2022**, *12*, 946. [CrossRef]
9. Druzhinina, O.; Gedminienė, L.; van den Berghe, K. Geochemical Study of the Iron Age Settlement Occupational Layer and the Early Roman Time Agricultural Layer at Voorthuizen, The Netherlands. *Minerals* **2022**, *12*, 373. [CrossRef]
10. Ghandour, I.M.; Aljahdali, M.H. Elemental Enrichment in Shallow Subsurface Red Sea Coastal Sediments, Al-Shuaiba, Saudi Arabia: Natural vs. Anthropogenic Controls. *Minerals* **2021**, *11*, 898. [CrossRef]

minerals

MDPI

Article

Geochemical Records of the Late Glacial and Holocene Paleoenvironmental Changes from the Lake Kaskadnoe-1 Sediments (East Sayan Mountains, South Siberia)

Elena V. Bezrukova [1,*], Alena A. Amosova [2] and Victor M. Chubarov [2]

[1] Laboratory of Ecological Geochemistry and Evolution of Geosystems, Vinogradov Institute of Geochemistry, Siberian Branch of the Russian Academy of Sciences, 664033 Irkutsk, Russia

[2] Laboratory of X-ray Analysis, Vinogradov Institute of Geochemistry, Siberian Branch of the Russian Academy of Sciences, 664033 Irkutsk, Russia; amosova@igc.irk.ru (A.A.A.); chubarov@igc.irk.ru (V.M.C.)

* Correspondence: bezrukova@igc.irk.ru; Tel.: +7-3-952-511-092

Abstract: Long-term and continuous lake sedimentary records offer enormous potential for interpreting paleoenvironmental histories and for understanding how terrestrial environments might respond to current global warming conditions. However, sedimentary records that contain the Late Glacial and Holocene epochs are scarce in deep continental high-mountain regions. A 150 cm sediment core was obtained from Lake Kaskadnoe-1 in the East Sayan Mountains (South Siberia, Russia, 2080 m above sea level), containing a unique record of the last 13,200 calibrated years (cal yr). Chronological control was obtained by AMS 14C dating. Here, we show the first detailed X-ray fluorescence (XRF) geochemical record, with the goal of broadening our knowledge of the paleoenvironmental history of the East Sayan Mountains in the past. The determination of major compounds and trace elements (Sr, Zr) was performed from each centimeter of the Lake Kaskadnoe-1 sediment core. The inorganic geochemistry indicates significant variations in elemental composition between two major lithological units of the sediment core: the Late Glacial dense grey silty clay (150–144 cm), and the upper interval (0–143 cm) mostly consisted of dark biogenic-terrigenous silt, accumulated during the Holocene. The Late Glacial sediments accumulated 13,200–12,800 cal yr BP are characterized by high values of CIA, Mg/Al, K/Al, and Mn/Fe, and are depleted in Si/Al, Fe/Al, and Ca/Al. During the Younger Dryas cold episode, LOI enrichment was probably caused by the presence of less oxic conditions, as seen in lower Mn/Fe values, due to a longer period of lake ice-cover. The Early Holocene (12,000–7500 cal yr BP) is associated with a decreasing trend of mineral matter with fluvial transport to Lake Kaskadnoe-1 (low K/Al, Mg/Al) and stronger chemical weathering in the lake basin. The increase in Ti/Al, K/Al and CIA values over the last 7500 years suggests an increase in the terrigenous input into the lake. Low LOI values can be possibly explained by the presence of less dense vegetation cover in the basin. In summary, our data indicate that the geochemical indices and selected elemental ratios mirror the sedimentation conditions that were triggered by environmental and climate changes during the Late Glacial and Holocene.

Keywords: geochemistry; lacustrine sediments; Late Glacial; Holocene; major element composition; Lake Kaskadnoe-1; South Siberia; environmental change

Citation: Bezrukova, E.V.; Amosova, A.A.; Chubarov, V.M. Geochemical Records of the Late Glacial and Holocene Paleoenvironmental Changes from the Lake Kaskadnoe-1 Sediments (East Sayan Mountains, South Siberia). *Minerals* **2023**, *13*, 449. https://doi.org/10.3390/min13030449

Academic Editors: Marianna Kulkova and Dmitry Subetto

Received: 13 February 2023
Revised: 16 March 2023
Accepted: 20 March 2023
Published: 22 March 2023

1. Introduction

Mountain lakes and their catchments are, due to their remoteness, the most sensitive to global climate changes [1–3]. In the high-altitude areas of the East Sayan Mountains, there are many lakes of various origins [4–8]. Lake basins can be combined into four groups: groups determined by volcanic activity, glacial processes, postglacial fluvial processes, and complex factors. The lakes of the first group are directly associated with the lava flows and volcanoes of the Jom-Bolok River Valley. They are located both at the bottom of the valleys and in the apical mountain belt. Lakes of fluvial origin are presented by floodplain water basins. Some lakes in the studied region were formed by a complex of relief-forming processes, and their basins have a polygenetic nature, mainly of the glacial-fluvial type [9]. The Oka Plateau, where most of the lakes are of volcanic and glacier origins, is one of the few existing regions in the world of active Late Glacial and Holocene volcanism [10–13]. To date, there have been several studies conducted to provide some data on the structure, sources, and possible mechanism of those eruptions [13–16]. It is known that the eruptions of lava were multiphase [13], began in post-glacial time, and probably have not ended yet [15,16]. The recent dendrochronological analysis has shown that the latest eruption occurred during the period 682–792 A.D. [16]. The first tephrochronological studies of the Lake Kaskadnoe-1 sediments provided the most complete record of volcanic activity in the Jom-Bolok region [6]. This record represents the longest of the currently known, reliably dated sequence of the Holocene eruptions in Northern Asia. Reconstructions have shown that the first phase can be traced back to 13,200 cal yr BP and lasted until 6000 cal yr BP. The dormant period spans 6000–1000 cal yr BP. The second phase of eruptions began 1000 cal yr BP and probably is not finished yet [6].

However, knowledge of the last glaciation, landscapes, and climate of this area is still very poor. Thus, it was found that in the Sartan period (marine isotope stage 2), the glaciers here occupied vast areas; however, the glacial processes occurred mainly within mountain valleys and corries [17]. The mean weighted exposure ages of glacial sediments localized within valleys of the Oka Plateau are 16,440 $\pm$ 380 cal yr BP and 22,800 $\pm$ 560 cal. yr BP [17].

However, high-resolution paleoenvironmental records in the East Sayan Mountains are still scarce and depend predominantly on pollen, diatoms, and a single geochemical record from several lakes [7]. Thus, a multiproxy study on an alpine lake ESM-1 [4] at the modern tree line showed that steppe and tundra biomes were extensive in the East Sayan landscapes during the early Holocene. Boreal forests quickly expanded by 9100 cal yr BP, and dominated the landscape until ca 700 cal yr BP, when the greatest period of compositional turnover occurred. At this time, the alpine meadow landscape expanded and *Picea obovata* colonised new habitats along river valleys and lake shorelines, because of the prevailing cool, moist conditions [4]. Previously published pollen records from Lake Kaskadnoe-1 [8], Lake Khikushka [18], and Lake Sagan-Nur [19] (Figure 1) show changes in vegetation and climate present since ca 13,400 cal yr BP [18]. Predominantly open steppe- and tundra-like vegetation dominated the area during the Allerod interstadial, with noticeable participation of boreal trees. A short-term reduction in the forest biome at ca. 12,600–12,500 cal yr BP could be a response of regional vegetation to climate deterioration during the Younger Dryas stadial. The strengthening of the forest biome between 12,500 and 11,200 cal yr BP occurred due to the expansion of *Picea*, *Abies*, and *Larix*. Climate warming and a decrease in effective moisture after 11,200 cal yr BP led to the degradation of dark coniferous forests in the study area and to a gradual expansion of *Pinus sylvestris* and *Pinus sibirica* in the East Sayan Mountains. The warmest climate existed during the Early Middle Holocene, ca. 11,200–6500 cal yr BP. The *Larix* stands may have been re-established in the study region soon after 5000 cal yr BP. This trend is parallel to a decrease in summer insolation and an increase in winter insolation. Although the study site is in an area with active volcanism, there is no clear evidence that volcanic eruptions influenced the local vegetation [8].

The above indicate that, to date, there has been little paleoecological research conducted on the Oka Plateau using a geochemical approach. Meanwhile, inorganic geochemical analyses have proved to be a powerful tool to study the evolution of lacustrine sediments and catchments [20–22]. Therefore, in this study, we present sediment stratigraphy and geochemical data including the geochemistry of major elements and some trace elements, elemental ratios, chemical indexes of alteration (CIA), and biogenic silica SiO2bio and pollen indices along a 150 cm sediment profile from Lake Kaskadnoe-1 with the objectives of understanding the geochemical processes and improving paleoecological information from the Late Glacial and Holocene about the Oka Plateau in the southern part of East Siberia.

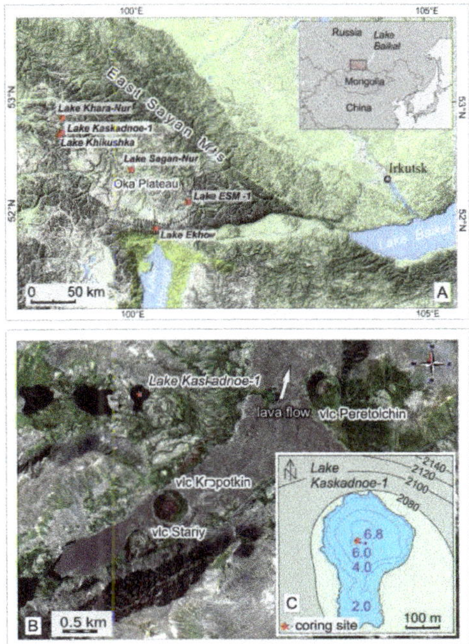

Figure 1. (**A**) Map based on Shuttle Radar Topography Mission (SRTM) v4.1 data [23], showing the position of the study area in the south of the Baikal region (upper right inset), and the position of Kaskadnoe-1 Lake on the Oka Plateau. Other red dots correspond to the previously studied sections of lake sediments of the Late Glacial–Holocene age in the East Sayan Mountains. (**B**) Position of Lake Kaskadnoe-1 near volcanic centers of Late Pleistocene-Holocene eruptions. The white arrow shows the direction of lava flows. (**C**) Bathymetric map with the coring site location. Numbers from 2080 to 2140 refer to elevations in meters above sea level. The contours of the Oka plateau are shown by a dotted line.

2. Regional Setting

The remoteness from the oceans and the high elevation of the Oka Plateau above sea level result in the dominance of a sharply continental climate with long and cold winters due to the effects of the Siberian anticyclone. In spring and autumn, west-north-west cyclonic activity increases. Summers are cool, and precipitation is brought mainly by cyclones from the south-southwest. The amplitude of the average temperatures between the warmest and coldest seasons is 50 °C [24]. The mean annual precipitation does not exceed 430 mm, and the mean annual near-surface air temperature is −4 °C. Because of the purity and sparsity of the atmosphere, the amount of summer insolation is high. This leads to significant surface heating during the day and rapid cooling at night, thus activating the processes of physical weathering of rocks and soils [25].

Lake Kaskadnoe-1 lies at 2080 m above sea level (Figure 1). The lake has a maximum depth of 7 m and a surface area of 2.5 km^2. The mountains, surrounding the lake, are composed of Paleozoic intrusive rocks (plagiogranite, granodiorite, pegmatite, diorite, gabbro-diorite, gabbro, and gabbro-norites) of the Tannuol complex [6]. Loose sediments occurring on slopes of the lake basin include Late Pleistocene–Holocene glacial, mudflow, and slope formations. Young volcanic rocks of the Jom-Bolok lava field are absent in the lake catchment and appear only 1.35 km SE of the lake (Figure 1B).

The lake is a tarn lake and originated as a result of glaciers retreating due to their melting. In winter, frozen soil and sediments around the lake are not deep due to the early lying snow cover in autumn, the considerable snow cover thickness, and the late melting of the snow cover in spring. The lake is fed mainly by atmospheric precipitation, snow melt water; the water may also come from the catchment area in the form of groundwater flow through permeable slope sediments. The excess water flows out through a stream and sediments on the southeastern shore of the lake.

The lake catchment has discontinuous vegetation cover; meadows predominate with shrubby birch (*Betula nana* ssp. *rotundifolia*), and there are rare occurrences of Siberian larch (*Larix sibirica*), willow (*Salix* sp.), and prostrate fir (*Abies sibirica*). In the lake basin, the vegetation consists of different shrubs and scattered larch trees. Locally, the lake shores and shallow water are occupied by coastal-aquatic vegetation dominated by sedges (Cyperaceae) in association with grasses (Poaceae).

3. Material and Methods

3.1. Sample Collection and Sediment Lithology

In 2015, we conducted a bathymetric survey using the three-dimensional 6-beam digital depth sounder Humminbird Matrix 748x3D (Johnson Outdoors Marine Electronics, Seongnam, Korea). As a result of that survey, a bathymetric map was obtained (Figure 1C). The coring operation was performed using rope-operated UWITEC Piston Corer (Niederreiter, Mondsee, Austria) with PVC liners of a 63 mm inner diameter. The sedimentary sequence was continuously cored, including penetration into the underlying glacial sediments. A 150 cm long core was retrieved. The core sediments were described in 2 cm intervals using the smear slide method in three replicates. On the basis of the dominant constituents, the sediment was classified and the dominant and minor lithology was indicated. The particles identified under a microscope were highlighted in the following groups: <10 μm (clay), 10–50 μm (silt), and >50 μm (sand).

3.2. AMS 14C Dating

As the terrestrial plant remnants and shells of aquatic mollusks are absent in the sediments of Lake Kaskadnoe-1, a bulk organic sediment fraction was used for radiocarbon dating. Seven AMS^{14}C ages were obtained in the Poznan Radiocarbon Laboratory (Table 1). The radiocarbon ages were then calibrated using R package clam [26] and the IntCal20 calibration curve [27].

Table 1. AMS 14C dates and calibrated ages for the Lake Kaskadnoe-1 core. Calibration was performed using R package version 2.5.0. https://CRAN.R-project.org/package=clam (accessed on 27 February 2023) [27].

Sample ID	Core Depth, cm	^{14}C yr BP	^{14}C Ages Corrected on Reservoir Effect of 980 Years	Calibrated 95.4% Range (cal yr BP)	Modeled Age, cal yr BP
Poz-76417	11	2070 ± 30	1090 ± 30	933–1057	1000
Poz-76459	67	7060 ± 50	6080 ± 50	6801–7147	6950
Poz-76460	91	8960 ± 50	7980 ± 50	8649–8991	8840
Poz-76461	113	9820 ± 60	8840 ± 60	9650–10,163	9920
Poz-76462	128	11,160 ± 60	10,180 ± 60	11,513–12,076	11,830
Poz-76463	142	11,820 ± 70	10,840 ± 70	12,711–12,957	12,790
Poz-76604	149	12,310 ± 70	11,330 ± 70	13,114–13,348	13,220

3.3. Determination of Major and Trace Element Composition

Samples for the determination of major compounds and trace elements (Sr, Zr) were selected from each centimeter of the Lake Kaskadnoe-1 sediment core. Extra pure lithium metaborate was dried at 450 °C for 4 h. The samples (110 mg) were carefully mixed with 1.1 g of extra pure lithium metaborate and 7 drops of a 40 mg/mL LiBr solution as a releasing agent in a platinum crucible and fused in the automatic electric multi-position furnace TheOX (Claisse, Québec, QC, Canada)) at 1050 °C for 19 min. The dilution factor was 1:10. This procedure allows the formation of fused glasses of a diameter of 10–12 mm [28]. The XRF spectra were measured by wavelength-dispersive XRF spectrometer S4 Pioneer (Bruker AXS, Karlsruhe, Germany). The calibration curves were constructed using the certified reference materials of sedimentary rocks [28].

The principal component analysis (PCA) of major elements was accomplished using the correlation matrix of major and trace elements in the "factoextra" package of the R software version 1.0.5 [29].

To evaluate the conditions of sedimentation in Lake Kaskadnoe-1, the loss on ignition (LOI) method was applied. LOI was analyzed following standard procedures [30]; sediments at 1 cm intervals were dried at 105 °C for 24 h and combusted at 550 °C for 4 h. This allowed the determination of the organic matter (LOI) in 149 sediment samples. The lithology of core sediments was described in 2 cm intervals using the smear slide method in three replicates. The previously published results from the Lake Kaskadnoe-1 sediment core, including biogenic silica SiO_{2bio}, sediment physical properties, and pollen [8], were utilized for interpretation in the present study.

4. Results

4.1. Lithology and Chronology

The visual inspection of the sediments showed two major lithological units: between 150 and 144 cm, the Lake Kaskadnoe-1 core is characterized by dense grey silty clay with a minor admixture of sand; the upper interval (0–143 cm) mostly consists of dark-olive biogenic-terrigenous silt (Figure 2). The results of the smear slide method demonstrated the more complicated lithological composition of the Lake Kaskadnoe-1 sediments and revealed that the sand content, in general, does not exceed 8%. The content of sand is rather high, at 13,200–12,800 cal yr BP, and then it decreases from 12,800 to 12,000 cal. yr BP. In sediments accumulated during the last 7500 cal. years, sand shows a gradually decreasing trend.

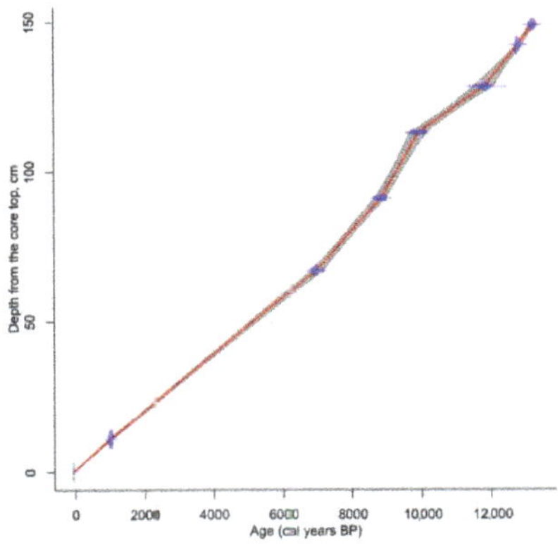

Figure 2. Results of applying the CLAM age-depth model of the Lake Kaskadnoe-1 core based on seven calibrated AMS radiocarbon dates (weighted mean age, dotted red line), overlaying the calibrated distributions of individual 14C ages (blue), with 95% probability intervals (gray shaded).

The obtained dates suggest the accumulation of the recovered core sediments during the Late Glacial and Holocene. The results of the radiocarbon dating are given in Table 1, and the ^{14}C age-depth model is shown in Figure 3. An almost linear age-depth relationship suggests continuous sedimentation. However, the obtained results show that the radiocarbon age of the uppermost sediments (11–12 cm) turned out to be older than expected (Table 1). Therefore, a reservoir effect could be an issue in Lake Kaskadnoe-1. To overcome this problem, a lake reservoir effect of 980 ^{14}C years was determined as the intercept of the Lake Kaskadnoe-1 age-depth model at the modern sediment surface. Such a method is commonly in use [31,32].

Therefore, when constructing the age model, we subtracted the supposed reservoir age of 980 years from all the radiocarbon dates prior to their calibration to calendar ages as is usually performed [33]. The sedimentation rate was calculated between adjacent dates by the median values.

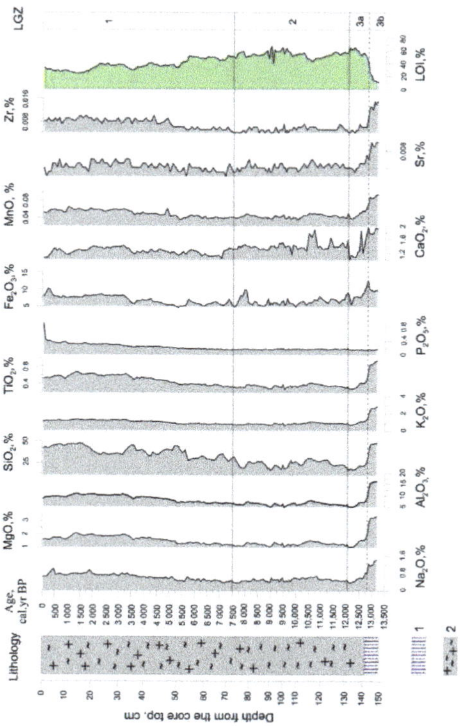

Figure 3. Core lithology and variations in major elements and some trace elements and LOI content in the Lake Kaskadnoe-1 sediments. Following the geochemical analysis, the percentages of 2 microelements and 10 oxides of each of the 150 samples were estimated. Three local geochemical zones (LGZ 1–3) with differing percentages of elements and oxides were identified. The CONISS program [34] for stratigraphically constrained cluster analysis by the method of incremental sum of squares was used to draw LGZ boundaries. Symbols for lithological description: 1—gray dense clay; 2—dark-olive biogenic-terrigenous silt with different proportions of diatom valves.

4.2. *Variations in Major Elements and Trace Elements*

Variations in the major elements and some trace elements including Na_2O (Na, %), MgO (Mg, %), total Fe_2O_3 (FeO + Fe_2O_3) (TFe, %), Al_2O_3 (Al, %), TiO_2 (Ti, %), SiO_2 (Si, %), MnO (Mn, %), K_2O (K, %), P_2O_5 (P, %), CaO (Ca, %), Sr (%), and Zr (%) are illustrated in Figure 3.

The contents of all major elements, except for P, and trace elements show their maxima between 13,200–12,800 cal yr BP (local geochemical zone 3b, LGZ) in the gray glacial clay layer. Between ca 12,800–12,000 cal yr BP (biogenic-terrigenous silt layer, LGZ 3a), the values of Na, K, Ca, Mg, Al, Si, Ti, Mn, Sr, and Zr decline, and LOI increases. Between 12,000 and 7500 cal yr BP (LGZ 2), the contents of all major elements remain almost constant, and concentrations of only Na, Mg, Al and Si slightly increase. After 7500 cal yr BP (LGZ 1), the contents of all elements increase apparently, whereas the LOI value tends to decrease gradually and markedly. The value of P reaches its maximum in the uppermost 2 cm of the core.

A number of studies have demonstrated that Al normalization can correct the problem of variable dilution in element records because of the input of total organic carbon. Therefore, we normalized the elements to Al, which is the most insoluble (under both oxic and anoxic conditions) and common terrestrially derived element [35] for the sake of evaluating the chemical solution and migration relative to Al. When normalized to Al, the contents of all the major elements in the Lake Kaskadnoe-1 record, except for P, and of the trace elements, show their maxima between 13,200–12,800 cal yr BP (local geochemical zone 3b, LGZ) in the gray glacial clay layer. Between ca 12,800–12,000 cal yr BP (biogenic-terrigenous silt layer, LGZ 3a), the values of Na, K, Ca, Mg, Al, Si, Ti, Mn, Sr, and Zr decline, and LOI increases. Between 12,000 and 7500 cal yr BP (LGZ 2), the contents of all major elements remain almost constant, and concentrations of only Na, Mg, Al and Si slightly increase. After 7500 cal yr BP (LGZ 1), the contents of all elements increase apparently, whereas the LOI value tends to decrease gradually and markedly. The value of P reaches its maximum in the uppermost 2 cm of the core (Figure 4).

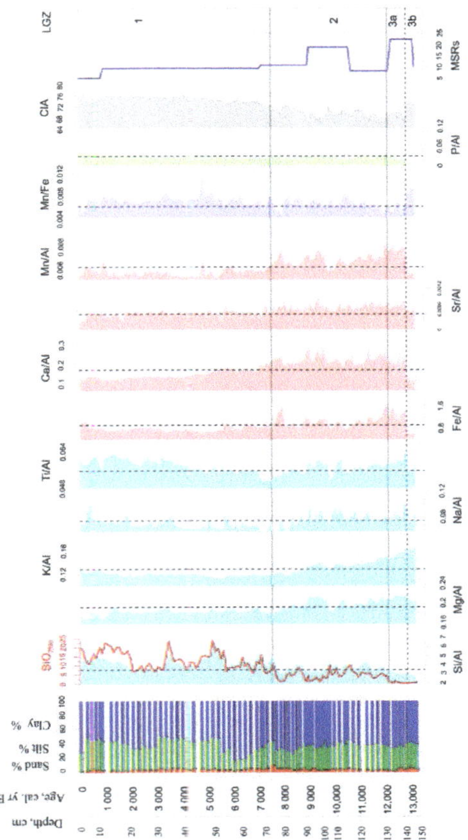

Figure 4. Changes in selected sediment characteristics and geochemical indices of the Lake Kaskadnoe-1 core. Vertical dotted lines indicate mean ratio values. Mean sedimentation rates (MSRs) were calculated between adjacent dates and are expressed in cm per 1000 years. Three local geochemical zones (LGZ 1–3) with differing geochemical records were identified based on the statistical evaluation of samples by principal component analysis (PCA) and demonstrated good agreement with LGZ 1–3 boundaries identified by the CONISS program. Subzones within LGZ -3 were identified by visual inspection.

Principal component analysis (PCA) was used to reduce the dimensionality of the elemental dataset, in order to try to identify the main factors controlling the elemental distributions and to aid the interpretation of the geochemical behavior of specific elements.

The results are summarized in the variable loadings on the first two principal components as illustrated in Figure 5. Two axes explain 57% of the total variance. PCA axis-1 accounts for 40% of the total and is characterized by high positive loadings of K/Al, Mg/Al, Mn/Al, Fe/Al, Ca/Al, and LOI, and high negative loadings of CIA, Si/Al and Mn/Fe ratios. PCA axis-2 (17% of the total variance) is characterized by a high positive loading of clay, and a negative loading of a silt fraction and the Si/Al ratio. The elements with high loadings on PCA axis-1 (K, Mg, Fe, Ca, Na) occur mainly in oxides and siliciclastic material, which are typically land-derived through the erosion of lithogenic material and are indicative of surface runoff processes. The second component is dominated by clay (positive loading) and silt (negative loading). In general, a correlation biplot reveals that elements distributed in the sedimentary strata could be divided into three distinct element clusters, and these clusters were considered as the local geochemical zones (LGZ 1–3).

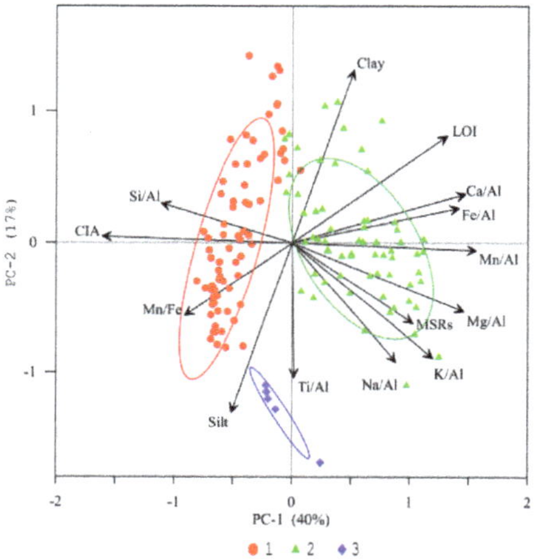

Figure 5. Summary of the principal component analysis of element ratios and distribution of selected indices along the studied core; 1—the group of ratios and indices related to silicate weathering and the supply of clastic material; 2—the group of ratios and indices associated with the supply of clastic material and weathering, leaching, and soil development; 3—the group related to the variations in the sediment grain size. Red circles —LGZ -1, green triangles—LGZ -3, and violet diamonds—LGZ -3.

5. Discussion

5.1. Potential Climatic Implication of Major Elements and Some Trace Elements

Following [36,37], bottom sediments retrieved from glacial Lake Kaskadnoe-1 are regarded as the sedimentary archive from the watershed. The point is that the continental climate of East Siberia causes minor chemical weathering during warm periods and strong physical weathering during cold periods. Thus, the elemental composition of the cores primarily depended on physical weathering, during which the elements were mainly supplied into the proglacial lakes as clastic material [7]. This is particularly true for the sediments of lakes located near glaciers [7]. The geochemical composition of sediments is primarily influenced by erosion and biotic processes in the watershed, as well as by the sediment grain size, and only slightly depends on sediment chemical alteration in the lake's fresh cold water [38]. These processes reflect changes in the paleoecological environment and respond to those changes. At the same time, Fe and Mn variations in sediments could be used as indicators of the past redox conditions in the soils within the catchment area or in the sediments themselves [39,40].

Climatic conditions are known to be responsible for the chemical composition of dissolved substances and play a major role in determining the intensity of weathering processes in watersheds [41]. It is also accepted that geochemical weathering processes in lake catchments are related to moisture and thermal regimes [37]. In the case of Lake Kaskadnoe-1, which is a semi-closed-basin lake that is fed by precipitation and groundwater, the materials, derived from various sources in the watershed due to weathering, are directly delivered to the lake. Therefore, characteristics of chemical elements in lake sediments can serve as indicators of the weathering history in the lake catchment. The measured and calculated major oxides were used to calculate the CIA index, which reflects the intensity of silicate weathering. Accordingly, CIA attains values of about 50 in non-weathered rocks, and attains values close to 100 in weathered varieties. A CIA index of less than 70 is an indicator of an incipient chemical weathering stage. A CIA index of >80 indicates high chemical weathering [42]. The analysis of the CIA relationships with indicators characterizing grain size (Al/Si) made it possible to determine the suitability of the CIA indicator as a proxy for the intensity of chemical weathering and climatic variations in the lake sediments [43].

Higher Ti/Al values can reflect the transportation of more terrigenous materials to the lake due to water runoff to the lake and can serve as an indicator of precipitation increase [44]. However, this ratio can be also indicative of the eolian input [45] as well as grain size changes [46]. The Mg/Al and K/Al ratios are used to reflect changes in fluvial transport to the lake [46,47].

Further, we use the above information on the paleoecological significance of geochemical indices and elemental ratios from the Lake Kaskadnoe-1 sediments to reconstruct the natural environment in the lake's basin.

5.2. Variations in Catchment CIA and Climate in the Past 13,200 cal Years

According to the age model, between 13,200 and 12,800 cal yr BP, dense silty clay accumulated in Lake Kaskadnoe-1. Higher sand content, Mg/Al, K/Al, Si/Al values and the lowest amount of SiO_{2bio} are indicative of siliciclastic matter input from the watershed with the meltwater of local glaciers. Elevated CIA and Sr/Al values may imply still-weak chemical weathering processes and/or intensive silicate weathering in the Lake Kaskadnoe-1 basin. At this time, the glacier was likely located around or very close to the lake, thus suggesting the maximum influx in the mineral matter produced by glacier erosion into the lake. Probably, when the glacier was close to the lake and was actively melting, there was an intensive formation of new material due to the physical weathering of the glacier bed. We assume that higher sand content and higher Mg/Al, K/Al, and Si/Al values describe the inflow of fine-grained material formed as a result of glacier erosion into the lake. Similar changes in the "mobile" state of the glacier were also observed in other mountainous regions of East Siberia [7]. The large amount of suspended minerogenic particles probably had a negative effect on the algal taxa, which found expression in very low values of biogenic silica. The mean sedimentation rate was high at this time (Figure 4), coinciding with the maximum mineral influx.

The clay layer formation coincided with a fast rate of glacier degradation possibly due to warming during the Bølling-Allerod interstadial. An intensive glacier melting on the Oka Plateau during the warming of the Bølling-Allerod interstadial correlates well with the retreat of the Grigoriev Glacier in the Tien Shan Mountains due to a wetter and warmer climate [48], glacier degradation in the Russian Altai mountains [49], and the air temperature increase in Greenland by 14,000 cal yr BP by almost 6 °C [50]. Moreover, the clay layer formation (fluvioglacial deposits) was produced almost simultaneously with the formation of fluvioglacial deposits of the Tompuda end moraine (located on the east shoreline of Lake Baikal) until 12,000 cal yr BP [7]. The warming of the Bølling-Allerod interstadial seems to be the cause of regional deglaciation both around Lake Baikal and in the Oka Plateau. For instance, cosmogenic ^{10}Be ages of erratic boulders in the southern coastal area of Lake Baikal are concentrated in the period 15,000–13,000 cal yr BP [51].

As demonstrated by the pollen indices, abundances of pollen, indicative of tundra shrubs, such as *Betula nana* (Figure 6), were at the high values around Lake Kaskadnoe-1, similarly to the present case. High pollen accumulation rates (PARs) (Figure 6) and the abundance of spruce *Picea* and *Larix* pollen suggest their distribution in the Lake Kaskadnoe-1 basin shortly after deglaciation, which coincided with their expansions in the adjacent areas of the Altai Mountains, Lake Baikal [48,52].

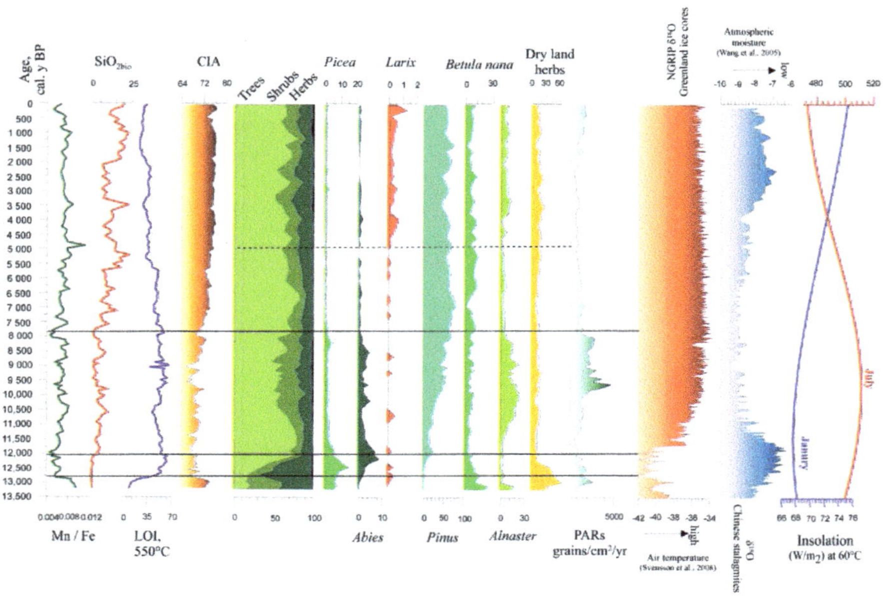

Figure 6. Comparison of palaeoenvironmental indicators from Lake Kaskadnoe-1 and the $\delta^{18}O$ records from Greenland, as an indicator of the northern hemisphere (NH) air temperature [53], and from Chinese stalagmites [54], as an indicator of the Pacific moisture-bearing monsoon intensity, and the NH June insolation at 55°N [55], plotted on their respective time scales. The black horizontal lines delineate the identified local geochemical zones (LGZ 1–3) shown in Figures 3 and 4. The uppermost dotted horizontal line highlights the appearance of larch trees near the Lake Kaskadnoe basin due to changes in the Late Holocene environment. (For interpretation of the references to color in this figure legend, the reader is referred to the web version of this article.)

A higher abundance of dry land herb pollen at that time (Figure 6) implies the occurrence of relatively warm summers, leading to the heating of the south-facing slopes, which favored the development of steppe vegetation. The sum of the pollen indices indicates a fairly dense vegetation cover and developed soils, which may have prevented the inflow of coarse mineral matter into the lake from the slopes.

Later, ca 12,800–12,000 cal yr BP, the biogenic-terrigenous silt started to accumulate. The share of organic matter increased. Decreases in dry bulk density values, sand content, and the K/Al ratio (Figures 3 and 4) are indicative of a weaker influx of large terrigenous matter with the glacial meltwater. At that time, the glacier was likely already in retreat. On the other hand, an increase in so-called terrigenous elements (Si/Al, Mg/Al, Fe/Al, Ti/Al, and Na/Al) and lower CIA values may suggest the more intensive input of fine terrigenous matter to the lake as the glacier flour. We also suggest that the increasing trend of the Ca/Al values could be a consequence of the input of the pelitomorphic carbonate of the glacial flour from the moraine sediments of the lake catchment.

The local glaciers likely decreased in area again, and a finer mineral fraction entered the lake as a result of moraine sediment erosion. Higher Ti/Al values could indicate both high soil moisture in the lake basin and the intensification of aeolian transport under windy conditions. However, the pollen indices indicate thick snow cover and plentiful soil moisture (Figure 5), which provided a favorable environment for *Picea*, *Larix*, and the maximum presence of *Abies* ca. 12,000 cal yr BP. The high soil moisture supports the conclusion that titanium arrived within the clastic fraction mostly from the basin slopes rather than via aeolian transport.

The increase in LOI could have resulted from both the influx of plant organic matter into the lake and/or a reducing environment in the internal conditions of the lake. Lower Mn/Fe ratios also suggest a reducing environment. Since the 12,800–12,000 cal yr BP interval corresponds to the first half of the Younger Dryas cooling [56], a reducing environment in Lake Kaskadnoe-1 could have been likely be due to the longer ice-covered period.

The next time interval, 12,000–7500 cal yr BP, is associated with a decreasing trend of mineral matter with fluvial transport to Lake Kaskadnoe-1 (Ti/Al, K/Al, Mg/Al, Fe/Al) and weak chemical weathering in the lake basin (low CIA index). An increase in the Mn/Fe ratio suggests that the increased mixing of the water column was likely due to a longer period of open water in a warmer than previous climate. Minor changes in Ca/Al and Sr/Al ratios reflect the continued input of fine pelitomorphic carbonate as a result of moraine sediment erosion. The authors believe that this process was facilitated not only by the meltwater of glaciers and snowfields but also by a higher amount of precipitation, which is suggested by the pollen indices (Figure 6). A decrease in the amount of *Picea* along with an increase in that of *Abies* and mesophytic shrub alder *Alnaster fruticosa* pollen indicate mild winters, cool summers, high soil humidity, a thick snow cover, and the lack of late-spring/summer frosts [57]. The dense vegetation cover (highest PARs) likely prevented the influx of coarse-grained material into the lake and contributed to the input of nutrients, thus leading to an increase in the productivity of the lake system (higher SiO_{2bio}).

A continuous increase in the pollen percentages of pines (Figure 6, Pinus curve), reflects the progressive expansion of pine forests and the strengthening position of the taiga biome in East Sayan and is in line with many other pollen records from across Eurasia [58]. Between 12,000 and 7500 cal yr BP, higher-than-present summer insolation in the middle latitudes of the Northern Hemisphere (Figure 6) promoted a stronger-than-present summer monsoon (Figure 6) and intensified south-easterly moisture transport to South Siberia. Warm summers contributed to the full degradation of the glacier in the Lake Kaskadnoe-1 basin since there is a high correlation between glacier mass changes and regional summer temperatures for the glaciers of East Siberia [59].

Over the time interval between 7500 and 5500 cal yr BP, further decreases in Sr/Al, K/A, Mg/Al, Na/Al, and Fe/Al values (Figure 5) suggest a minimum fluvial input. Ti/Al values may point to the increased eolian contribution in the lacustrine sediments during a more arid climate within the Lake Kaskadnoe-1 basin. These conclusions can be supported by the pollen indices and lower LOI values. Noticeable decreases in the amount of *Abies*, *Picea*, *Alnaster* pollen and low PSRs values imply lower soil moisture and/or annual precipitation, and they imply the disappearance of arboreal vegetation from the basin of Lake Kaskadnoe-1.

The decrease in humidity was likely the major reason of the lower input of mineral matter into the bottom sediments, resulting in low MSRs. The maximum distribution of *Pinus* (Figure 6) supports the conclusion that the changes in geochemical indices were due to a dramatically reduced meltwater inflow and/or a lower amount of atmospheric precipitation. This conclusion is consistent with the maximum distribution of both pines in the Altai-Sayan Region and with the highest reconstructed temperature index for sites located above the upper forest limit from 9000 to 6000 cal yr BP [60,61]. A warmer-than-earlier climate resulted in the stronger weathering of surrounding rocks and aquatic productivity, as seen in the increased CIA and SiO_{2bio} values (Figure 6).

The geochemical and pollen indicators show that the Late Pleistocene glaciers cardinally shrunk or fully disappeared from the Lake Kaskadnoe-1 basin by 7500 cal yr BP, which is in line with their disappearance in other locations in the south of East Siberia [7].

The increase in Ti/Al and K/Al values over the last 5500 years suggests an increase in the terrigenous input into the lake. The decreasing Mg/Al, Ca/Al, and Sr/Al ratios imply the very weak erosion of moraine sediments in the basin of the lake due to a noticeable reduction in the water flows that drained the slopes. The low LOI values can be possibly explained by the less dense vegetation cover in the basin. The increase in Mn/Fe and Ti/Al ratios implies the more intensive mixing of the water column due to increased wind activity and/or the shorter periods of ice cover.

The pollen indices indicate that scarce arboreal vegetation (larch) existed close to the Lake Kaskadnoe-1 basin (Figure 6), suggesting a colder and more humid climate in the Lake Kaskadnoe-1 watershed based on the ecological and climatic preferences of larch [62]. The general climate cooling in the Altai-Sayan region later than 5500 cal yr BP led to the glacial advances in the Russian Altai ca 5000 cal yr BP (the most extensive Holocene advance) [63] and to the permafrost aggradation and reconstructed changes in the local and regional vegetation [64]. However, the higher CIA values indicate the occurrence of slightly intensified chemical weathering in the lake basin, thus suggesting that heat and moisture are sufficient to maintain the present equivalent weathering intensity.

6. Conclusions

The first inorganic geochemistry record of Lake Kaskadnoe-1 sediments indicates downcore variations in elemental composition over the past 13,200 cal years. In sum, with the pollen, biogenic silica, LOI records, and the lithological description of the lake sediments, the inorganic geochemistry shows that the Lake Kaskadnoe-1 sediments resulted from a combination of fluvioglacial and climate interactions throughout the Late Glacial–Holocene. The fine clay sediments formed in 13,200–12,800 cal yr BP shows that Lake Kaskadnoe-1 already existed at the core location. At that time, the glacier was likely located around or very close to the lake, thus suggesting a maximum influx of the mineral matter produced by glacier erosion into the lake. The formation of Lake Kaskadnoe-1 at 13,200 cal yr BP may indicate that the Bølling-Allerød warming was most likely a trigger for deglaciation and the formation of glacial lakes in East Siberia.

The onset of biogenic-terrigenous silt accumulation at 12,800–12,000 cal yr BP indicates the glacier retreat from the lake basin. However, minerogenic matter sedimentation still dominated.

The higher aquatic productivity and presence of the most productive terrestrial landscapes, reconstructed for the period 12,000–7500 cal yr BP, were probably related to the wettest climate conditions throughout the whole period of sediment accumulation in Lake Kaskadnoe-1. Geochemical indices reflect the input of chemically weakly altered, relatively coarse-grained sediments brought from the basin slopes by the water flows.

Since 7500 cal yr BP, an increase in the rate of chemical weathering has been seen. There was a maximum increase in aquatic productivity, suggesting there has been the highest nutrient input into the lake over the last 13,500 years.

As the volcanic rocks are lacking in the basin and catchment area of Lake Kaskadnoe-1, the authors could not find a clear association between the geochemical composition of bottom sediments from Lake Kaskadnoe-1 and volcanic activity.

Author Contributions: Conceptualization, E.V.B.; data curation, E.V.B., A.A.A. and V.M.C.; formal analysis, E.V.B., A.A.A. and V.M.C.; funding acquisition, E.V.B.; methodology, E.V.B. and A.A.A.; supervision, E.V.B.; visualization, E.V.B. and V.M.C.; writing—review and editing, E.V.B. and A.A.A. All authors have read and agreed to the published version of the manuscript.

Funding: This research received no external funding.

Data Availability Statement: The data presented in this study are available on request from the corresponding author. The data are not publicly available due to the incompleteness of the multiproxy studies of the Lake kaskadnoe-1 sediments.

Acknowledgments: The research was performed in accordance with the state assignment of the Vinogradov Institute of Geochemistry SB RAS (project number: N 0284-2021-0003) using equipment of the Center of isotopic and geochemical Research. The authors are grateful to A.A. Shchetnikov, I.A. Filinov, and M.A. Krainov for their invaluable help in the field work. Special thanks go to Marina Khomutova for the initial translation of the manuscript and to anonymous reviewers for their useful comments and suggestions.

Conflicts of Interest: The authors have no conflict of interest to declare.

References

1. Catalan, J.; Bartons, M.; Camarero, L.; Grimalt, J.C. Mountain waters as witnesses of global pollution. In *Living with Water: Targeting Quality in a Dynamic World*; Pechan, P., de Vries, G.E., Eds.; Springer: New York, NY, USA, 2013; pp. 31–67.
2. Woolway, R.; Merchant, C. Worldwide alteration of lake mixing regimes in response to climate change. *Nat. Geosci.* **2019**, *12*, 271–276. [CrossRef]
3. Zhang, G.; Yao, T.; Xie, H.; Yang, K.; Zhu, L.; Shum, C.K. Response of Tibetan Plateau lakes to climate change: Trends, patterns, and mechanisms. *Earth Sci. Rev.* **2020**, *208*, 103269. [CrossRef]
4. Mackay, A.W.; Bezrukova, E.V.; Leng, M.J.; Meaney, M.; Nunes, A.; Piotrowska, N.; Self, A.; Shchetnikov, A.A.; Shilland, E.; Tarasov, P.E.; et al. Aquatic ecosystem responses to Holocene climate change and biome development in boreal, central Asia. *Quat. Sci. Rev.* **2012**, *41*, 119–131. [CrossRef]
5. Bezrukova, E.V.; Shchetnikov, A.A.; Kuzmin, M.I.; Sharova, O.G.; Kulagina, N.V.; Letunova, P.P.; Ivanov, E.V.; Kraynov, M.A.; Kerber, E.V.; Filinov, I.A.; et al. First data on the environment and climate change within the Zhom-Bolok volcanic field (Eastern Sayan Mountains) in the Middle–Late Holocene. *Dokl. Earth Sci.* **2016**, *468*, 527–531. [CrossRef]
6. Shchetnikov, A.A.; Bezrukova, E.V.; Krivonogov, S.K. Late Glacial to Holocene volcanism of Jom-Bolok Valley (East Sayan Mountains, Siberia) recorded by microtephra layers of the Lake Kaskadnoe-1 sediments. *J. Asian Earth Sci.* **2019**, *173*, 291–303. [CrossRef]
7. Stepanova, O.G.; Trunova, V.A.; Osipov, E.Y.; Kononov, E.E.; Vorobyeva, S.S.; Parkhomchuk, E.V.; Kalinkin, P.N.; Vorobyeva, E.E.; Vershinin, K.E.; Rastigeev, S.A.; et al. Glacier dynamics in the southern part of East Siberia (Russia) from the final part of the LGM to the present based on from biogeochemical proxies from bottom sediments of proglacial lakes. *Quat. Int.* **2019**, *524*, 4–12. [CrossRef]
8. Bezrukova, E.V.; Shchetnikov, A.A.; Kulagina, N.V.; Amosova, A.A. Lateglacial and Holocene vegetation and environmental change in the Jom-Bolok volcanic region, East Sayan Mountains, South Siberia, Russia. *Boreas* **2021**, *50*, 935–947. [CrossRef]
9. Shchetnikov, A.; Bezrukova, E.V. Lakes of the Jom-Bolok Volcanoes Valley in the East Sayan Mts., Baikal region: Morphogenesis and potential for regional paleoenvironmental studies. *J. Geogr. Sci.* **2019**, *29*, 1823–1840. [CrossRef]
10. Atkinson, T.W. *Oriental and Western Siberia: A Narrative of Seven Years' Explorations and Adventures in Siberia, Mongolia, the Kirghis Steppes, Chinese Tartary and Part of Central Asia*; Bradley, J.W., Ed.; J.W. Bradley: Philadelphia, PA, USA, 1859.
11. Kropotkin, P.A. *A Trip to the Oka Guard Post. Memoirs of the Russian Geographical Society, Siberian Division*; Tipografija Okruzhnogo Shtaba: Irkutsk, Russia, 1867; pp. 9–10. (In Russian)
12. Obruchev, S.V.; Lurye, M.L. Kropotkin and Peretolchin volcanoes in East Sayan. In *Transactions of Laboratory of Volcanology*; Publisher of the AN USSR: Moscow, Russia, 1954. (In Russian)
13. Yarmolyuk, V.V.; Nikiforov, A.V.; Ivanov, V.G. The structure, composition, sources and mechanism of Zhom-Bolok valley lava flows (Holocene, South-Baikal volcanic area). *Vulkanol. Seismol.* **2003**, *5*, 41–59. (In Russian)
14. Kiselev, A.I.; Medvedev, M.E.; Golovko, G.A. *Volcanism of the Baikal Rift Zone and Problems of Deep Magma Genesis*; Nauka: Novosibirsk, Russia, 1979. (In Russian)
15. Ivanov, A.V.; Arzhannikov, S.G.; Demonterova, E.I.; Arzhannikova, A.V.; Orlova, L.A. Jombolok Holocene volcanic field in the East Sayan Mts., Siberia, Russia: Structure, style of eruptions, magma compositions, and radiocarbon dating. *Bull. Volcanol.* **2011**, *73*, 1279–1294. [CrossRef]
16. Arzhannikov, S.G.; Ivanov, A.V.; Arzhannikova, A.V.; Demonterova, E.I.; Jolivet, M.; Voronin, V.I.; Buyantuev, V.A.; Oskolkov, V.A. Age of the Jombolok lava field (East Sayan): Evidence from dendrochronology and radiocarbon dating. *Russ. Geol. Geophys.* **2017**, *58*, 20–36. [CrossRef]
17. Arzhannikov, S.G.; Braucher, R.; Jolivet, M.; Arzhannikova, A.V. Late Pleistocene glaciations in southern East Sayan and detection of MIS 2 terminal moraines based on beryllium (10Be) dating of glacier complexes. Russ. *Geol. Geophys.* **2015**, *56*, 1509–1521. [CrossRef]
18. Bezrukova, E.V.; Kulagina, N.V.; Reshetova, S.A.; Shchetnikov, A.A.; Krainov, M.A.; Filinov, I.A. Environment of the Oka Plateau (East Sayan Mountains) in the Late glacial and Holocene: A case study of a complex record from the Lake Khikushka sediments. *Geomorphology* **2022**, *3*, 61–73. [CrossRef]
19. Bezrukova, E.V.; Reshetova, S.A.; Volchatova, E.V.; Kuzmin, M.I. First reconstructions of vegetation and climate changes in the central part of the Oka Plateau (East Sayan Mountains) in the Middle-Late Holocene. *Dokl. Earth Sci.* **2022**, *506*, 687–692. [CrossRef]

20. Koinig, K.A.; Shotyk, W.; Lotter, A.F.; Ohlendorf, C.; Sturm, M. 9000 years of geochemical evolution of lithogenic major and trace elements in the sediment of an alpine lake-the role of climate, vegetation, and land-use history. *J. Paleolimnol.* **2003**, *30*, 307–320. [CrossRef]
21. Schmidt, R.; Roth, M.; Tessadri, R.; Weckström, K. Disentangling late-Holocene climate and land-use impacts on an Austrian alpine lake using seasonal temperature anomalies, ice-cover, sedimentology, and pollen tracers. *J. Paleolimnol.* **2008**, *40*, 453–469. [CrossRef]
22. Minyuk, P.S.; Borkhodoev, V.Y.; Wennrich, V. Inorganic geochemistry data from Lake El'gygytgyn sediments: Marine isotope stages 6–11. *Clim. Past.* **2014**, *10*, 467–485. [CrossRef]
23. Strelkov, S.A.; Vdovin, V.V. *Altai-Sayan Mountain Region (Evolution of Topography of Siberia and the Far East)*; Publishing House Nauka: Moscow, Russia, 1969. (In Russian)
24. Plyusnin, V.M.; Kitov, A.D.; Ivanov, E.N.; Sheinkman, V.S. Distinctive characteristics of formation and dynamics of Nival-Glacial Ge-osystems in the South of East Siberia and on Mongolian Altai. *Geogr. Nat. Resour.* **2013**, *34*, 5–18. [CrossRef]
25. Jarvis, A.; Reuter, H.I.; Nelson, A.; Guevara, E. Hole-Filled SRTM for the Globe Version 4, Available from the CGIAR-CSI SRTM 90m Database. 2008. Available online: http://srtm.csi.cgiar.org (accessed on 10 January 2023).
26. Blaauw, M.; Christen, J.A.; Vazquez, J.E.; Goring, S. Classical Age-Depth Modelling of Cores from Deposits. R Package Version 2.5.0. 2022. Available online: https://CRAN.R-project.org/package=clam (accessed on 27 February 2023).
27. Reimer, P.J.; Austin, W.E.N.; Bard, E.; Bayliss, A.; Blackwell, P.G.; Ramsey, C.B.; Butzin, M.; Cheng, H.; Edwards, R.L.; Hogg, M.G.; et al. The IntCal20 Northern Hemisphere radiocarbon age calibration curve (0-55 kcal BP). *Radiocarbon* **2020**, *62*, 725–757. [CrossRef]
28. Amosova, A.A.; Chubarov, V.M.; Pashkova, G.V.; Finkelshtein, A.L.; Bezrukova, E.V. Wavelength dispersive X-ray fluorescence determi-nation of major oxides in bottom and peat sediments for paleoclimatic studies. *Appl. Radiat. Isot.* **2019**, *144*, 118–123. [CrossRef]
29. Kassambara, A.; Mundt, F. Package 'Factoextra': Extract and Visualize the Results of Multivariate Data Analyses. R Package Version 1.0.5. 2017. Available online: https://CRAN.R-project.org/package=factoextra (accessed on 3 March 2023).
30. Heiri, O.; Lotter, A.F.; Lemcke, G. Loss on ignition as a method for estimating organic and carbonate content in sediments: Reproducibility and comparability of results. *J. Paleolimnol.* **2001**, *25*, 101–110. [CrossRef]
31. Shen, J.; Liu, X.; Wang, S.; Matsumoto, R. Paleoclimatic changes in the Qinghai lake area during the last 18,000 years. *Quat. Int.* **2005**, *136*, 131–140. [CrossRef]
32. Zhou, W.; Cheng, P.; Wang, H.; Zhu, Y.; Jull, A.J.T.; Wu, Z. ^{14}C Chronostratigraphy for Qinghai Lake in China. *Radiocarbon* **2014**, *56*, 143–155. [CrossRef]
33. Kobe, F.; Bezrukova, E.V.; Leipe, C.; Shchetnikov, A.A.; Goslar, T.; Wagner, M.; Kostrova, S.S.; Tarasov, P.E. Holocene vegetation and climate history in Baikal Siberia reconstructed from pollen records and its implications for archaeology. *Archaeol. Res. Asia* **2020**, *23*, 100209. [CrossRef]
34. Grimm, E.C. *Tilia 1.7.16 Software*; Illinois State Museum Research and Collection Center: Springfield, IL, USA, 2011.
35. Zhong, W.; Pen, Z.; Xue, J.; Ouyang, J.; Tang, X.; Cao, J. Geochemistry of sediments from Barkol Lake in the westerly influenced northeast Xinjiang: Implications for catchment weathering intensity during the Holocene. *J. Asian Earth Sci.* **2012**, *50*, 7–13. [CrossRef]
36. Zhang, C.; Zhang, W.; Feng, Z.; Mischke, S.; Gao, X.; Gao, D.; Sun, F. Holocene hydrological and climatic change on the northern Mongolian Plateau based on multi-proxy records from Lake Gun Nuur. *Palaeogeogr. Palaeoclimatol. Palaeoecol.* **2012**, *325*, 75–86. [CrossRef]
37. Brown, E.T.; Le Callonnec, L.; German, C.R. Geochemical cycling of redox sensitive metals in sediments from Lake Malawi, a diagnostic paleotracer for episodic changes in mixing depth. *Geochim. Cosmochim. Acta.* **2000**, *64*, 3515–3523. [CrossRef]
38. Sabatier, P.; Moernaut, J.; Bertrand, S.; Van Daele, M.; Kremer, K.; Chaumillon, E.; Arnaud, F. A Review of Event Deposits in Lake Sediments. *Quaternary* **2022**, *5*, 34. [CrossRef]
39. Ulrich, M.; Matthes, H.; Schmidt, J.; Fedorov, A.; Schirmeister, L.; Siegert, C.; Schneider, B.; Strauss, J. Holocene thermokarst dynamics in Central Yakutia e A multi-core and robust grain-size endmember modeling approach. *Quat. Sci. Rev.* **2019**, *218*, 10–33. [CrossRef]
40. Ortega-Guerrero, B.; Avendano, D.; Caballero, M.; Lozano-García, S.; Brown, E.T.; Rodríguez, A.; García, B.; Barceinas, H.; Soler, A.M.; Albarran, A. Climatic control on magnetic mineralogy during the late MIS 6—Early MIS 3 in Lake Chalco, central Mexico. *Quat. Sci. Rev.* **2020**, *230*, 106163. [CrossRef]
41. Baumer, M.M.; Wagner, B.; Meyer, H.; Leicher, N.; Lenz, M.; Fedorov, G.; Pestryakova, L.A.; Melles, M. Climatic and environmental changes in the Yana Highlands of north-eastern Siberia over the last c. 57 000 years, derived from a sediment core from Lake Emanda. *Boreas* **2021**, *50*, 114–133. [CrossRef]
42. Guo, L.; Zhang, B.; Xiong, S.; Wu, J.; Chen, Z.; Cui, J.; Chen, Y.; Ye, W.; Zhu, L. Shifts in the silicate weathering regime in South China during the Meso-Cenozoic linked to Asian summer monsoon evolution. *Glob. Planet. Change* **2022**, *212*, 103809. [CrossRef]
43. Yusupova, A.R.; Nourgalieva, N.G. Geochemical basis of climate change indication in the Holocene sediments of Lake Bannoe (Southern Urals, Russia). *Uchenye Zap. Kazan. Universiteta. Seriya Estestv. Nauk.* **2021**, *163*, 514–526. (In Russian) [CrossRef]
44. Alongi, D.M. Macro- and Micronutrient Cycling and Crucial Linkages to Geochemical Processes in Mangrove Ecosystems. *J. Mar. Sci. Eng.* **2021**, *9*, 456. [CrossRef]

45. Martinez-Ruiz, F.; Kastner, M.; Gallego-Torres, D.; Rodrigo-Gamiz, M.; Nieto-Moreno, V.; Ortega-Huertas, M. Paleoclimate and paleoceanography over the past 20,000 yr in the Mediterranean Sea Basins as indicated by sediment elemental proxies. *Quat. Sci. Rev.* **2015**, *107*, 25–46. [CrossRef]
46. Takeuchi, N.; Fujita, K.; Aizen, V.B.; Narama, C.; Yokoyama, Y.; Okamoto, S.; Naoki, K.; Kubota, J. The disappearance of glaciers in the Tien Shan Mountains in Central Asia at the end of Pleistocene. *Quat. Sci. Rev.* **2014**, *103*, 26–33. [CrossRef]
47. Bueno, C.; Figueira, R.C.L.; Ivanoff, M.D.; Toldo, E.E.; Ferreira, P.A.L.; Fornaro, L.; Garcıa-Rodrıguez, F. Inferring centennial terrigenous input for Patos Lagoon, Brazil: The world's largest choked coastal lagoon. *J. Paleolimnol.* **2021**, *66*, 157–169. [CrossRef]
48. Blyakharchuk, T.A.; Wright, H.E.; Borodavko, P.S.; van der Knaap, W.O.; Amman, B. Late Glacial and Holocene vegetational changes on the Ulagan high-mountain plateau, Altai Mountains, southern Siberia. *Palaeogeogr. Palaeoclimatol. Palaeoecol.* **2004**, *209*, 259–279. [CrossRef]
49. Blyakharchuk, T.A.; Wright, H.E.; Borodavko, P.S.; van der Knaap, W.O.; Ammann, B. Late Glacial and Holocene vegetational history of the Altai Mountains (southwestern Tuva Republic, Siberia). *Palaeogeogr. Palaeoclimatol. Palaeoecol.* **2007**, *245*, 518–534. [CrossRef]
50. Liu, Z.; Otto-Bliesner, B.L.; He, F.; Brady, E.C.; Tomas, R.; Clark, P.U.; Carlson, A.E.; Lynch-Stieglitz, J.; Curry, W.; Brook, E.; et al. Transient Simulation of Last Deglaciation with a New Mechanism for Bølling-Allerød Warming. *Science* **2009**, *325*, 310–314. [CrossRef]
51. Horiuchi, K.; Matsuzaki, H.; Osipov, E.; Khlystov, O.; Fujii, S. Cosmogenic Be-10 and Al-26 dating of erratic boulders in the southern coastal area of Lake Baikal, Siberia. *Nucl. Instrum. Methods Phys. Res. B* **2004**, *223–224*, 633–638. [CrossRef]
52. Shichi, K.; Takahara, H.; Krivonogov, S.K.; Bezrukova, E.V.; Kashiwaya, K.; Takehara, A.; Nakamura, T. Late Pleistocene and Holocene vegetation and climate records from Lake Kotokel, central Baikal region. *Quat. Int.* **2009**, *205*, 98–110. [CrossRef]
53. Svensson, A.; Andersen, K.K.; Bigler, M.; Clausen, H.B.; Dahl-Jensen, D.; Davies, S.M.; Johnsen, S.J.; Muscheler, R.; Parrenin, F.; Ras-mussen, S.O.; et al. A 60,000 year Greenland stratigraphic ice core chronology. *Clim. Past.* **2008**, *4*, 47–57. [CrossRef]
54. Wang, Y.; Cheng, H.; Edwards, R.L.; He, Y.; Kong, X.; An, Z.; Wu, J.; Kelly, M.J.; Dykoski, C.A.; Li, X. The Holocene Asian monsoon: Links to solar changes and North Atlantic climate. *Science* **2005**, *308*, 854–857. [CrossRef]
55. Laskar, J.; Robutel, P.; Joutel, F.; Gastineau, M.; Correia, A.C.M.; Levrard, B. A long-term numerical solution for the insolation quantities of the Earth. *Astron. Astrophys.* **2004**, *428*, 261–285. [CrossRef]
56. Rasmussen, S.O.; Bigler, M.; Blockley, S.P.; Blunier, T.; Buchardt, S.L.; Clausen, H.B.; Cvijanovic, I.; Dahl-Jensen, D.; Johnsen, S.J.; Fischer, H.; et al. A stratigraphic framework for abrupt climatic changes during the Last Glacial period based on three synchronized Greenland ice-core records: Refining and extending the INTIMATE event stratigraphy. *Quat. Sci. Rev.* **2014**, *106*, 14–28. [CrossRef]
57. Tinner, W.; Lotter, A.F. Holocene expansions of Fagus sylvatica and Abies albain Central Europe: Where are we after eight decades of debate? *Quat. Sci. Rev.* **2006**, *25*, 526–549. [CrossRef]
58. Binney, H.; Edwards, M.; Macias-Fauria, M.; Lozhkin, A.; Anderson, P.; Kaplan, J.O.; Andreev, A.; Bezrukova, E.; Blyakharchuk, T.; Jankovska, V.; et al. Vegetation of Eurasia from the last glacial maximum to present: Key biogeographic patterns. *Quat. Sci. Rev.* **2017**, *157*, 80–97. [CrossRef]
59. Shahgedanova, M.; Popovnin, V.; Aleynikov, A.; Stokes, C.R. Geodetic mass balance of Azarova Glacier, Kodar mountains, eastern Siberia, and its links to observed and projected climatic change. *Ann. Glaciol.* **2011**, *52*, 129–137. [CrossRef]
60. Blyakharchuk, T.A.; Chernova, N.A. Vegetation and climate in the Western Sayan Mts according to pollen data from Lugovoe Mire as a background for prehistoric cultural change in southern Middle Siberia. *Quat. Sci. Rev.* **2013**, *75*, 22–42. [CrossRef]
61. Zhang, D.; Feng, Z. Holocene climate variations in the Altai Mountains and the surrounding areas: A synthesis of pollen records. *Earth Sci. Rev.* **2018**, *185*, 847–869. [CrossRef]
62. Kloeppel, B.D.; Gower, S.T.; Trechel, I.W.; Kharuk, V. Foliar carbon isotope discrimination in *Larix* species and sympatric evergreen conifers: A global comparison. *Oecologia* **1998**, *14*, 153–159. [CrossRef] [PubMed]
63. Agatova, A.R.; Nazarov, A.N.; Nepop, R.K.; Rodnight, H. Holocene glacier fluctuations and climate changes in the southeastern part of the Russian Altai (South Siberia) based on a radiocarbon chronology. *Quat. Sci. Rev.* **2012**, *43*, 74–93. [CrossRef]
64. Kharuk, V.I.; Im, S.T.; Petrov, I.A.; Yagunov, M.N. Decline of Dark Coniferous Stands in Baikal Region. Contemp. *Probl. Ecol.* **2016**, *9*, 617–625. [CrossRef]

minerals MDPI

Article

Distribution of Mercury in the Water-Suspended Matter-Bottom Sediments System of the Lake Onego Water Area

Victor Malov [1,*], Dmitry Subetto [2,3], Vera Strakhovenko [1,3], Ekaterina Ovdina [1,3] and Natalia Belkina [3]

1 V.S. Sobolev Institute of Geology and Mineralogy, Siberian Branch of the Russian Academy of Sciences (RAS), 3 Akad. Koptyug Ave., 630090 Novosibirsk, Russia
2 Department of Geography, Herzen State Pedagogical University of Russia, 48 Moika Emb., 191186 Saint-Petersburg, Russia
3 Northern Water Problems Institute, The Karelian Research Centre, Russian Academy of Sciences, 50 Alexander Nevsky Ave., 185030 Petrozavodsk, Russia
* Correspondence: malov@igm.nsc.ru

Abstract: This paper presents for the first time data on the lateral distribution of total mercury in the water-suspended matter-bottom sediments system for the entire water area of Lake Onego, the second largest lake in Europe. The patterns of the total mercury vertical distribution in two types of bottom sediments stratification, have been established. The total mercury content in the Lake Onego water averages 0.32 ± 0.07 μg/L. The most common form of mercury in water is dissolved+colloid, with the exception of water samples from the Kondopoga and Povenetsky Bays. In the material of the sedimentation traps, the mercury content is 0.5 ± 0.3 μg/g, and in the upper and lower parts of the bottom sediments section is 0.067 ± 0.003 and 0.041 ± 0.001 μg/g, respectively. The paper considers the factors responsible for the mercury content increasing from the bottom sediments up through the sections. These factors include anthropogenic pollution, migration of Hg and its redeposition at the geochemical barrier together with Fe and Mn.

Keywords: mercury; Hg; bottom sediments; Lake Onego

Citation: Malov, V.; Subetto, D.; Strakhovenko, V.; Ovdina, E.; Belkina, N. Distribution of Mercury in the Water-Suspended Matter-Bottom Sediments System of the Lake Onego Water Area. *Minerals* **2022**, *12*, 1410. https://doi.org/10.3390/min12111410

Academic Editor: Pierfranco Lattanzi

Received: 22 September 2022
Accepted: 4 November 2022
Published: 6 November 2022

1. Introduction

Mercury is a toxic element. According to the World Health Organization, Hg is among the ten most dangerous chemicals [1], due to its high mobility and bioaccumulative ability [2]. Mercury represents the greatest danger when it enters aquatic ecosystems, where it can be transformed into a more toxic form - methylmercury, which accumulates in aquatic food chains [3].

In aquatic ecosystems, most toxicants accumulate in the bottom sediments [4] and organisms living in close contact with these deposits are exposed to chemicals either directly or through food chains.

The formation of methylmercury (MeHg) is one of the main processes occurring in aquatic ecosystems. The methylation process takes place mainly in the water column, as well as in the sediments of fresh and marine waters, because of interrelated biochemical, chemical and photochemical processes. The methylation of Hg(II) under natural conditions is primarily a biological process, mediated by sulfate-reducing and iron-reducing bacteria [5].

Methylmercury easily accumulates in organisms and is bio-amplified in food chains to concentrations significantly higher than concentrations in surface waters [6].

Methylmercury is a widespread and neurotoxic pollutant that can cause a deficiency of essential elements as a result of competition for the active sites of biologically important molecules affecting the central nervous system of living organisms [7].

One of the main research objects of freshwater ecosystems are lakes, which play an important role in the processes of chemical elements concentration, as they are the final reservoirs of runoff, along with seas and oceans. In the large lakes of Europe, the processes of chemical elements redistribution are comparable to the inland seas. The transformation of inorganic mercury to monomethylmercury in the aquatic environment can lead to high concentrations of monomethylmercury in fish and have a negative impact on the health of wild animals and people who consume fish [8].

In freshwater ecosystems, the predominant sources of mercury are the direct atmospheric deposition, wastewater and river runoff. The atmospheric deposition of mercury is especially important in lakes with a large ratio of the water surface area to the volume of lake water and small catchment areas [9].

Mercury enters freshwater systems from various sources and undergoes complex transport pathways [10]. These pathways correspond to their characteristic processes of mercury form transformation. As a result, for each of its entering ways, the freshwater ecosystems, mercury can be in its own form, which, in turn, greatly affects its future condition in the freshwater ecosystem [11].

As is the case for almost all global trace element balances, there are uncertainties regarding the estimation of the anthropogenic and natural mercury amount "stored" in various environmental components, flows between mercury reservoirs and the rate of mercury removal from the biosphere. At the moment, there is a large number of studies devoted to this problem [10,12–18]. However, the interpolation of such data on unique objects will give a large number of errors in relation to estimates. This is due to the fact that when studying the ecosystem of a very large lake (water area >1000 km^2), it is necessary to take into account the uneven development of the large- and small-scale hydrodynamic processes, as well as the mosaic nature of anthropogenic impact, which does not manifest itself immediately, often with a long time break. Such unique objects include Lake Onego.

The aim of this work is to assess the mercury distribution in the components of the water-suspended matter-bottom sediments system in the Lake Onego water area.

2. Materials and Methods

2.1. Study Area

The object of the study is Lake Onego. Lake Onego (61.69° north latitude, 35.66° east longitude, 33 m above sea level), with a catchment area of 62.8×10^3 km^2, a volume of water mass (291.7 km^3) and water area (9720 km^2) is the second largest waterbody in Europe. It is located in the area of the two large geological structures junction - the southeastern slope of the Baltic Crystal Shield and the Russian Platform. The lake basin was formed in the period between the Archean and Proterozoic and experienced a complex path of tectonic transformations in continental and marine environments. The last glaciation deepened the depression and blocked the relief with a cover of glacial, fluvio- and limnoglacial deposits. The lake basin has a complex structure, which determines the difference in the coastline outlines and the bottom relief form. A complex relief with large depth differences, an abundance of bays and islands, characterizes the northern part of the lake [19].

Lake Onego has 1152 tributaries of which 52 are more than 10 km long. The main tributaries are the Vodla, Shuya and Suna Rivers, the average annual flow rate which is 4.36, 3.1 and 2.5 km^3/year, respectively. According to published data, the main source of mineral and organic substances entering Lake Onego is river runoff, which provides 60% of the runoff into the lake [19,20].

The territory of the Lake Onego basin is the most economically developed part of the Republic of Karelia, Russia. The main environmental risk factor is the anthropogenic impact of industrial centers located on the shores of the lake. Such centers include the cities of Petrozavodsk (mechanical engineering, construction, food industry, transport and utilities) and Kondopoga (pulp and paper mill).

In addition, the uniqueness of Lake Onego lies in the fact that shungite rocks come to the daytime surface in the catchment area. Shungite rocks (widespread in the northern part of the Lake Onego catchment area) are a complex of volcanogenic-sedimentary rocks of Precambrian age (~2 billion years) with a large amount of carbonaceous matter [21]. In a broader sense, they can be attributed to "black shales", but taking into account that in this term for carbonaceous matter, there is no specificity of organic matter, both in composition and in genesis [22]. Black shales, and shungite-bearing rocks, are characterized by high concentrations of Hg and a number of other rare elements. For example, the Hg content in maksovites (a type of shungite-bearing rocks) is on average 0.79 ± 0.41 g/t [22], which significantly exceeds the Clarke number of mercury for crustal rocks—0.08 g/t [23].

2.2. Sampling

The actual material was sampled in 2016–2021 from the "Ecologist" research vessel, throughout the entire water area of Lake Onego: Povenetsky Bay, Zaonezhsky Bay, Small Onego, Lizhemskaya Bay, Unitskaya Bay, Kondopoga Bay, Petrozavodsk Bay, Big Onego, Central Onego, South Onego (Figure 1). A comprehensive sampling of the lake system components was made, water samples were taken in conjunction with the suspended matter, the sedimentary material from the sedimentation traps and samples of bottom sediments.

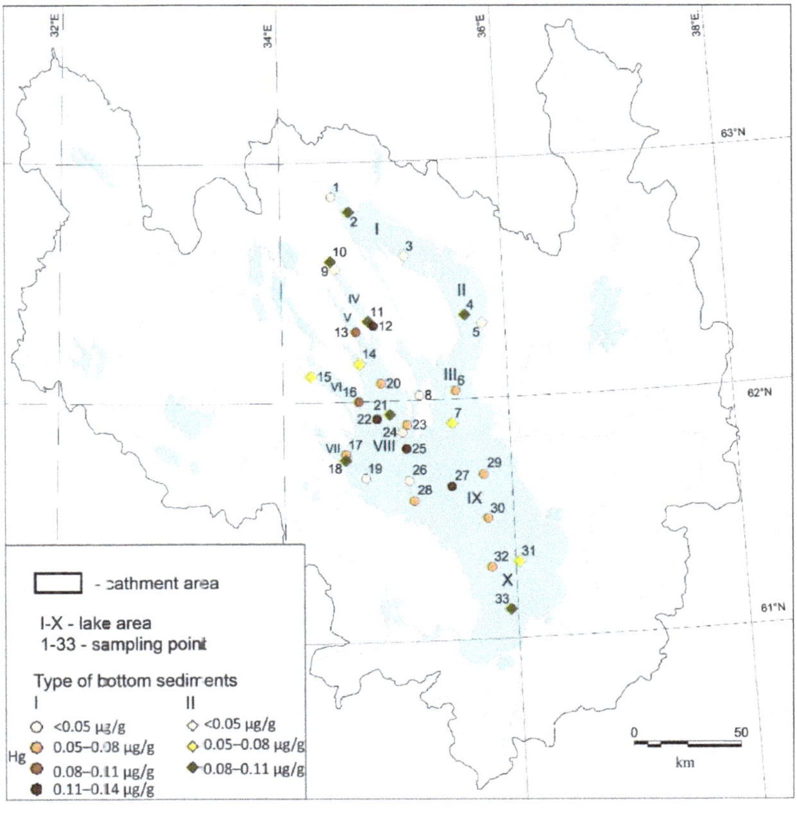

Figure 1. The Lake Onego ecosystem components sampling scheme: I—Povenetsky Bay; II—Zaonezhsky Bay; III—Small Onego; IV—Unitskaya Bay; V—Lizhemskaya Bay; VI—Kondopoga Bay; VII—Petrozavodsk Bay; VIII—Big Onego; IX—Central Onego; X—South Onego.

Water sampling was carried out using a Ruther RT bathometer (Hydrometpribor, Riga, Latvia). The separation of the suspended matter in the water was carried out by vacuum filtration on specially prepared pre-weighted membrane filters with a pore diameter of 0.45 μm. The aqueous filtrates were preserved with concentrated HNO_3 at the rate of 1 mL of acid per 100 mL of the sample.

Since suspended matter from lake water entering the bottom sediments is complex, having a different particle size, multicomponent system of terrigenous, biogenic and chemical materials, its study required the installation of sedimentation traps at more than two dozen points in various areas of Lake Onego. For this purpose, a simplified version of the UST–100 sedimentation trap (ST) was made in the Laboratory of Paleolimnology, NWPI KarRS RAS, Petrozavodsk, Russia. The installation of the equipment was carried out, in accordance with the recommendations given in [24]. The exposure time of the sedimentation trap was one year. The detailed information about the sedimentation traps and the method of their installation is described in [25].

Bottom sediment cores were sampled using Limnos, GOIN, Perfiliev samplers and heavy gravity core samplers. In total, 34 cores of bottom sediments up to 3.2 m long and with sampling steps from 1 cm to 10 cm (260 samples) were sampled.

2.3. Analytical Methods

Analytical studies of the lake components were conducted in the Analytical Center for multi-elemental and isotope research SB RAS, Novosibirsk, Russia.

The suspended matter from the water surface horizon of Lake Onego was isolated by filtering a water sample through a 0.45 μm membrane filter before they it was clogged. The filters with suspended matter were dried in a drying cabinet at 105 °C to a constant mass. The decomposition of the filters was carried out in an acidic condition (in 10 mL of concentrated nitric acid) in a microwave system SpeedWave®four (Berghof, Berlin, Germany). Then, the resulting solution was brought to a volume of 20–25 mL. The Hg content in the obtained solutions and filtered water was determined by "cold vapor" atomic absorption spectroscopy on an analyzer (Perkin Elmer 3030B, Waltham, MA, USA). The detection limit is 0.02 μg/L; the relative measurement error is 20%.

The total mercury content in the sedimentary material samples of the sedimentation traps and bottom sediments was determined, in accordance with an accredited methodology M 03-09-2013 by the Russian Federal Environmental Regulations, using the analyzer "RA-915M" (Lumex, St. Petersburg, Russia) with the device "RP-91C" (Lumex, St. Petersburg, Russia). The detection limit is 1.0 μg/kg.

Major (Ca, Na, K, Al, Fe, Mg, Mn) and trace (Sb, Sr, Ba, Be, Cd, Co, Ni, Cr, Zn, Pb etc.) elements in the samples of the suspended matter, sedimentation traps material and bottom sediments were determined via atomic absorption using a Solaar M6 instrument, equipped with a Zeeman and deuterium background corrector (Thermo Electron, Waltham, MA, USA). Two versions of the atomic absorption were used: flame atomization (acetylene-air and nitrous oxide-acetylene) to quantify the content of a wide range of chemical elements, the content of which, in the samples was >0.0001 mass%, and the electrothermal atomization for the quantitative determination of the lower contents (less than 0.000001 mass%).

The laboratory has a set of more than 40 standard samples. Specifically, we used the Russian standard sample BIL1 (standard sample composition of the bottom silt of Lake Baikal) to control the reliability of the results. The standard samples of the appropriate composition were analyzed with each batch of samples, which allowed us to control the accuracy of the analysis. The laboratory regularly participates in Russian Interlaboratory comparison tests that confirmed a high quality assessment of the atomic absorption analysis.

The measurement range of the mass fraction of elements (μg/g) is presented in Table 1. The limits of the total relative measurement error (±δ at P = 0.95) are ±30%. The normalization of the error characteristics was carried out taking into account the possible dilution of the analyzed solution, but not more than 1000 times.

Table 1. Measurement range of the mass fraction of elements (μg/g).

Element	Electrothermal Atomization	Flame Atomization
Al	-	5.0–5.0 × 10^4
K	-	5.0–5.0 × 10^5
Na	-	5.0–5.0 × 10^5
Mg	-	5.0–5.0 × 10^5
Ca	-	5.0–5.0 × 10^3
Fe	-	5.0–5.0 × 10^3
Be	0.50–1.0 × 10^3	5.0–1.0 × 10^3
V	-	5.0–1.0 × 10^3
Cr	-	1.0–5.0 × 10^3
Mn	-	1.0–5.0 × 10^3
Co	0.5–1.0 × 10^3	1.0–5.0 × 10^3
Ni	0.5–1.0 × 10^3	1.0–5.0 × 10^3
Cu	-	1.0–5.0 × 10^3
Zn	-	1.0–5.0 × 10^3
Sr	-	5.0–5.0 × 10^3
Cd	0.05–1.0 × 10^3	1.0–5.0 × 10^3
Sb	5.0–1.0 × 10^3	1.0–5.0 × 10^3
Ba	-	5.0–5.0 × 10^3
Pb	0.5–1.0 × 10^3	1.0–5.0 × 10^3

Note: "-" means that the method was not used to analyze the element.

2.4. Data Analysis

To explore the correlations between the data, the concentration values of the chemical elements were subjected to a hierarchical cluster analysis. Based on the indexes of the correlation coefficients, similar paired groups of chemical elements were connected, then the next most similar paired groups, and so on, until all chemical elements were grouped on the dendrogram by averaging [26].

Analytical data processing was carried out using Microsoft Office Excel software package and Statistica 6.0 (6.0, Dell, Novosibirsk, Russian).

3. Results

3.1. Mercury in the Water–Suspended Matter System

The total mercury content in the Lake Onego water averages 0.32 ± 0.07 μg/L. The minimum values were obtained in water samples of Povenetsky Bay (0.022 μg/L). The maximum values are typical for water samples taken in South Onego (0.852 μg/L) and Big Onego (0.552 μg/L). These values exceed the maximum permissible concentrations (0.5 μg/L) [27].

The values obtained earlier by other researchers, on average, are lower than those obtained by us and do not exceed the maximum permissible concentrations. For instance, the concentration of mercury in the water of Petrozavodsk Bay is similar to the concentrations of mercury for this area obtained in 2016 by [28]. Similar values were reported in [29]. According to these, the average mercury content in water is 0.028 μg/L (0.017–0.039 μg/L).

The mercury concentration in suspended matter in the Lake Onego water varies from 0.002 μg/L (Lizhemskaya Bay) to 0.073 μg/L (Zaonezhsky Bay), the average content is 0.024 ± 0.006 μg/L. The mercury content in the dissolved+colloid form is higher than the mercury content in the suspended form, and varies in a wide range: from 0.01 μg/L in Kondopoga Bay and Povenetsky Bay to 0.85 μg/L in South Onego. Thus, the predominant form of mercury in water is dissolved + colloid (Figure 2). Mercury, in suspended form prevails only in the Kondopoga and Povenetsky Bays.

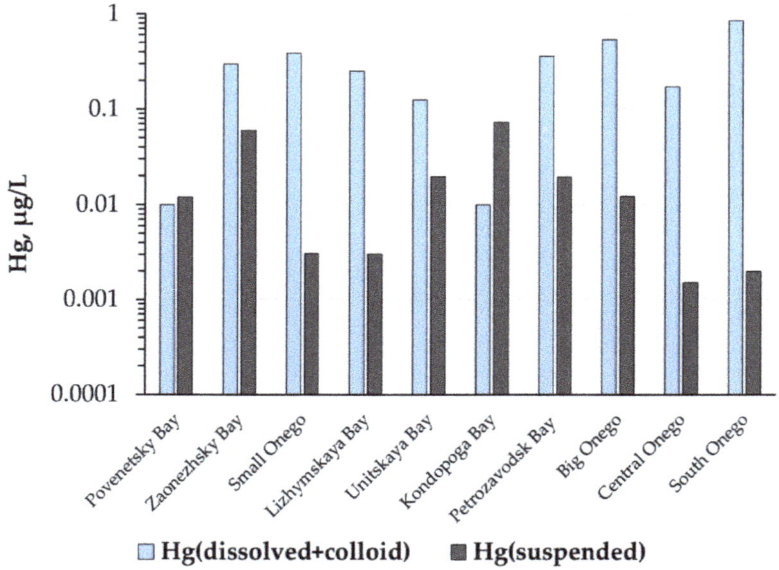

Figure 2. The mercury form in the water in the different areas of Lake Onego.

We found that the mercury content in the sedimentation traps material varies over a wide range of 0.062–4.37 µg/g. It is worth noting that high values are observed only in two areas of Lake Onego: in the Lizhemskaya (4.37 µg/g) and in Povenetsky Bays (0.76 µg/g). The obtained values exceed the values of the maximum permissible concentration (0.5 µg/g) and require further study and identification of the mercury source [27]. For the remaining areas, the average mercury concentration in the sedimentation trap material is 0.12 µg/g.

To confirm the results obtained, sedimentation traps were re-installed in the same areas of the lake. In 2021, similar values of mercury concentrations were obtained in the material of the sedimentation traps (Table 2, Figure 3). The sedimentation traps located in the Lizhemskaya and Povenetsky Bays could not be recovered in 2021.

Table 2. Mercury content in the components of the Lake Onego ecosystem.

Lake Area	Hg in Sedimentation Traps Material (2020/2021), µg/g	Hg in Water		Hg in Bottom Sediments	
		$Hg_{dissolved}$, µg/L	$Hg_{suspended}$, µg/L	0–20 cm, µg/g	20–200 cm, µg/g
Povenetsky Bay	0.76/ND	0.01	0.012	0.041	0.023
Zaonezhsky Bay	0.09/0.07	0.3	0.06	0.05	0.019
Small Onego	0.19/0.09	0.385	0.003	0.053	0.033
Lizhemskaya Bay	4.37/ND	0.25	0.003	0.073	0.044
Unitskaya Bay	0.18/0.08	0.125	0.02	0.07	0.016
Kondopoga Bay	0.10/0.15	0.01	0.073	0.089	0.031
Petrozavodsk Bay	0.11/0.16	0.36	0.019	0.052	0.027
Big Onego	0.12/0.12	0.54	0.012	0.076	0.058
Central Onego	ND/0.10	0.173	0.002	0.069	0.044
South Onego	0.06/0.07	0.85	0.002	0.08	0.044

Note: ND—no data available.

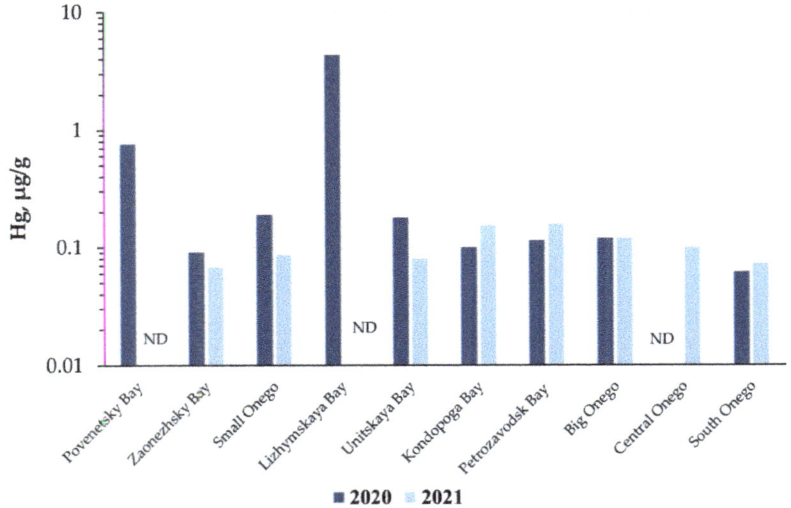

Figure 3. Mercury content in the sedimentation traps material from the different areas of Lake Onego in 2020 and 2021. ND—no data available.

3.2. Mercury in the Bottom Sediments

According to [30], the lake bottom is composed (from top to bottom): Holocene lacustrine deposits—silt and sand (lH, where l is lacustrine), Upper Pleistocene deposits—varved clays (lgIII, where lg—limno-glacial), glacial and fluvial glacial deposits of the Upper Pleistocene—coarse sand with pebbles, clays with boulders, boulder loams (gIII; fIII, where g is glacial, f is fluvioglacial).

The geochemical and mineral compositions of the Lake Onego bottom sediments are described in detail in [31–33] The authors suggest that the upper part (0–20 cm) of the Holocene bottom sediments can be divided into two stratigraphic types, based on the mineralogical composition and geochemical features. The first type of bottom sediments is characterized by the presence of enriched Fe-Mn layers formed near the water-bottom sediment boundary. In the second type, such layers are absent, and the upper part of the bottom sediments is represented by a small zone of oxidized silts. Below the redox barrier, there are homogeneous grayish-green silts, which are present in both types of the Lake Onego bottom sediments.

From the analysis of mercury concentrations in the bottom sediments section, it was found that in the range of 20–200 cm, the distribution is uniform, does not change along the entire interval and averages 0.041 ± 0.001 µg/g (Figure 4).

In the upper part of the section (above 20 cm), a sharp increase in Hg concentrations is observed, the average value is 0.067 ± 0.003 µg/g. The highest concentrations are characteristic of oxidized layers and layers enriched with Fe-Mn.

If we consider the distribution of mercury by two types of bottom sediments stratification, we can conclude that the distribution of mercury in both types of bottom sediments sections are similar. The only difference is that the first type is characterized by higher extreme values, compared to the second type of sediments (0.19 µg/g for the first type and 0.12 µg/g for the second type of bottom sediments) (Figure 5).

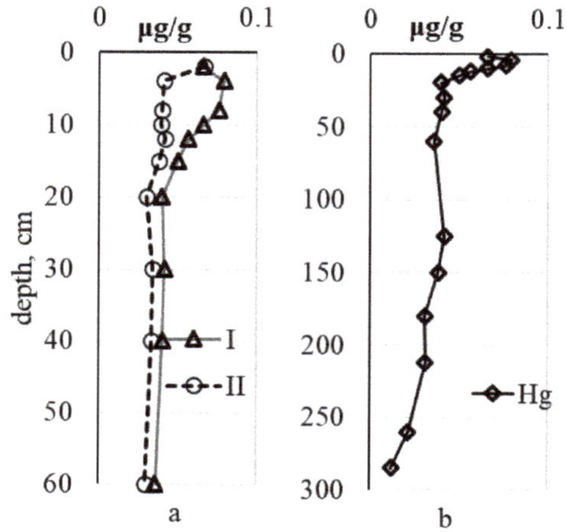

Figure 4. Distribution of mercury: (**a**) in the first (I) and second (II) type of the Lake Onego bottom sediments, (**b**) the average concentration in the Lake Onego bottom sediments.

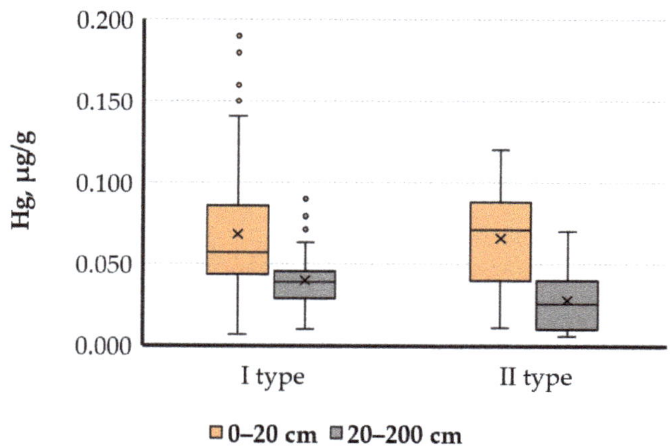

Figure 5. Mercury content in the first (I) and second (II) type of bottom sediments.

In the course of the work, the lateral distribution of Hg in the bottom sediments over the entire Lake Onego water area was studied for the first time. For the upper part of the bottom sediments, it was found that the mercury content has a uniform distribution and varies in the range of 0.041–0.089 µg/g (Figure 1). At the same time, lower values are characteristic for the northeastern part of Lake Onego (Povenetsky and Zaonezhsky Bays, Small Onego) than for the rest of the Lake Onego water area. The maximum values of the Hg content in the upper part of the bottom sediments, were established in the Kondopoga Bay (0.089 µg/g) (Figure 2).

As already mentioned above, throughout the entire water area, the mercury content in the lower horizons of the bottom sediments has lower concentrations than in the upper ones (Figure 6). The lowest values in the bottom sediments lower part are observed in Unitskaya Bay—0.016 µg/g, with the maximum values (0.058 µg/g) in the area of Big Onego.

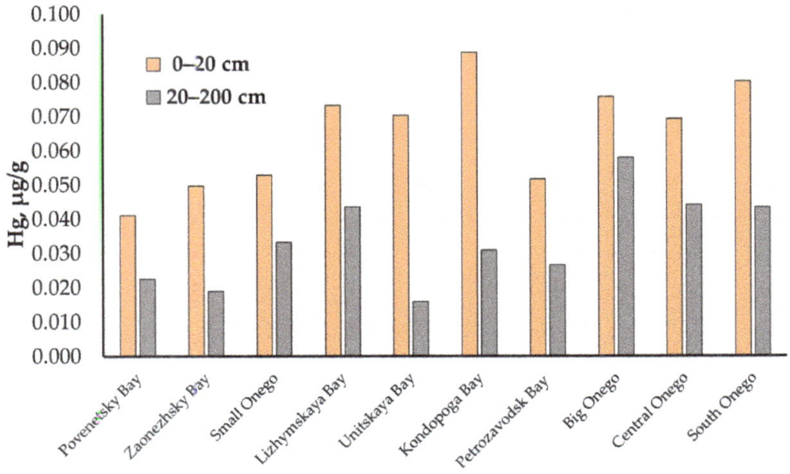

Figure 6. Mercury content in the bottom sediments of the different areas of Lake Onego.

Higher concentrations of mercury in the upper part of the bottom sediments may indicate either an increase in the flow of mercury intake, or a redistribution of mercury contents in the bottom sediments during diagenesis (more details in the "Discussion").

3.3. Results of the Cluster Analysis

The correlation between the chemical elements and their groups, calculated for the sedimentation trap material, water, and suspended material of Lake Onego, as well as for the bottom sediments divided into three groups, according to the depth and belonging to the first or second type of stratification (presence or absence of an oxidizing-reducing geochemical barrier in the bottom sediment), is shown on the dendrogram of the R-type cluster analysis (Figure 7).

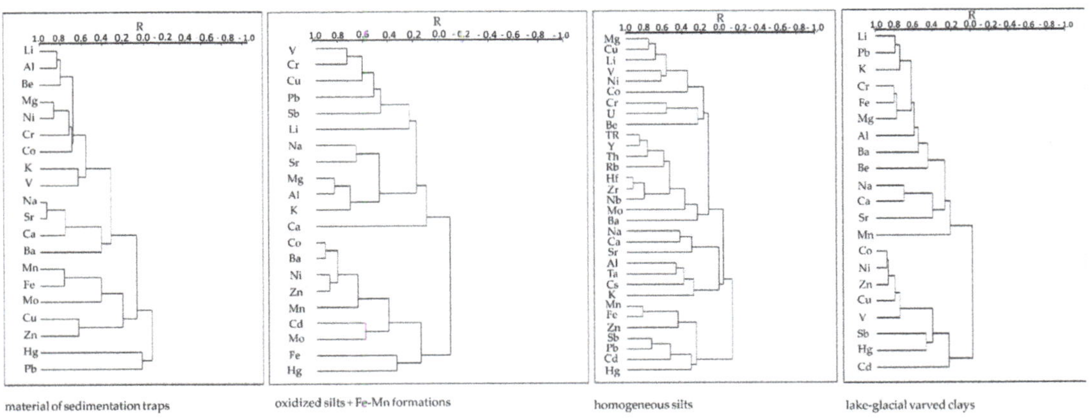

Figure 7. Dendrogram of the R-type cluster analysis calculated, based on the chemical element concentrations in the material of the sedimentation traps and the bottom sediments.

The separation of the chemical element groups is determined by the presence of several main mineral phases, which act as their concentrators. Such mineral phases for the upper part of the bottom sediments (which include oxidized silts + Fe-Mn formations) are: quartz, feldspar, layered silicates (chlorite, illite, mica) and Fe and Mn hydroxides [34]. In our case, we determined that the lithophilic elements form their own group, the isolation of which is due to the entry of these elements into silicate minerals, and a group of minerals closely related to Fe and Mn hydroxides. It is important to note that mercury, for this part of the section, is included in the group of elements associated with Fe and Mn.

For the homogeneous silts below, the main mineral phases are the same, but their quantitative ratio is different. In the homogeneous silts, Mn hydroxides are present in smaller quantities, while Fe hydroxides are completely absent, because they are transformed into phosphates and carbonates of Fe in the diagenesis under reducing conditions [30]. The different ratio of the main mineral phases leads to various combinations of chemical elements by groups (Figure 7). It is important to note that mercury stands apart and, apparently, is not associated with certain mineral phases.

For the lower part of the bottom sediments, which are represented by lake-glacial varved clays, a different picture is observed. Unlike the upper part of the section, chalcophilic elements together with Hg form an independent group, and Fe and Mn form a group with lithophilic elements.

As for the bottom sediments, for the material of the sedimentation traps, the combination of elements into groups is determined by the presence of several main mineral phases. From the point of view of the material composition, the bottom sediments and the material of sedimentation traps are identical, with the exception of the cationic composition of illite and chlorite. As part of the bottom sediments, the main part of illite and chlorite is represented by ferruginous varieties that were formed directly in the lake [25,30]. On the dendrogram of the R-type cluster analysis for the sedimentation trap material, as well as for the bottom sediments, lithophilic elements form their own groups, the isolation of which is due to the occurrence of these elements in the silicate minerals. Fe and Mn form a separate group, due to the presence of Fe-Mn hydroxides in the sedimentation trap material. Mercury is not included in any of the groups and stands alone, as well as Pb.

4. Discussion

In the works of the predecessors, detections of mercury concentrations exceeding the maximum permissible concentration were reported [27]. For example, the article [34] provides data on the mercury concentration in the muscles of fish consumed by residents of the cities of Petrozavodsk and Medvezhegorsk. The maximum values of mercury concentration in the muscles of perch, walleye and pike caught in Lake Onego, either exceed the values of the maximum permissible concentration (0.5 μg/g), or are close to these values.

In [35], it is stated that mercury concentrations exceeding the maximum permissible concentration (0.01 μg/L) were detected in 2013, in the waters of the Shuya, Andoma, Svir (Svirstroy), Polist, Veronda, Volkhov, Morye rivers, as well as in the water of Lake Onego near the settlements of Kondopoga, Medvezhegorsk and Povenets (up to 0.048 μg/L). In 2014, the mercury content up to 0.060 μg/L at different observation periods was noted in Kondopoga Bay, near the village of Peschanoe, in the rivers of Vytegra, Pola, Lovat, Psizha, Perehoda, Shelon, Veronda, Volkhov, Vuoksa, Tulema, Vidlitsa, Oyat, Pasha, Syas, Lava and Lake Ilmen. At the same time, mercury was found in higher concentrations—0.14–0.27 μg/L in the Lososinka River, Ilmen tributaries - Nisha, Polisti, Psizhe, Peregod, Sheloni, Veronda, Veryazh as well as in Sviritsa and Nazia.

According to [20] the main inflow of lithophilic elements and heavy metals into Lake Onego is associated with river runoff (68%–97% of the total intake of each element in the region). These data fully correspond to the materials published by [36]. However, the mercury content was not studied in these works.

However, Gorbunov et al. [34] show that the levels of mercury accumulation in soil and vegetation in the cities of Karelia do not exceed the standards approved in Russia and are close to the background.

As can be seen from the publications mentioned above, on the territory of Lake Onego, the increased concentrations of mercury are observed only in water and related environmental objects.

The atmospheric precipitation is the main source of mercury for the regions where there are no outcrops of mercury-containing rocks. Hydrology and mobility of soluble organic compounds are the main controlling factors of the Hg delivery to such aquatic ecosystems. The prevalence of mercury in suspended form in some areas indicates that there is at least one other source of mercury-anthropogenic activity. Most enterprises discharge wastewater containing trace elements, including mercury, directly into the water [37]. Chlorine-alkaline plants and pulp and paper mills were large industrial sources discharging mercury with wastewater into reservoirs [38,39]. This is consistent with the location of industrial plants on the shore of the Kondopoga and Povenetsky Bays, which may be the source of such mercury-containing particles.

The source of mercury can be the electrolysis process in the production of chlorine and caustic soda. Such production is geographically always located next to consumers, among which are manufacturers of pulp and paper products. The most dangerous production from the point of view of the impact on the Lake Onego ecosystem is the Kondopoga Pulp and Paper Mill. Over the 90 years of its existence, since 1929, the volumes (from 2.6 in 1948 to 64.5 million m^3 in 1990) and the qualitative composition of the wastewater entering Kondopoga Bay have changed, in accordance with the technologies used. Currently, wastewater discharged by the Kondopoga Pulp and Paper Mill (~40 million m^3 per year) carries a significant amount of pollutants into the reservoir. Therefore, on average, ~2500 tons of suspended solids enter the bay during the year: 50 tons of phosphorus, 60 tons of nitrogen, etc. [20]. It is in the bottom sediments of Kondopoga Bay that high concentrations of mercury in the bottom sediments (0.19 μg/g) were detected, which is an order of value higher than in other parts of the lake. High concentrations of mercury in the sedimentation traps material in almost all bays, for example, Povenetsky Bay (0.76 μg/g), and especially in Lizhemskaya Bay (4.3 μg/g), may indicate the wind dispersion of the polluted dust. Similar concentration ranges are typical for sedimentary material/bottom sediments bearing traces of anthropogenic pollution [40–42]. Dust may come from the sump pond (which may dry out) used by The Kondopoga Pulp and Paper Mill.

It was observed that in the entire water area of Lake Onego, the mercury concentrations in the sedimentation traps material are higher than the mercury content in the corresponding cores of the bottom sediments. The difference in mercury concentrations in the material of the sedimentation traps at different times is presumably due to single anthropogenic releases of mercury into the aquatic ecosystem of the lake.

The material composition of the sedimentation traps material is similar to the upper part of the bottom sediments, except for the presence of the same illites and chlorites for the bottom sediments and for much more organic remains in the sedimentation traps material. The second fact may also be associated with the higher mercury concentrations in the material of the sedimentation traps, since mercury has a high bioaccumulative ability.

According to the vertical distribution graphs of the mercury contents in the bottom sediments, it can be concluded that in the Late Pleistocene and Holocene, the flow of mercury into the bottom sediments was constant. Only in modern times (about 200 years) there has been an increase in mercury concentrations in the bottom sediments. At the same time, the highest concentrations are observed in the oxidized part of the bottom sediments. According to radiometric dating using nonequilibrium ^{210}Pb [30], an increase in mercury concentrations is observed in the bottom sediments with an age up to 150 years ago, while a sharp increase has been observed since 80 years ago.

The close relationship of mercury with Fe and Mn, which is established for the upper part of the bottom sediments, is explained by the sorption of Hg by Fe and Mn hydroxides. Fe and Mn hydroxides play an important role in the circulation and transport of trace elements, due to their large surface area and high ability to sorb and co-deposit metals [43]. An increase in the intake of Fe and Mn hydroxides in the Holocene, caused by climate humidification may be the reason for an increase in the intake of mercury together with Fe and Mn hydroxides. The source of mercury in this case may be the shungite rocks, for which the predominance of physical weathering in the Late Pleistocene was replaced by chemical weathering in the Holocene. Atmospheric precipitation is also a source of mercury. The increase in mercury intake from the atmosphere is associated with a global increase in the mercury intake into the environment caused by human industrial activity [44].

Another explanation for the close connection of Fe and Mn can be the mechanism by which Fe-Mn crusts (layers) are formed. Such crusts are formed, according to the classical scheme, referred to in the literature as the "manganese trap" [45]. In post-sedimentation processes, organic matter reduces Fe and Mn. As a result of this process, soluble divalent ions of these metals are formed. These ions enter the pore water, and then migrate up the section to the previously formed oxidized layer, in which free O_2 has not yet been used up for the oxidation of the organic matter. Here, the ions are oxidized, passing into insoluble forms and forming layers with increased concentrations of Fe and Mn. Apparently, at the time of the reduction of Fe and Mn and the dissolution of their hydroxides, Hg goes into pore solutions. Mercury migrates up the section together with Fe and Mn, where it is re-sorbed by newly formed Fe-Mn hydroxides.

5. Conclusions

For the first time concerning Lake Onego, the second largest freshwater lake in Europe, the lateral distribution of the total mercury in the water-suspended matter-bottom sediments system is considered. The analysis of the total mercury content distribution was carried out by taking into account two types of bottom sediment cores (determined by the presence or absence of an oxidation-reduction barrier) and various areas of the Lake Onego water area.

The study of the lateral mercury distribution in the bottom sediments over the entire Lake Onego area, showed that the mercury content in the bottom sediments does not significantly differ and varies in the range of 0.041–0.089 μg/g, with an average value in the upper and lower parts of the bottom sediments—0.067 ± 0.003 and 0.041 ± 0.001 μg/g, respectively.

An increase in mercury content in the bottom sediments vertically to the water-bottom sediment boundary has been established, which is mainly due to two factors: an increase in the global background of mercury associated with anthropogenic activity; migration, redistribution and concentration of mercury in the bottom sediments on the geochemical barrier, jointly with Fe and Mn.

The total mercury content in the water of Lake Onego is on average 0.32 ± 0.07 μg/L. The predominant form of mercury in water is dissolved+colloid, with the exception of the water samples from the Kondopoga and Povenetsky Bays.

It was found that the mercury content in the sedimentation traps material varies in a wide range from 0.062 μg/g to 4.37 μg/g, which is significantly higher than the mercury content in the upper part of the corresponding bottom sediments cores. This difference may be due to the atmospheric input of contaminated dust from paper mills.

Author Contributions: Conceptualization, V.M., V.S. and D.S.; methodology, V.S.; software, V.M.; validation, N.B., V.S. and D.S.; formal analysis, V.M. and V.S.; investigation, V.M., V.S., D.S., E.O. and N.B.; resources, V.M., V.S., D.S., E.O. and N.B.; data curation, V.S., E.O. and D.S.; writing—original draft preparation, V.M. and V.S.; writing—review and editing, V.S., E.O. and D.S.; visualization, V.M.; supervision, V.S. and D.S.; project administration V.S. and D.S.; funding acquisition, V.S., N.B. and D.S. All authors have read and agreed to the published version of the manuscript.

Funding: The main part of the work, concerned with field research, sample preparation for the chemical analytical work and results interpretation, was financially supported by a grant from the Russian Science Foundation No. 18-17-00176; studies and analytical work and results interpretation were carried out with the financial support from the Russian Foundation for Basic Research, research project No. 19-05-50014.

Conflicts of Interest: The authors declare no conflict of interest.

References

1. Raj, D.; Maiti, S.K. Sources, toxicity, and remediation of mercury: An essence review. *Environ. Monit. Assess.* **2019**, *191*, 1–22. [CrossRef] [PubMed]
2. Selin, N. Global Biogeochemical Cycling of Mercury: A Review. *Annu. Rev. Environ. Resour.* **2009**, *34*, 43–63. [CrossRef]
3. Hsu-Kim, H.; Kucharzyk, K.H.; Zhang, T.; Deshusses, M.A. Mechanisms Regulating Mercury Bioavailability for Methylating Microorganisms in the Aquatic Environment: A Critical Review. *Environ. Sci. Technol.* **2013**, *47*, 2441–2456. [CrossRef] [PubMed]
4. Burton, G.A.; Landrum, P.F. Toxicity of sediments. In *Encyclopedia of Sediments and Sedimentary Rocks*; Kluwer Academic Publishers: Amsterdam, The Netherlands, 2003; pp. 748–751.
5. Klapstein, S.; Nelson, J. Methylmercury biogeochemistry in freshwater ecosystems: A review focusing on DOM and photodemethylation. *Bull. Environ. Contam. Toxicol.* **2018**, *100*, 14–25. [CrossRef] [PubMed]
6. Walters, D.; Jardine, T.; Cade, B.; Kidd, K.; Muir, D.; Leipzig-Scott, P. Trophic Magnification of Organic Chemicals: A Global Synthesis. *Environ. Sci. Technol.* **2016**, *50*, 4650–4658 [CrossRef] [PubMed]
7. Wolfe, M.; Schwarzbach, S.; Sulaiman, R. Effects of mercury on wildlife: A comprehensive review. *Environ. Toxicol. Chem.* **1998**, *17*, 146–160. [CrossRef]
8. Mergler, D.; Anderson, H.; Chan, L.; Mahaffey, K.; Murray, M.; Sakamoto, M.; Stern, A. Methylmercury Exposure and Health Effects in Humans: A Worldwide Concern. *Ambio* **2007**, *36*, 3–11. [CrossRef]
9. Kocman, D.; Wilson, S.; Amos, H.; Telmer, K.; Steenhuisen, F.; Sunderland, E.; Mason, R.; Outridge, P.; Horvat, M. Toward an Assessment of the Global Inventory of Present-Day Mercury Releases to Freshwater Environments. *Int. J. Environ. Res. Public Health* **2017**, *14*, 138. [CrossRef]
10. Driscoll, C.T.; Mason, R.P.; Chan, H.M.; Jacob, D.J. Pirrone, N. Mercury as a global pollutant: Sources, pathways, and effects. *Environ. Sci. Technol.* **2013**, *47*, 4967–4983. [CrossRef]
11. Obrist, D.; Kirk, J.; Zhang, L.; Sunderland, E.; Jiskra, M.; Selin, N. A review of global environmental mercury processes in response to human and natural perturbations: Changes of emissions, climate, and land use. *Ambio* **2018**, *47*, 116–140. [CrossRef]
12. Streets, D.G.; Devane, M.K.; Lu, Z.; Bond, T.C.; Sunderland, E.M.; Jacob, D.J. All-time releases of mercury to the atmosphere from human activities. *Environ. Sci. Technol.* **2011**, *45*, 10485–10491. [CrossRef]
13. Amos, H.M.; Jacob, D.J.; Streets, D.G.; Sunderland, E.M. Legacy impacts of all-time anthropogenic emissions on the global mercury cycle. *Glob. Biogeochem. Cycles* **2013**, *27*, 410–421. [CrossRef]
14. Horowitz, H.M.; Jacob, D.J.; Amos, H.M.; Streets, D G.; Sunderland, E.M. Historical mercury releases from commercial products: Global environmental implications. *Environ. Sci. Technol.* **2014**, *48*, 10242–10250. [CrossRef] [PubMed]
15. Lamborg, C.H.; Hammerschmidt, C.R.; Bowman, K.L.; Swarr, G.J.; Munson, K.M.; Ohnemus, D.C.; Lam, P.J.; Heimbürger, L.-E.; Rijkenberg, M.J.A.; Saito, M.A. A global ocean inventory of anthropogenic mercury based on water column measurements. *Nature* **2014**, *512*, 65–68. [CrossRef] [PubMed]
16. Zhang, Y.; Jaeglé, L.; Thompson, L. Natural biogeochemical cycle of mercury in a global three-dimensional ocean tracer model. *Glob. Biogeochem. Cycles* **2014**, *28*, 553–570. [CrossRef]
17. Amos, H.M.; Sonke, J.E.; Obrist, D.; Robins, N.; Hagan, N.; Horowitz, H.M.; Mason, R.P.; Witt, M.; Hedgecock, I.M.; Corbitt, E.S.; et al. Observational and modeling constraints on global anthropogenic enrichment of mercury. *Environ. Sci. Technol.* **2015**, *49*, 4036–4047. [CrossRef] [PubMed]
18. Streets, D.; Horowitz, H.; Jacob, D.; Lu, Z.; Levin, L. ter Schure, A.; Sunderland, E. Total Mercury Released to the Environment by Human Activities. *Environ. Sci. Technol.* **2017**, *51*, 5969–5977. [CrossRef]
19. Filatov, N. *Onego Lake. Atlas*; KarNTS RAN: Petrozavodsk, Russia, 2010; p. 151.
20. Kulik, N.; Efremenko, N.; Belkina, N.; Strakhovenko, V.; Gatalskaya, E.; Orlov, A. Fe, Mn, Al, Cu, Zn, and Cr in the sedimentary matter of Lake Onego. *Quat. Int.* **2022**. [CrossRef]
21. Sokolov, V.; Kalinin, Y. *Shungites of Karelia and Ways of their Complex Use*; KarNTS RAN: Petrozavodsk, Russia, 1975; p. 240.
22. Filippov, M. *Shungitonosnye Porody Onezhskoi Strucury*; Fillipov, M.M., Ed.; KarNTS RAN: Petrozavodsk, Russia, 2002; p. 280.
23. Taylor, S. Abundance of chemical elements in the continental crust: A new table. *Geochim. Cosmochim. Acta* **1964**, *28*, 1273–1285. [CrossRef]
24. Lukashin, V.; Klyuvitkin, A.; Lisitzin, A.; Novigatsky, A. The MSL-110 small sediment trap. *Oceanology* **2011**, *51*, 699–703. [CrossRef]
25. Strakhovenko, V.; Belkina, N.; Efremenko, N.; Potakhin, M.; Subetto, D.; Frolova, L.; Nigamatzyanova, G.; Ludikova, A.; Ovdina, E. The First Data on the Mineralogy and Geochemistry of the Suspension of Lake Onego. *Russ. Geol. Geophys.* **2022**, *63*, 55–71. [CrossRef]
26. Davis, J.C.; Sampson, R.J. *Statistics and Data Analysis in Geology*, 3rd ed.; Wiley: New York, NY, USA, 1986; p. 646.
27. Feng, C.; Wu, F.; Zhao, X.; Li, H.; Chang, H. Water quality criteria research and progress. *Sci. China Earth Sci.* **2012**, *55*, 882–891. [CrossRef]
28. Efremova, T.; Sabylina, A.; Lozovik, P.; Slaveykova, V.; Zobkova, M.; Pasche, N. Seasonal and spatial variation in hydrochemical parameters of Lake Onego (Russia): Insights from 2016 field monitoring. *Inland Waters* **2019**, *9*, 227–238. [CrossRef]
29. Gorbunov, A.; Lyapunov, S.; Ermolaev, B.; Okina, O.; Frontas'eva, M.; Pavlov, S. Macro- and microelements distribution in urbanized ecosystems of karelia. *Hum. Ecol.* **2020**, *27*, 4–14. [CrossRef]

30. Subetto, D.; Rybalko, A.; Strakhovenko, V.; Belkina, N.; Tokarev, M.; Potakhin, M.; Aleshin, M.; Belyaev, P.; Dubois, N.; Kuznetzov, V.; et al. Structure of Late Pleistocene and Holocene Sediments in the Petrozavodsk Bay, Lake Onego (NW Russia). *Minerals* **2020**, *10*, 964. [CrossRef]
31. Strakhovenko, V.; Subetto, D.; Ovdina, E.; Danilenko, I.; Belkina, N.; Efremenko, N.; Maslov, A. Mineralogical and geochemical composition of Late Holocene bottom sediments of Lake Onego. *J. Great Lakes Res.* **2020**, *46*, 443–455. [CrossRef]
32. Strakhovenko, V.; Subetto, D.; Ovdina, E.; Belkina, N.; Efremenko, N. Distribution of Elements in Iron-Manganese Formations in Bottom Sediments of Lake Onego (NW Russia) and Small Lakes (Shotozero and Surgubskoe) of Adjacent Territories. *Minerals* **2020**, *10*, 440. [CrossRef]
33. Strakhovenko, V.; Belkina, N.; Subetto, D.; Rybalko, A.; Efremenko, N.; Kulik, N.; Potakhin, M.; Zobkov, M.; Ovdina, E.; Ludikova, A. Distribution of rare earth elements and yttrium in water, suspended matter and bottom sediments in Lake Onego: Evidence of the watershed transformation in the Late Pleistocene. *Quat. Int.* **2021**. [CrossRef]
34. Gorbunov, A.; Lyapunov, S.; Ermolaev, B. Distribution of mercury in natural and urban environments of karelia, northwest russia. *Hum. Ecol.* **2019**, *26*, 10–17. [CrossRef]
35. Ignatieva, N.; Petrova, T.; Guseva, M. Estimation of pollution of surface waters on the territory of the drainage basin of Lake Ladoga according to hydro-chemical indicators. *Proc. Samara Sci. Cent. Russ. Acad. Sci.* **2015**, *17*, 91–96.
36. Sanin, A.; Strokov, A.; Terskii, P. Assessment of natural processes impact on content of heavy metals in water of Lake Onego. Vestnik of Saint Petersburg University. *Earth Sci.* **2020**, *65*, 146–171. [CrossRef]
37. Nriagu, J.; Pacyna, J. Quantitative assessment of worldwide contamination of air, water and soils by trace metals. *Nature* **1988**, *333*, 134–139. [CrossRef] [PubMed]
38. Clarkson, T. The three modern faces of mercury. *Environ. Health Perspect.* **2002**, *110*, 11–23. [CrossRef] [PubMed]
39. Hylander, L.; Meili, M. The Rise and Fall of Mercury: Converting a Resource to Refuse After 500 Years of Mining and Pollution. *Crit. Rev. Environ. Sci. Technol.* **2005**, *35*, 1–36. [CrossRef]
40. Wu, T. Distribution of methylmercury in a mercury-contaminated ecosystem. *Pract. Period. Hazard. Toxic Radioact. Waste Manag.* **2006**, *10*, 192–197. [CrossRef]
41. Yan, H.; Li, Q.; Meng, B.; Wang, C.; Feng, X.; He, T.; Dominik, J. Spatial distribution and methylation of mercury in a eutrophic reservoir heav-ily contaminated by mercury in Southwest China. *Appl. Geochem.* **2013**, *33*, 182–190. [CrossRef]
42. Wang, Y.; Fang, M.; Chien, L.; Lin, C.; Hsi, H. Distribution of mercury and methylmercury in surface water and surface sediment of river, irrigation canal, reservoir, and wetland in Taiwan. *Environ. Sci. Pollut. Res.* **2019**, *26*, 17762–17773. [CrossRef]
43. Nelson, Y.; Lion, L.; Shuler, M.; Ghiorse, W. Effect of Oxide Formation Mechanisms on Lead Adsorption by Biogenic Manganese (Hydr)oxides, Iron (Hydr)oxides, and Their Mixtures. Environ. *Sci. Technol.* **2002**, *36*, 421–425. [CrossRef]
44. Beal, S.; Osterberg, E.; Zdanowicz, C.; Fisher, D. Ice Core Perspective on Mercury Pollution during the Past 600 Years. *Environ. Sci. Technol.* **2015**, *49*, 7641–7647. [CrossRef]
45. Yudovich, Y. *Geochemistry of Manganese*; IG Komi NTs URO RAN: Syktyvkar, Russia, 2014; p. 540.

minerals

MDPI

Article

Lipid Biomarker and Stable Isotopic Profiles through Late Carboniferous–Early Triassic of the Deepest Well MS-1 in the Junggar Basin, Northwest China

Shuncun Zhang [1,3], Tao Wang [1,3], Hui Guo [1,3], Shengyin Zhang [1,3] and Bo Chen [2,*]

1 Northwest Institute of Eco-Environment and Resources, Chinese Academy of Sciences, Lanzhou 730000, China
2 Guangxi Key Laboratory of Green Chemical Materials and Safety Technology, Beibu Gulf University, Qinzhou 515000, China
3 Gansu Key Laboratory of Petroleum Resources, Lanzhou 730000, China
* Correspondence: cbo-11@163.com

Abstract: The Carboniferous–Triassic period was an important stage of global sea–land transformation, with coal formation in the Carboniferous, biological extinction at the end of the Permian, and global drought in the Triassic. The MS-1 well in the Mosuowan High of the Junggar Basin is the deepest well drilled in Northwestern China. In this paper, we investigate the sedimentary environment and climate evolution of the Mosuowan area in the central Junggar Basin during the Late Carboniferous–Early Permian by the petrothermal, lipid biomarker, and isotopic composition of mud shale core samples, and explore the tectonic–climatic events and Central Asian orogenic belt evolution driving the sedimentary environment. The study shows that the organic matter from the Upper Carboniferous to the Lower Permian is at a mature stage, but biomarkers maintained the primary information although the organic matter was subjected to thermal evolution. In the late Carboniferous period (Tamugan Formation), the study area was a closed remnant sea with a relatively humid climate, triggering lush terrestrial vegetation and high organic carbon content in the sediments, which had the potential to evolve into natural gas. During the Xiazijie Formation of the Middle Permian, tectonic activity shifted to the subsidence period, and the salinity of the water decreased after a large input of fresh water. The lake basin area expanded, and the content of aquatic organisms continued to increase. As the Lower Permian stratigraphy is missing, this sea–land transition seems to jump. The low and upper Urho Formations of the Middle–Upper Permian are a deltaic foreland deposit, and geochemical indicators show an overall lake retreat process with a continuous increase in organic matter content of terrestrial origin. The lithologic assemblage of the Triassic Baikouquan Formation is braided river deltaic sedimentation with migration of deposition centers of the lake basin. In conclusion, the Late Carboniferous–Early Permian period was influenced by global changes, Paleo-Asian Ocean subduction, and continental splicing, which resulted in a continuous increase in terrestrial organic matter, water desalination, and oxidation-rich sediments in the Mosuowan region, but the P–T biological mass extinction event was not recorded.

Keywords: geochemistry; environment; tectonic evolution; Carboniferous–Triassic; Junggar Basin

Citation: Zhang, S.; Wang, T.; Guo, H.; Zhang, S.; Chen, B. Lipid Biomarker and Stable Isotopic Profiles through Late Carboniferous–Early Triassic of the Deepest Well MS-1 in the Junggar Basin, Northwest China. *Minerals* **2022**, *12*, 1299. https://doi.org/10.3390/min12101299

Academic Editors: Marianna Kulkova and Dmitry Subetto

Received: 14 September 2022
Accepted: 14 October 2022
Published: 15 October 2022

Publisher's Note: MDPI stays neutral with regard to jurisdictional claims in published maps and institutional affiliations.

1. Introduction

Between the Carboniferous and Triassic periods, north China experienced global events such as coal-forming movements, paleo-oceanic closures and land emergence, and the Permian–Triassic biological mass extinction [1,2]. The Junggar Basin is located at the southwest edge of the Central Asian orogenic belt, where it is the key part of the intersection of the Siberian plate, the Kazakhstan plate, and the Tarim plate. The Carboniferous of Junggar was characterized by the subduction of the ancient Asian Ocean and regional volcanic activity. The Paleozoic–Mesozoic period experienced collisional basin formation, faulting, subsidence, and regenerated foreland basin stages. Meanwhile, the Paleozoic–Mesozoic was also an important stage for hydrocarbon source rock and reservoir development in the Junggar Basin, and several Permian–Triassic hydrocarbon-bearing sags are located in the central part of the basin, such as Mahu Sag, Jimsar Sag, Shawan Sag, etc. [3–5]. At present, sand and conglomerate, dense sandstone, and oil shale of the Permian–Triassic in the northwest margin of the Junggar Basin have become an important, prospective oil and gas reservoir [6,7].

The sedimentation evolution of the Junggar Basin was not only influenced by global tectonic–climatic events but also deeply constrained by the accretionary and extension of the Central Asian orogenic belt. The evolution of the Junggar Basin from the Paleozoic to the Mesozoic was closely related to the subduction and subsidence of the Paleo-Asian Ocean in the northern Xinjiang region. During the Carboniferous period, the Junggar Basin developed several ophiolite belts and regional volcanic clastic rocks. The central part of the basin mainly experienced mechanical subsidence in the Early Permian, thermal subsidence with post-fracture in the Middle Permian, and tectonic inversion during the Late Permian–Early Triassic [8–10]. Paleozoic calc-alkaline volcanic rocks on the perimeter of the basin are the main source of sediments, and the braided, river-fan, deltaic, shallow lake phase sediments were developed from the mountain front region to the lake. The paleontological data show that the northern part of the Junggar Basin experienced four important paleoclimatic events from the Carboniferous to the Triassic period: the Late Carboniferous cooling event, the Middle Permian greenhouse event, the Late Permian–Early Triassic drought event, and the Middle to Late Triassic humid event [11–14].

A large number of geochemical studies have been carried out on volcanic and sedimentary rocks in the Junggar Basin, mainly focusing on tectonic environment discrimination, hydrocarbon source rock evaluation, and sedimentary environment reconstruction in the Carboniferous and Permian systems [15–17]. Geophysical data and sedimentary rock records from the drilled strata have also been used to map the depositional environment [18–20]. However, there is still a lack of regional, continuous, geochemical data for reconstructing the environmental changes from the Carboniferous to Triassic period. In this study, core samples from the deepest well, MS-1, in the Junggar Basin were analyzed for pyrolysis, lipid biomarkers, and carbon isotopic compositions. We studied the composition of the sedimentary organic matter, depositional environment, and climatic evolution of Mosuowan High in the central Junggar Basin during the Carboniferous–Triassic period. The global, tectonic–climatic events and the evolution of the Central Asian orogenic belt which drove the depositional environment of the Junggar Basin are further discussed.

2. Geological Setting

The Junggar Basin is located between the Altai and Tian Shan mountains in Xinjiang, China (92°00′–95°30′ E, 43°20′–45°00′ N) (Figure 1). It is a Late Carboniferous–Quaternary sedimentary basin with a total area of about 130,000 km^2. The Junggar Basin is one of the largest, superimposed, hydrocarbon-bearing basins in northwest China, and its geotectonic position is located at the intersection of three major plates: the Siberian Plate, the Kazakhstan Plate, and the Tarim Plate [21,22]. The formation and evolution of the Junggar Basin were closely related to the subduction and subsidence of the ancient Asian Ocean in the northern Xinjiang region [23–25]. Numerous ophiolite belts are exposed in the West Junggar region, including many ophiolite belts such as Dalabut, Baijiantan, Barrek, Tangbal, and Mayil, which are key evidence of the extinction of the multi-island arc oceanic crust suture [22,26–28]. Based on the Permian tectonic features and late tectonic transformation in the basin, the Junggar Basin is divided into six primary tectonic units, including the Western Uplift, the Eastern Uplift, the Luliang Uplift, the Tianshan Foreland Depression, the Central Depression, and the Ulungu Depression (Figure 1) [29,30].

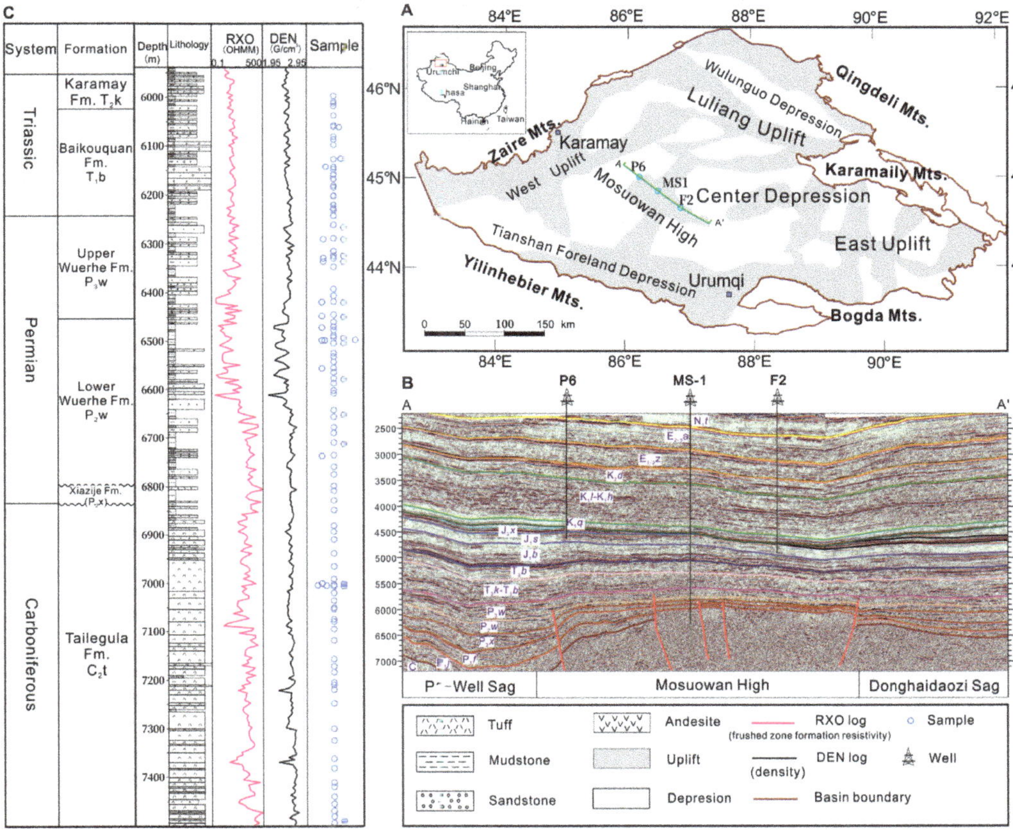

Figure 1. (**A**) Map of sampling location in Junggar Basin, China. (**B**) Seismic profile of well P6, MS-1, F2. (**C**) Detailed stratigraphic column of the sampled interval.

The MS-1 well is located in the Mosuowan Uplift in the Junggar Basin Central Depression, which is a typical basement uplift. The two flanks in the middle of and to the east of the Mosuowan Uplift have developed northwest-trending, geodesic, spreading reverse faults (Figure 1), which have opposite tendencies and make the Mosuowan Uplift have the morphology of a back-breaking block [31]. The stratigraphy drilled in the Carboniferous–Triassic MS-1 well is the Baikouquan Formation (T_1b), Permian Upper Urho Formation (P_3w), Lower Urho Formation (P_2w), Xiazijie Formation (P_2x), and Carboniferous Tamugang (C_2t) [32]. In the Pre-Carboniferous–Carboniferous period, the central and western part of the Junggar Basin showed an island arc uplift area with calc-alkaline volcanic eruptions. At the beginning of the Permian, it was an island arc uplift zone with a northwest–west trend. With the intensification of denudation and sea level change, the island arc uplift was finally covered during the deposition of the Lower Permian Fengcheng Formation (P_1f). At the end of deposition of the P_2x, fold uplift occurred under the influence of north–south extrusion, and the P_2x was locally erased, and the magnitude of the Mosuowan Uplift increased. The P_2w–P_3w were buried again during the depositional period. The Triassic was the period of stable deposition of the Mosuowan High [33–35].

3. Sample and Experiment

3.1. Sample Collection

In this study, 73 samples were collected from the MS-1 well, including 32 samples from the C_2t, 9 samples from the P_2x, 16 samples from the P_2w, 7 samples from the P_3w, and 9 samples from the T_1b. All of the samples were prepared for the determination of total organic carbon and rock pyrolysis parameters. A total of 64 samples were prepared for gas chromatography mass spectrometry (GC-MS) analysis, and 59 samples were selected for chloroform asphalt "A" stable carbon isotope analysis (Tables S1 and S2).

3.2. Experimental Analysis

Extraction and analysis of saturated hydrocarbons: Powder samples were placed in a Soxhlet extractor and extracted with chloroform for 72 h. The extract was concentrated and purified by passing through a silica gel/alumina (1:1) chromatographic column, and the saturated hydrocarbon fractions were collected by washing with 20 mL of n-hexane four times. The saturated hydrocarbon fractions were run in full scan and selected ion monitoring (SIM) mode by a GC-MS Thermal Scientific DSQ II quadrupole mass spectrometer equipped with a 30 m × 0.25 mm × 0.25 μm HP-5 MS column with helium as the carrier gas. The oven temperature was initially set to increase from 80 °C to 320 °C at a rate of 10 °C/ min and held at 320 °C for 20 min. The oven was held at 35 °C for 1 min, then the temperature was increased from 35 °C to 120 °C at a rate of 10 °C/min, then from 120 °C to 300 °C at a rate of 3 °C/min, and, finally, held at 300 °C for 30 min. Detection of the biomarkers, hopanes and steranes, was carried out by monitoring the respective, typical ions: m/z 191 for hopanes, m/z 217 for 5α,14α,17α-steranes, and m/z 218 for 5α,14β,17β-steranes.

Total organic carbon (TOC) and rock pyrolysis: The samples (about 100 mg) were crushed, sieved through a 120 mesh sieve, pretreated with 1 mol/L HCL to remove carbonates from the samples, and then heated to 1200 °C in a pyrolysis oven to obtain the TOC. The sample pyrolysis data were collected using a Rock-Eval II instrument according to the theory of Peter et al. [36]. Where S_1 represents the number of liquid hydrocarbons per unit mass of raw oil rock detected at 300 °C, S_2 represents the yield of caseinate cracked per unit mass of raw oil rock detected at 300–600 °C by heating. If the sediment is still in the mature evolution stage in the buried state, the S_2 peak represents the existing potential of the rock to produce oil. S_1 + S_2 characterizes the sum of the number of hydrocarbons already produced and potentially able to be produced in the hydrocarbon source rock, which is called hydrocarbon generation potential, i.e., the total number of hydrocarbons that may be produced by the rock, and it gradually decreases with the depletion of the organic matter hydrocarbon generation potential and the hydrocarbon removal process. The parameter T_{max} (°C) is the Rock-Eval pyrolyzer temperature at which the hydrocarbon compounds are observed to reach their maximum degree of pyrolysis, and this parameter can also indicate the maturity of the hydrocarbon compounds.

Chloroform asphalt "A" isotopic analysis: The carbon isotopic composition of chloroform asphalt "A" was determined offline by MAT 253 gas stable isotope mass spectrometer. The results of the carbon isotope analysis are reported as δ sign relative to the Pee Dee Belemnite (PDB) standard, and the analytical precision of the method was better than ±0.2‰ with reproducibility better than 0.2‰.

4. Results and Discussion

4.1. TOC and Rock Pyrolysis Parameters

The TOC of the gray mudstone, tuff, and dark-gray tuff of C_2t was almost less than 0.50% for all, while the TOC of gray-black tuff and coal was relatively high, with ranges of 0.61%–4.64% and 3.26%–5.53%, respectively. The hydrocarbon potential (S_1 + S_2) of the samples in C_2t was 0.25–10.18 mg/g, and the maximum pyrolysis temperature (T_{max}) was 441–502 °C. The TOC of P_2x ranged from 0.05% to 1.16%, with S_1 + S_2 and T_{max} of 0.14–0.96 mg/g and 425–468 °C, respectively. The TOC of P_2w and P_3w mudstone ranged from 0.44% to 1.21%, and $S_1 + S_2$ from 0.29 to 3.41 mg/g, while the T_{max} values were 425–471 °C. The T_1b samples were divided into two groups, the brown-gray, mud-bearing siltstone and the gray mudstone. The TOC of the brown-gray siltstone of the T_1b was very low, 0.15%–0.17%, and the TOC of the gray mudstone section was 0.50%–0.77%, with $S_1 + S_2$ ranging from 1.26 to 2.83 mg/g and T_{max} values of 430–441 °C (Table S1, Figure 2). In general, only samples from the Late Carboniferous had high organic matter content, hydrocarbon generation potential, and thermal evolution degree. If considered on a regional scale, they can be used as a set of effective hydrocarbon source rocks.

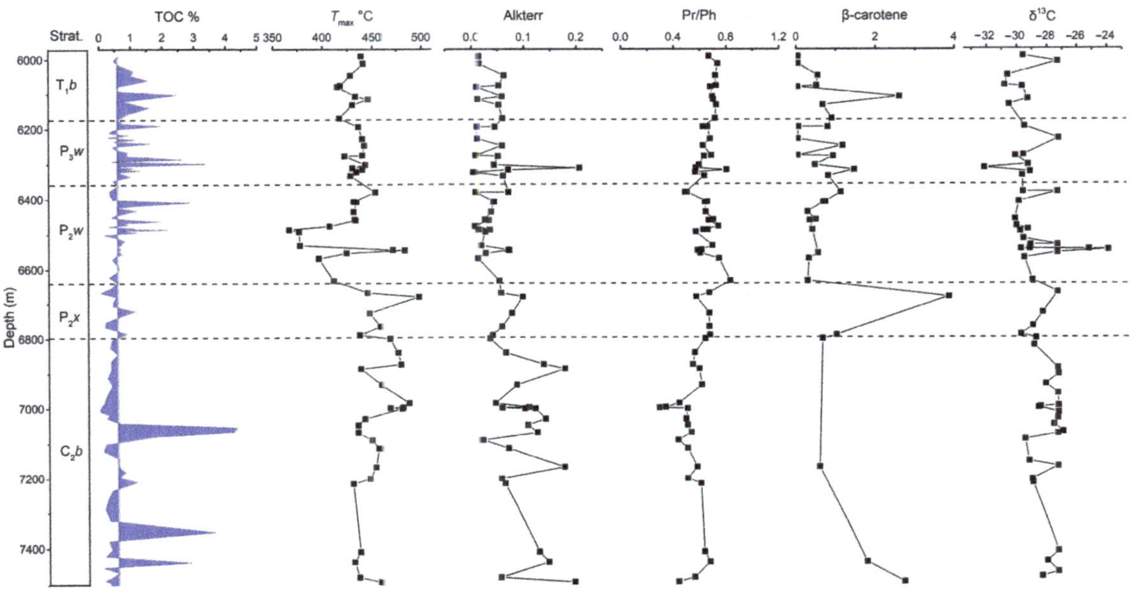

Figure 2. Profiles of rock pyrolysis and organic geochemical parameters in MS-1 well.

4.2. Molecular Geochemistry of Organic Matter

4.2.1. n-Alkanes, Pr/Ph, β-Carotene, and $\delta^{13}C$ of Chloroform Asphalts

The contents and variations of n-alkanes and acyclic isoprene hydrocarbons in the core samples from the C_2t, P_2x, P_2w, P_3w, and T_1b of the MS-1 well were analyzed by gas chromatography. Table 1 summarizes the parameters such as carbon dominance index (CPI_1 and CPI_2) [37], Pr, Ph, β-carotene, and $\delta^{13}C$ of the chloroform asphalt. The n-alkanes of the MS-1 well have single-peak and double-peak distribution patterns, in which the main carbon numbers of the front carbon peak are C_{17} and C_{19}, and the main carbon numbers of the back carbon peak are C_{23} and C_{27}. The n-alkanes of the C_2t are mainly bimodal, with the main carbon peak being C_{19} and C_{23}. The n-alkanes of the P_2x, P_2w, P_3w, and T_1b are mainly single-peaked, with the main carbon peaks of C_{17} and C_{19} (Figure 2). The LMW/HMW of samples in C_2t ranged from 0.4 to 2.36, with a mean value of 1.04. The LMW/HMW of the P_2x, P_2w, and P_3w samples ranged from 1.29 to 1.81, 1.15 to 11.64, and 0.37 to 11.43, respectively, and the LMW/HMW of T_1b ranged from 1.39 to 8.69 with a mean value of 1.04. The CPI_1 and CPI_2 of the C_2t were 0.95–1.31 and 1.04–2.0, respectively, while the CPI_1 and CPI_2 of the P_2x were 1.01–1.1 and 0.82–1.65, respectively, and those of the T_1b were 1.05–1.06 and 1.14–1.19, respectively. The ratio of terrestrial organic hydrocarbons (C_{27}, C_{29}, C_{31}, C_{33}) from higher vascular plant waxes and conifer-derived lipid detritus to the total sum of organic matter is known as the terrestrial index (Alkterr) [38–40], and the Alkterr values of Upper Carboniferous, Permian, and Lower Triassic samples were 0.02–0.19, 0.01–0.21, and 0.01–0.06, respectively.

Table 1. Parameters of δ^{13C} values of n-alkanes, Pr, Ph, β-carotene, and chloroform asphalts from MS-1 well of the Late Carboniferous–Early Triassic.

Strata		β-Carotene /$\Sigma C_{12\text{-}35}$	C_{max} [a]	CPI_1 [b]	CPI_2 [c]	Alkterr [d]	LMW/HMW [e]	Pr/Ph	$\delta^{13}C$ (PDB) [f]
T_1b	Min.	0.04	17	1.05	1.14	0.01	1.39	0.67	−30.81
	Max.	0.9	19	1.06	1.19	0.06	8.69	0.74	−29.28
	Mean	0.53		1.05	1.17	0.05	1.74	0.72	−30.085
P_3w	Min.	0.03	17	1.03	1.11	0	0.37	0.56	−32.17
	Max.	1.46	19	1.06	1.65	0.21	11.43	0.8	−29.15
	Mean	0.79		1.05	1.16	0.04	2.01	0.63	−29.58
P_2w	Min.	0.28	17	1.02	0.98	0.01	1.15	0.49	−30.14
	Max.	1.12	19	1.08	1.25	0.07	11.64	0.83	−23.93
	Mean	0.445		1.07	1.16	0.03	2.29	0.65	−29.3
P_2x	Min.	1	17	1.01	0.82	0.04	1.29	0.57	−29.76
	Max.	3.85	19	1.1	1.21	0.1	1.81	0.67	−28.32
	Mean	2.425		1.05	1.09	0.06	1.57	0.67	−28.705
C_2t	Min.	0.56	17	0.95	1.04	0.02	0.4	0.29	−29.52
	Max.	2.7	27	1.33	2	0.19	2.36	0.67	−27
	Mean	1.2		1.05	1.215	0.105	0.86	0.5	−28.405

[a] C_{max}: maximum peak carbon of n-alkanes; [b] CPI_1: $1/2[\Sigma C_{15-21}$ (odd carbon)/ΣC_{14-20} (even carbon) + ΣC_{15-21} (odd carbon)/ΣC_{16-22} (even carbon)]; [c] CPI_2: $1/2[\Sigma C_{25-35}$ (odd carbon)/ΣC_{24-34} (even carbon) + ΣC_{25-35} (odd carbon)/ΣC_{26-36} (even carbon)]; [d] Alkterr: $(C_{27} + C_{29} + C_{31} + C_{33})/\Sigma C_{14-38}$; [e] LMW/HMW: $\Sigma nC_{21-}/\Sigma nC_{22+}$; [f] $\delta^{13}C = \delta^{13}C$ of chloroform asphalts.

Organisms have different carbon isotope compositions, and the $\delta^{13}C$ values range from −22‰ to −31‰ for C3 plants and from −10‰ to −16‰ for C4 plants [1,2]. HCO_3^- in lake water provides carbon for photosynthesis of aquatic plants, and, hence, they have high $\delta^{13}C$ values. The $\delta^{13}C$ of phytoplankton in warm seawater is close to −20‰, while that of phytoplankton in cold seawater is around −30‰ [1,3]. Seagrasses have $\delta^{13}C$ values ranging from −10‰ to −16‰, and marine algae have $\delta^{13}C$ values around −18‰ [4]. The carbon isotopic composition of non-marine aquatic plants and algae is generally in the range of −12‰ to −26‰ [5]. In general, high $\delta^{13}C$ values correspond to a warm period and low values to a cold period [6,7]. The $\delta^{13}C$ (PDB) of the chloroform asphalts of the samples from the C_2t, P_2x, P_2w, P_3w, and T_1b ranged from −29.52 to −27‰, −29.76 to −28.32‰, −30.14 to −23.93‰, −32.17 to −29.15‰, and −30.81 to approximately −29.28‰, respectively (Table 1 and Table S2, Figures 2 and 3).

The Pr/Ph ratio is a useful indicator of the oxidation-reduction conditions of the sediment environment [41–44]. The Pr/Ph values of the samples from the C_2t, P_2x, P_2w, P_3w, and T_1b ranged from 0.29 to 0.67, 0.57 to 0.67, 0.49 to 0.83, 0.56 to 0.8, and 0.67 to 0.74 (Table 1 and Table S2, Figures 2–4). Β-carotene is the most prominent compound in the carbon skeleton of carotenoids preserved in lakes and highly restricted marine environments, and its higher content indicates a higher reduction in waters [45,46]. Β-carotene was detected in only a few samples from the C_2t, with a highly variable content. The ratio of β-carotene/$\Sigma C_{12\text{-}35}$ values ranged from 0.56 to 2.7 and had a mean value of 1.42. The two samples from the P_2x contained β-carotene, with β-carotene/$\Sigma C_{12\text{-}35}$ values ranging from 1.0 to 3.85 and a mean value of 2.42. The mean values of β-carotene/$\Sigma C_{12\text{-}35}$ in the P_2x, P_2w, P_3w, and T_1b were 0.51, 0.64, and 0.42. (Table 1 and Table S2, Figures 2 and 3).

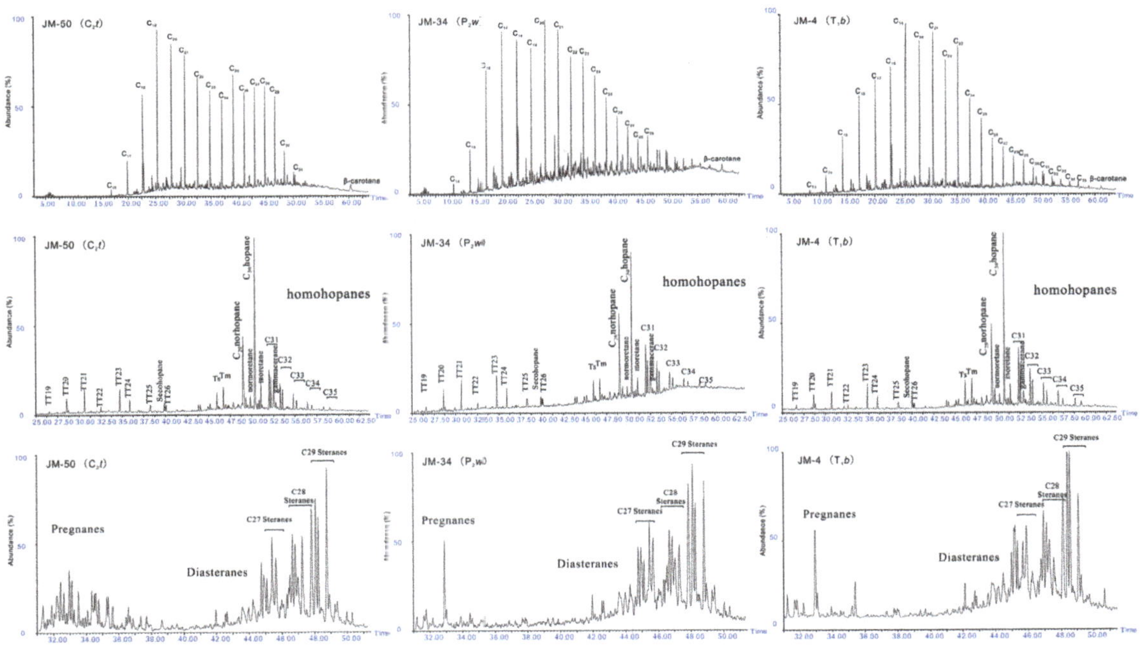

Figure 3. Chromatograms and mass fragmentogram of m/z = 191, 217, in the samples from MS-1 well of the Late Carboniferous–Early Triassic.

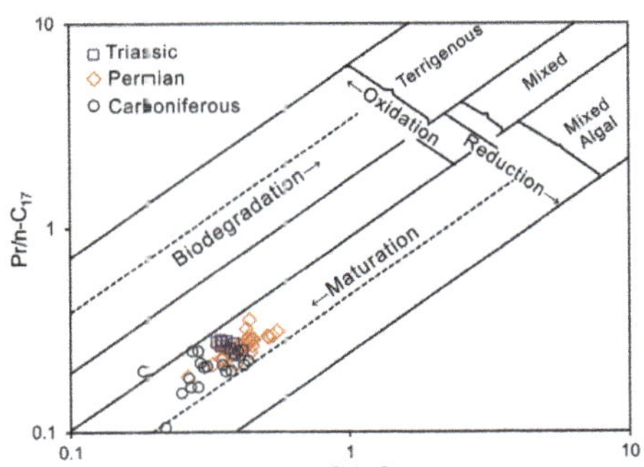

Figure 4. The correlation diagram of Ph/n-C_{18} and Pr/n-C_{17} in MS-1 well of the Late Carboniferous–Early Triassic.

4.2.2. Terpenoids and Steroids

Terpenes (including hopanes) are biomarker compounds, including tricyclic terpenes and pentacyclic triterpenes. The tricyclic terpenes of these samples in terms of tricyclic terpene abundance were distributed $C_{19} < C_{20} < C_{21} < C_{23}$ in an "ascending" pattern, and the tetracyclic C_{24} terpene abundance was higher. The C_{31}17β,21α(H)-hopane αβ-22S/(S + R) values for the C_2t, P_2x, P_2w, P_3w, and T_1b ranged from 0.38 to 0.67, 0.58 to 0.64, 0.51 to 0.64, 0.53 to 0.6, and 0.57 to 0.65 (Table 2). The values of C_{29}/C_{30}-hopane in the Carboniferous, Permian, and Triassic samples were 0.34–0.72, 0.43–0.89, and 0.42–0.82, respectively. Gammacerane is C_{30} pentacyclic triterpenoid, and high levels of gammacerane represent strongly reducing hypersaline environments [47–49], which were detected in all samples from the MS-1 well (Figure 3). The gammacerane indices (gammacerane/C_{31}-hopane) of the C_2t, P_2x, P_2w, P_3w, and T_1b ranged from 0.28 to 2.47, 0.51 to 0.73, 0.42 to

1.3, 0.21 to 0.67, and 0.21 to 0.83, respectively. The ratios of C_{27}-17α(H)-tridecane (Tm) to C_{27}-18α(H)-tridecane (Ts) effectively reflect the depositional environment and organic matter maturity [50–53]. Ts/Tm values in the Carboniferous, Permian, and Triassic samples ranged from 0.27 to 1.18, 0.28 to 0.91, and 0.22 to 0.79 (Tables 2 and S2, Figures 3 and 5). Steroids are another group of biomarker compounds, which are tetracyclic compounds with alkyl side chains, and the tetracyclic structures are collectively called steroid nuclei. The MS-1 well samples were rich in steroids, including C_{27}–C_{29} regular steroids and rearranged steroid series, mainly regular steroids (Figure 4). The C_{29}αααα-20S/20(S + R) values in the samples of C_2t, P_2x, P_2w, P_3w, and T_1b ranged from 0.19 to 0.49, 0.42 to 0.49, 0.42 to 0.52, 0.42 to 0.49, and 0.43 to 0.5, respectively, while the C_{29}-αββ/(αββ + αααα) values ranged from 0.23 to 0.51, 0.47 to 0.51, 0.44 to 0.53, 0.47 to 0.53, 0.47 to 0.53, and 0.47 to 0.57, respectively (Tables 2 and S2, Figures 3, 5 and 6).

Table 2. Parameters of hopane and sterane in Late Carboniferous–Early Triassic of MS-1 well.

Strata		C31αβ-22S/(S + R)	Gama [a]	C29/C30	Ts/Tm	C29-S [b]	C29-α [c]	ααα20R		
								C27	C28	C29
T_1b	Min.	0.57	0.21	0.42	0.22	0.43	0.47	11.37	19.33	43.04
	Max.	0.65	0.83	0.82	0.79	0.5	0.57	34.27	33.16	55.47
	Mean	0.59	0.41	0.63	0.54	0.46	0.53	27.5	21.15	50.57
P_3w	Min.	0.53	0.21	0.54	0.28	0.42	0.47	16.74	24.04	43.11
	Max.	0.6	0.67	0.89	0.69	0.49	0.53	27.4	32.35	55.28
	Mean	0.58	0.56	0.59	0.49	0.46	0.48	21.76	28.49	50.14
P_2w	Min.	0.51	0.42	0.48	0.35	0.42	0.44	22.47	23.25	40.79
	Max.	0.64	1.3	0.78	0.7	0.52	0.53	31.34	32.94	49.78
	Mean	0.57	0.51	0.68	0.53	0.46	0.51	27.44	26.52	46.87
P_2x	Min.	0.56	0.43	0.31	0.42	0.47	17.95	22.7	41.07	0.58
	Max.	0.61	0.76	0.89	0.49	0.51	34.92	30.91	52.75	0.64
	Mean	0.6	0.59	0.81	0.45	0.5	26.6	24.16	47.16	0.59
C_2t	Min.	0.38	0.28	0.34	0.27	0.19	0.23	18.25	17.76	37.14
	Max.	0.67	2.47	0.72	1.18	0.49	0.51	39.27	27.67	63.54
	Mean	0.585	0.635	0.57	0.725	0.43	0.45	28.92	23.63	46.57

[a] Gama. = gammacerane/C31-hopane. [b] C29-S = ααα-20S/20(S + R). [c] C29-α = C29[c] αββ/(αββ + ααα).

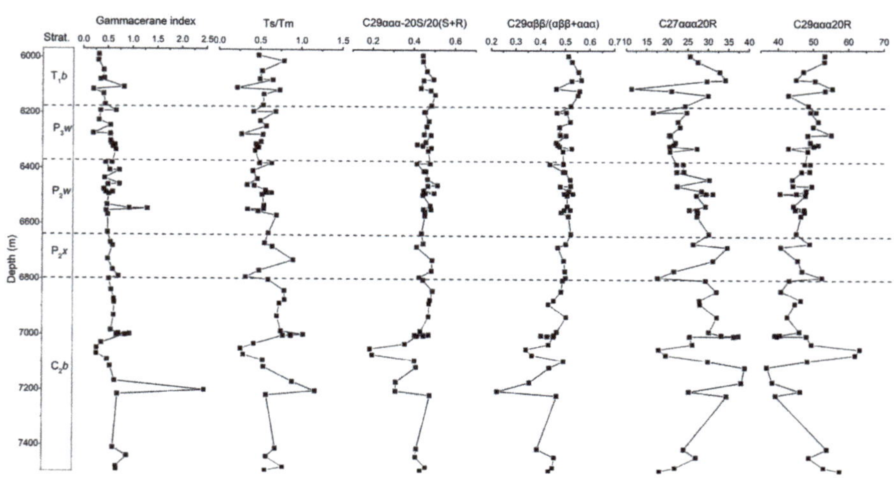

Figure 5. Profile for parameters from hopanoids and steranes of samples from the MS-1 well of the Late Carboniferous–Early Triassic.

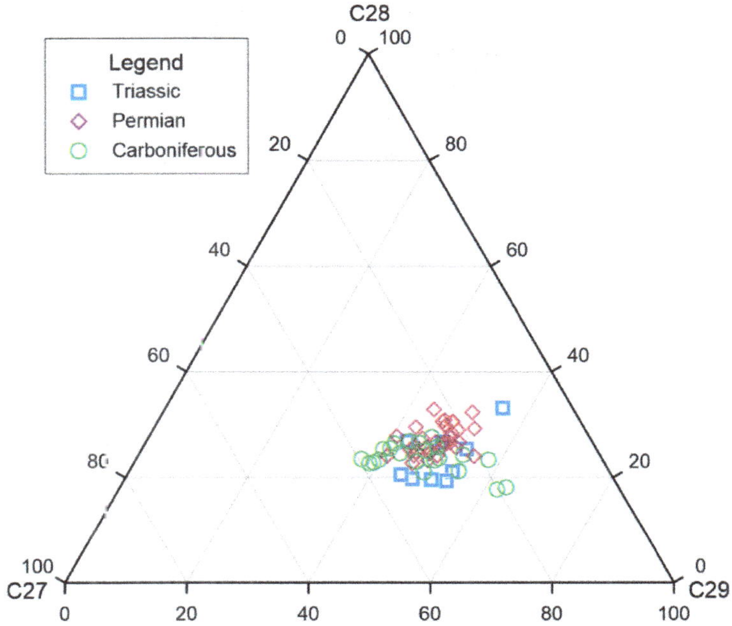

Figure 6. Ternary plot showing the distribution of regular steranes (C27–C28–C29) of samples from the MS-1 well of the Late Carboniferous–Early Triassic.

4.2.3. Organic Matter Maturity

The maximum pyrolysis temperature (T_{max}) in the Carboniferous samples from the MS-1 well was mainly 441–502 °C, indicating that the hydrocarbon source rocks are in the stage of high maturity–overmaturity evolution. T_{max} values of 425–468 °C for hydrocarbon source rocks of the Xiazijie Formation from the Permian indicated that these rocks have also reached a high-maturity evolutionary stage. The samples from the Lower Permian Urho Formation had a T_{max} range of 425–471 °C, which shows that the hydrocarbon source rocks have reached the mature–high-maturity evolutionary stage. The samples from the Urho Formation of the Upper Permian also showed a mature evolutionary stage with T_{max} values of 428–443 °C. The T_{max} values of the Baikouquan Formation of the Triassic were mainly 430–441 °C, indicating that these rocks have just entered the mature evolutionary stage (Table 1, Figure 2).

Biomarker compound parameters are another tool for evaluating the maturity of organic matter in sedimental rocks [51,54,55]. During the heating of hopanes and steranes under geological conditions, the R conformation is converted to the S conformation, forming a mixed conformation of R + S. In general, a higher 22S/(S + R) ratio represents higher maturity [52]. In addition, $C_{29}\alpha\beta\beta$ has higher stability than $C_{29}\alpha\alpha\alpha$, and the value of C_{29} $\alpha\beta\beta/(\alpha\alpha\alpha + \alpha\beta\beta)$ increases abruptly in the first half of a raw oil zone. The C_{31} 17β, 21α(H)-hopane, and 31αβ-22S/(S + R) values, sterane $C_{29}\alpha\alpha\alpha$-20S/20(S + R) values, and C_{29}-$\alpha\beta\beta/(\alpha\beta\beta + \alpha\alpha\alpha)$ values from the MS-1 well all indicated that the Late Carboniferous–Early Triassic organic matter of MS-1 has reached the mature stage in general.

In general, the ratio of C_{27}-17α(H)-triacontane (Tm) to C_{27}-18α(H)-triacontane (Ts) can effectively reflect the organic matter maturity and depositional environment, and, in the post-rock stage, C_{27}-17α(H)-trisnorhopane (Tm) is less stable than C_{27}-18α(H)-trisnorhopane (Ts), and, as the maturity of crude oil increases, the Ts/Tm values also increase with increasing maturity of crude oil. and for immature samples. The Ts/Tm values of samples from the C_2t similarly indicated that the organic matter has reached the maturity stage. Ts/Tm can also reflect the redox conditions of the depositional environment, with low Ts/Tm values for oxidizing environments and high values for reducing environments [52]; however, the samples from the MS-1well did not exhibit significant environmental indicator features.

The Rock-Eval indicated that the organic matter of the MS-1 well is in the mature stage, but did not show an increase in the parameters of organic matter maturity with increasing depth of the formation. Ts/Tm, $C_{29}\alpha\alpha\alpha$-20S/20(S + R), and C_{29}-$\alpha\beta\beta/(\alpha\beta\beta + \alpha\alpha\alpha)$ even showed a decreasing trend with organic matter type and depositional environment, indicating that the biomarkers in the study area are not significantly affected by maturity, and, on the other hand, supporting that MS-1 organic geochemical parameters can be used to reconstruct the paleoclimate and paleoenvironment.

5. Sedimentary Environment

5.1. Carboniferous

The Telegula Formation (C_2t)

The C_2t stratigraphic logs can be divided into three lithologic zones (Figures 1 and 7): the depth of 6806.0–6952.0 m is mainly gray-brown, conglomeratic mudstone with gray, conglomeratic mud siltstone in unequal thickness interbedded and gray tuff with tuff siltstone interbedded; the depth of 6952.0–7265.0 m is dominated by gray and greenish-gray tuff, interbedded with black coal, gray-black sunken tuff, tuffaceous mudstone, and andesite; and the depth of 7265.0–7500.0 m is dominated by gray, dark-gray, and gray-black tuffs, interbedded with gray-black tuffaceous siltstone. The lithologic assemblage indicates that the C_2t was frequently volcanic, the climate was relatively warm and humid, terrestrial higher plants were relatively developed, and large amounts of terrestrial organic matter were usually deposited in nearshore or shallow water along with tuffaceous debris. $C_{29}\alpha\alpha\alpha\alpha$20R and isotopes of organic matter (−29.52 to −27‰) also indicated a high content of terrestrial organic matter mixed with mixed aquatic plants in the sediments of the MS-1 well during this period. Meanwhile, the abundance of short-chain carbon number n-alkanes, Pr/Ph, and gammacerane indices indicates high salinity and reduction in the water column [52]. Additionally, intermittent volcanic activities can carry the terrestrial organic matter to the water column, causing a dramatic increase in organic matter and a rise in the salinity and reducibility of the water column [56]. In terms of biomarker trends, the aquatic life content tends to increase while the salinity and oxidation of the water column increases. Considering that the geochemistry of the Carboniferous volcanic rocks of the MS-1 well shows an island arc setting, while the complete closure of the western ocean of the Junggar Basin was delayed until the Early Permian, we believe that the MS-1 well was in a relatively closed residual sea environment, and the relatively humid climate made terrestrial plants flourish. Two recessions and one flooding event can be identified by the chloroform asphalts of $\delta^{13}C$ (PDB), showing two decreasing trends of $\delta^{13}C$ and an abrupt rise at 7300 m depth.

5.2. Permian

5.2.1. The Xiazijie Formation (P_2x)

The P_2x is directly unconformable, overlying the Carboniferous (Figures 1 and 7); 6641.0–6655.0 m is a set of dark-gray mudstone. It is mainly gray-white, muddy siltstone from the depth of 6655.0 to 6700.0 m, interbedded with gray mudstone and siltstone; an area of gray-white, muddy siltstone from the depth of 6700.0 to 6800.0 m, interbedded with gray mudstone and siltstone with varying thickness; and gray, gravelly siltstone from a depth of 6800.0 to 6806.0 m. The lithology of the P_2x period is typical of lacustrine sedimentation. Since the stratigraphy of the Fengcheng Formation is missing [57–59], we speculate that the super-lake basin of the Fengcheng Formation gradually shrunk with stratigraphic uplift, and the P_2x turned into a period of subsidence. During P_2x, several alluvial fans developed in the western part of the Junggar Basin, and the sediments were characterized by coarse-grained sandstone and conglomerate. The Mosuowan area served as the sedimentary center, and the gray, fine sandstone and mudstone were mainly deposited there. Biomarkers and $\delta^{13}C$ (PDB) of the chloroform asphalts (−29.76 to −28.32‰) indicated a sustained increase in aquatic life content, a decrease in water column salinity, and an increase in water column oxidation.

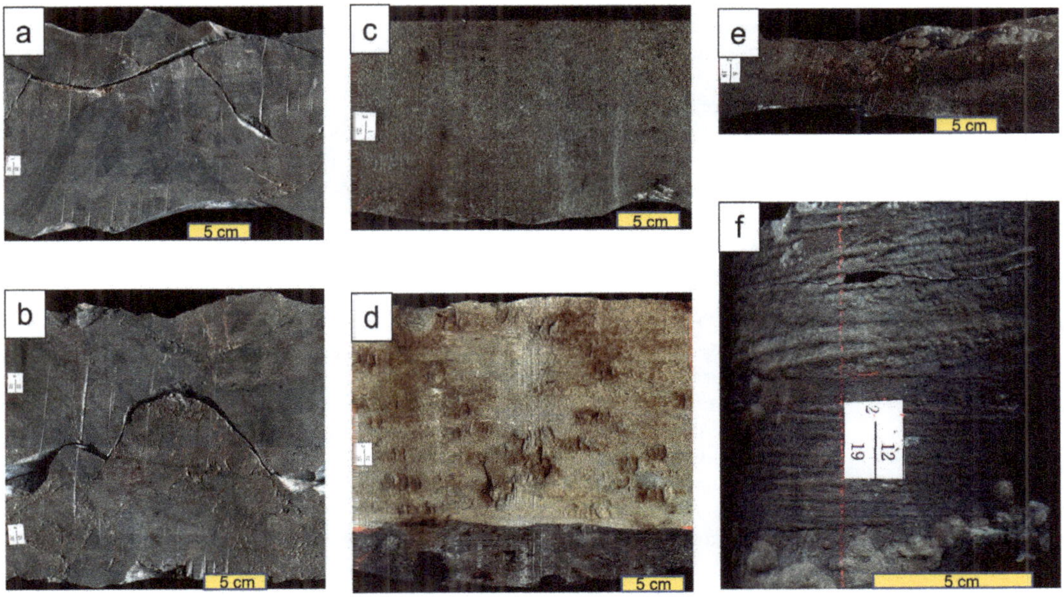

Figure 7. Scanning core photos of samples from the MS-1 well: (**a**,**b**), tuff in C_2t; (**c**), siltstone mudstone in P_2w; (**d**), grayish-white siltstone and black mudstone in P_3w, with scouring surface between them; (**e**,**f**), siltstone mudstone in T_1b.

5.2.2. The Lower and Upper Urho Formation (P_2w, P_3w)

The lithology of the Middle Permian–Upper Permian Urho Formation (Lower and Upper Urho Formation) is characterized as follows (Figures 1 and 7): It consists of grayish-white siltstone, muddy siltstone, and conglomeratic siltstone, interbedded with grayish-brown, brownish-gray, gray, and greenish-gray mudstone, platy mudstone, and siltstone; from the depth of 6455.0–6548.0 m, it consists of greenish-gray, gray, brownish-gray, and dark-gray mudstone and siltstone, interbedded with grayish-white and muddy siltstone and siltstone; from 6548.0 to 6641.0 m, it consists of gray siltstone, muddy siltstone, and conglomeratic siltstone, interbedded with gray and dark-gray mudstone and siltstone; an interval (depth of 6178.0–6256.0 m) consists of brownish-gray, grayish-brown, beige, and gray, muddy siltstone, siltstone, and conglomerate, interbedded with brownish-gray and grayish-brown mudstone in unequal thickness; and the depth of 6256.0–6374.0 m is gray siltstone, sandstone, and muddy siltstone, interbedded with brownish-gray, grayish-brown, brown mudstone, and silt mudstone. The lithological assemblage of the P_2w and P_3w is a deltaic foreland deposit. Combined with geochemical indicators, it shows an overall lake retreat process with a continuous increase in terrestrial organic matter content and small changes in water salinity and reducibility. The $\delta^{13}C$ (PDB) of the chloroform asphalts at P_2w ranged from −30.14 to −23.93‰, during which the maximum flooding surface occurred with $\delta^{13}C$ reaching −24‰. In contrast, P_3w had the largest lake retreat event during C_2t–T_1b, reflecting the climate transition from wet to dry. β-carotene/$\Sigma C_{12\text{-}35}$ showed that the salinity of P_3w was higher than that of P_2w, which might have been caused by the repeated lake level fluctuations.

5.2.3. The Baikouquan Formation (T_1b)

The lithology of T_1b is divided into two sections (Figures 1 and 7), the first section (depth of 5939.0–5991.0 m) is gray, brownish-gray, off-white, muddy siltstone, siltstone, sand conglomerate, and gravelly, fine sandstone interbedded with brownish-gray, gray mudstone, and coarse siltstone, while the second section (depth of 5991.0–6178.0 m) is brownish-gray, off-white, gray, muddy siltstone, siltstone, gravelly, muddy siltstone, conglomeratic siltstone, and sand conglomerate, interbedded with gray and brown-gray mudstone. The lithologic assemblage of the T_1b is braided river deltaic sedimentation with migration of deposition centers of the lake basin and reduced water salinity. A transient flooding event was found in the middle of the Baikouquan Formation, which increased the organic matter content of terrestrial sources and increased water salinity. The $\delta^{13}C$ (PDB) of the chloroform asphalts ranged from −30.81 to −29.28‰, with obvious higher plant carbon isotope characteristics. Pr/Ph had been increasing continuously since P_3w and reached its highest point, indicating a significant change in the depositional environment invention, i.e., from shallow lakeside to braided river deposition.

6. Global Tectonics and Environmental Change Drives

Large-scale sea retreat events during the Carboniferous period exposed the seafloor at the land margin, expanded the land area globally, and created fertile swamps and wetlands as freshwater rivers flowed through the new continental mass [57,60,61]. At the same time, the warm and humid climate made Lycopodium grow densely, and the tall ferns provided raw materials for coal formation, and the European continent and Siberia in the north, as well as the land of north China, which are the major coal-gathering areas in the world [62]. The C_2t coal formation and higher, plant-rich tuffs drilled in the Carboniferous MS-1 well also reflect the global climatic characteristics of this period, but the strong volcanic activity in the Late Carboniferous Junggar Basin may have affected the growth of Carboniferous fern plants and, thus, the regional development of the coal formation. The most important global event of the Permian was the Permian–Triassic extinction that occurred at the end of the Permian, an extinction event that can be divided into one to three stages [61,63–65]. The first mini-peak may have been due to a gradual change in the environment due to sea level changes, oceanic hypoxia, and the arid climate due to the formation of the Pangaea continent. The later peaks were rapid and dramatic, probably due to impact events, super volcanic eruptions, or abrupt sea level changes that caused a massive release of methane hydrates that lasted about 60,000 years [65–67]. However, this event has no obvious carbonate record in the Junggar Basin. Since the Permian sediments of the MS-1 well are dominated by terrestrial organic matter, the negative organic carbon isotope drift is not significant. The core samples show that the Urho Formation samples are darker in color than the Baikouquan Formation, indicating that its depositional environment may have dealt with a global anoxic background. Influenced by the global aridification triggered by the Pangaea, the drought in the Junggar Basin reached its peak in the Late Permian–Early Triassic. The drought in Junggar began in the Middle Permian, and the lakes on the southern edge of Junggar dried up in its late stage, with evaporation exceeding recharge, and the lake extent south of Karameri in the Early Triassic also gradually decreased until it was replaced by the red layer, further confirming the persistence and development of the arid climate, a fact also illustrated by the gradual increase in the oxidation of the sedimentary environment of the MS-1 well (Figure 8).

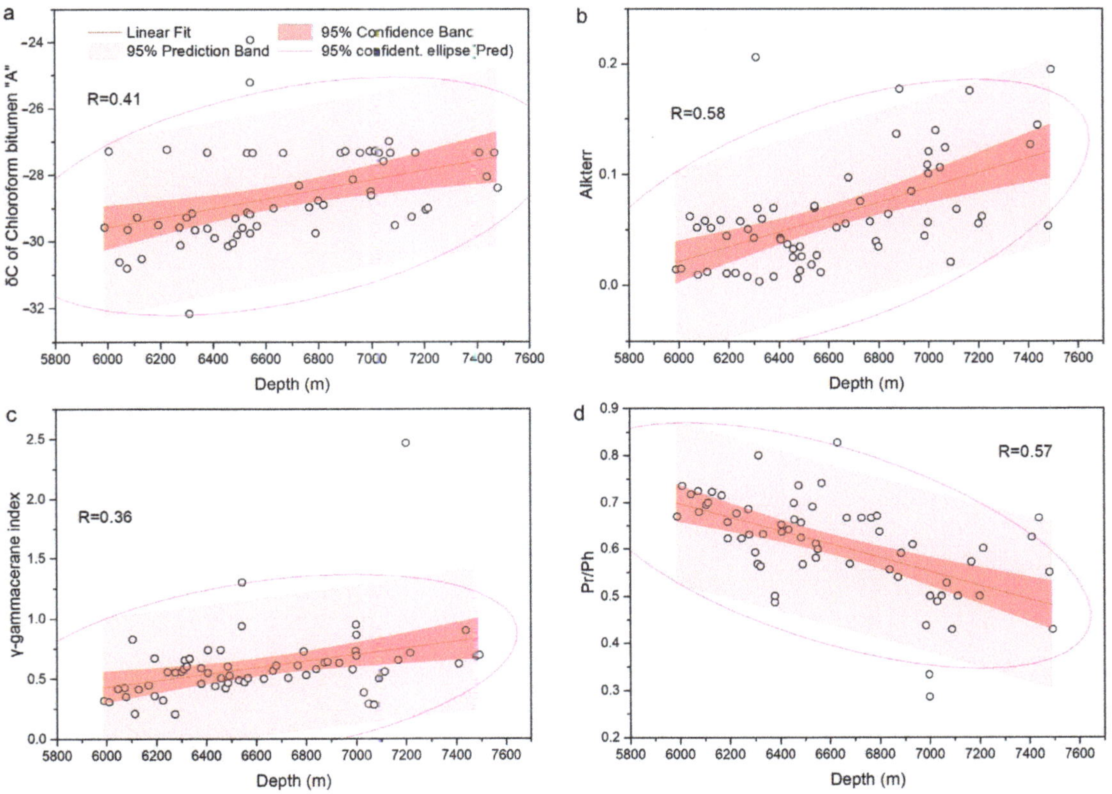

Figure 8. Correlation between depth and δC of chloroform bitumen "A" (**a**), Alkterr (**b**), γ-gammacerane index (**c**), Pr/Ph (**d**).

The Late Paleozoic orogenic belt in northern Xinjiang and adjacent areas developed a series of successive, southward accretionary belts along the broad South Siberian active margin, forming the Alaska–Aleutian arc system (Kokchetav–Morth Tianshan), the Japanese arc system (Altai, Chinese Middle Tianshan), and the Mariana arc system (Balkhash, West Junggar, East Junggar) [4,9] (Figure 9). Paleomagnetic reconstructions indicate that complex nappes occurred in the Early Paleozoic. In the Late Devonian, a complex Kokchetav–North Tianshan arc system with a straight extension of the main axis was formed under growth and fusion of the circum-microcontinental type. Approximately corresponding to the late Early Carboniferous, the Junggar paleo sheet formed a rift trough along the Bogda area, separating the Junggar-occupied sheet. The rift trough did not reach the level of the emergence of oceanic crust, so it is also called the Ora Valley. After the middle of the Early Carboniferous, the Siberian plate and the Junggar paleolithic sheet underwent continental collisional orogeny, and the ocean basin disappeared. The ocean between the Junggar paleo sheet and Kazakhstan finally closed at the end of the Carboniferous or the beginning of the Permian, while the ocean south of the basin closed in the Late Carboniferous and entered the intra-land orogenic stage, and the relatively closed residual marine environment in the Late Carboniferous study area coincided with it. In the Permian period, geosyncline closed, and folded mountain series were formed successively, and the paleo-plates were gradually linked to form a joint paleo-continent (Pangaea). With the further expansion of the land area, gymnosperms began to decline on land, true and seed ferns flourished, and the first gymnosperms appeared. From the Permian to the Middle Triassic, the Junggar Basin was in the foreland basin tectonic development stage, and the sediment evolution of the Xiazijie,

Urho, and Baikouquan Formations also reflected the gradual disappearance of marine fingerprints and the increasingly obvious characteristics of continental sedimentation; the organic matter of terrestrial origin continued to increase, and the salinity of water bodies became increasingly low. In conclusion, regional tectonic evolution constrained the variability of lake waters in the MS-1 well, while the Carboniferous global warming and wetting event, Permian anoxic event, and Late Permian–Early Triassic drought event influenced the organic matter composition and redox conditions of the lake waters.

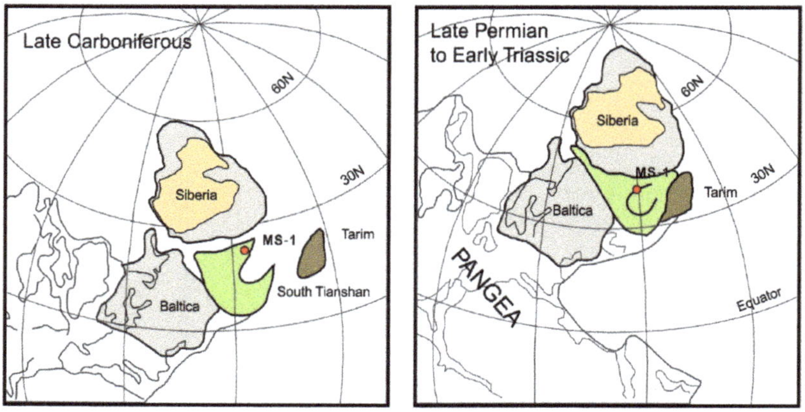

Figure 9. Paleogeographic reconstruction for the bending of the Kazakhstan oroclines. Modified after [68,69].

7. Conclusions

The organic carbon content in the Carboniferous–Lower Permian sedimentary rocks drilled from MS-1 in the Mosuowan area of the Junggar Basin is low, but the Carboniferous tuffs and coal-bearing rocks are high in organic matter and have potential hydrocarbon generation capabilities. The organic matter evolution is all at a mature stage, but biomarker parameters were not significantly subjected to thermal evolution. During the Late Carboniferous, the Mosuowan area was located in a relatively enclosed remnant marine environment with a relatively humid climate that allowed for vigorous terrestrial vegetation, abundant rivers flowing into the water, and reduced water salinity yet increased depth. The Xiazijie Formation of the Middle Permian was a period of basin subsidence during which the basin area expanded significantly, and organic geochemical information indicates a continuous increase in aquatic organisms, a decrease in lake water salinity, and an increase in water oxidation. The lithologic assemblage of the Urho Formation in Permian was developed on the deltaic foreland. The biomarkers and isotopes of organic matter show an overall lake retreat process with a continuous increase in terrestrial organic matter and little change in water salinity and reduction, during which significant flooding events showed a significant decrease in terrestrial organic matter, an increase in aquatic organisms, and an increase in water salinity and reduction. The Baikouquan Formation of the Early Triassic was deposited as a braided river delta, with a migration of deposition centers of the lake basin and a decrease in water salinity, during which there was a brief flooding event, an increase in organic matter of terrestrial origin, and an increase in water salinity. Paleo-Asian ocean subduction and continental collision constrained the changes of waters in the MS-1 well during the Late Carboniferous–Early Permian, while the continuous increase in sediment, terrestrial organic matter, water desalination, and oxidation enrichment in the Mosuowan area, and the global warming and wetting events of the Carboniferous, Permian anoxic events, and Late Permian–Early Triassic drought events affected the lipid biomarkers and stable isotopes of the organic matter composition; however, the P–T biological mass extinction events were not clearly recorded.

Supplementary Materials: The following are available online at https://www.mdpi.com/article/10.3390/min12101299/s1, Table S1: TOC and rock pyrolysis parameters of the Late Carboniferous-Early Triassic rocks in MS-1 well, Table S2: Parameters of δ^{13C} values of n-alkanes, Pr, Ph, β-carotene and chloroform extracts from MS1 wells of the Late Carboniferous-Early, Table S3: Parameters of hopance and sterane in Late Carboniferous-Early Triassic of MS1 well.

Author Contributions: Conceptualization: S.Z. (Shuncun Zhang). Data curation: T.W. and H.G. Investigation: T.W., H.G. and S.Z. (Shengyin Zhang). Methodology: B.C. Writing—original draft: S.Z. (Shuncun Zhang). Writing—review and editing: S.Z. (Shengyin Zhang). All authors have read and agreed to the published version of the manuscript.

Funding: This study was financially supported by the Guangxi Natural Science Foundation, China (2020GXNSFBA297128); the Special Talent Project of Guangxi Science and Technology Base, China (Guike AD20238041); the National Natural Science Foundation of China (grant no. 41872145).

Institutional Review Board Statement: Not applicable.

Informed Consent Statement: Not applicable.

Data Availability Statement: The data are contained within the article.

Conflicts of Interest: The authors declare no conflict of interest.

References

1. Shen, S.Z.; Shi, G.R. Late Paleozoic deep Gondwana and its peripheries: Stratigraphy, biological events, paleoclimate and paleogeography. *Gondwana Res.* **2013**, *24*, 1–4. [CrossRef]
2. Feng, Y.; Song, H.J.; Bond, D.P.G. Size variations in foraminifers from the early Permian to the Late Triassic: Implications for the Guadalupian-Lopingian and the Permian-Triassic mass extinctions. *Paleobiology* **2020**, *46*, 511–532. [CrossRef]
3. Wang, Y.J.; Jia, D.; Pan, J.G.; Wei, D.T.; Tang, Y.; Wang, G.D.; Wei, C.R.; Ma, D.L. Multiple-phase tectonic superposition and reworking in the Junggar Basin of northwestern China-Implications for deep-seated petroleum exploration. *Aapg Bull.* **2018**, *102*, 1489–1521. [CrossRef]
4. Xiao, W.J.; Windley, B.F.; Allen, M.B.; Han, C.M. Paleozoic multiple accretionary and collisional tectonics of the Chinese Tianshan orogenic collage. *Gondwana Res.* **2013**, *23*, 1316–1341. [CrossRef]
5. Ju, W.; Hou, G.T.; Li, L.; Xiao, F.F. End Late Paleozoic tectonic stress field in the southern edge of Junggar Basin. *Geosci. Front.* **2012**, *3*, 707–715. [CrossRef]
6. Hendrix, M.S.; Brassell, S.C.; Carroll, A.R.; Graham, S.A. Sedimentology, organic geochemistry, and petroleum potential of Jurassic coal measures: Tarim, Junggar, and Turpan basins, northwest China. *Aapg Bull.-Am. Assoc. Pet. Geol.* **1995**, *79*, 929–959.
7. Xiao, M.; Wu, S.T.; Yuan, X.J.; Cao, Z.L.; Xie, Z.R. Diagenesis effects on the conglomerate reservoir quality of the Baikouquan Formation, Junggar Basin, China. *J. Pet. Sci. Eng.* **2020**, *195*, 17. [CrossRef]
8. Han, Y.G.; Zhao, G.C. Final amalgamation of the Tianshan and Junggar orogenic collage in the southwestern Central Asian Orogenic Belt: Constraints on the closure of the Paleo-Asian Ocean. *Earth-Sci. Rev.* **2018**, *186*, 129–152. [CrossRef]
9. Xiao, W.J.; Windley, B.F.; Han, C.M.; Liu, W.; Wan, B.; Zhang, J.; Ao, S.J.; Zhang, Z.Y.; Song, D.F. Late Paleozoic to early Triassic multiple roll-back and oroclinal bending of the Mongolia collage in Central Asia. *Earth-Sci. Rev.* **2018**, *186*, 94–128. [CrossRef]
10. Han, B.F.; Ji, J.Q.; Song, B.; Chen, L.H.; Zhang, L. Late Paleozoic vertical growth of continental crust around the Junggar Basin, Xinjiang, China (Part I): Timing of post-collisional plutonism. *Acta Petrol. Sin.* **2006**, *22*, 1077–1086.
11. Yu, K.H.; Cao, Y.C.; Qiu, L.W.; Sun, P.P.; Jia, X.Y.; Wan, M. Geochemical characteristics and origin of sodium carbonates in a closed alkaline basin: The Lower Permian Fengcheng Formation in the Mahu Sag, northwestern Junggar Basin, China. *Palaeogeogr. Palaeoclimatol. Palaeoecol.* **2018**, *511*, 506–531. [CrossRef]
12. Yang, W.; Feng, Q.A.; Liu, Y.Q.; Tabor, N.; Miggins, D.; Crowley, J.L.; Lin, J.Y.; Thomas, S. Depositional environments and cyclo- and chronostratigraphy of uppermost Carboniferous-Lower Triassic fluvial-lacustrine deposits, southern Bogda Mountains, NW China-A terrestrial paleoclimatic record of mid-latitude NE Pangea. *Glob. Planet. Chang.* **2010**, *73*, 15–113. [CrossRef]
13. Metcalfe, I.; Foster, C.B.; Afonin, S.A.; Nicoll, R.S.; Mundil, R.; Wang, X.F.; Lucas, S.G. Stratigraphy, biostratigraphy and C-isotopes of the Permian-Triassic non-marine sequence at Dalongkou and Lucaogou, Xinjiang Province, China. *J. Asian Earth Sci.* **2009**, *36*, 503–520. [CrossRef]
14. Gastaldo, R.A.; Neveling, J.; Clark, C.K.; Newbury, S.S. The terrestrial Permian-Triassic boundary event bed is a nonevent. *Geology* **2009**, *37*, 199–202. [CrossRef]
15. Meng, Z.Y.; Liu, Y.Q.; Jiao, X.; Ma, L.T.; Zhou, D.W.; Li, H.; Cao, Q.; Zhao, M.R.; Yang, Y.Y. Petrological and organic geochemical characteristics of the Permian Lucaogou Formation in the Jimsar Sag, Junggar Basin, NW China: Implications on the relationship between hydrocarbon accumulation and volcanic-hydrothermal activities. *J. Pet. Sci. Eng.* **2022**, *210*, 17. [CrossRef]
16. Hu, D.; Rao, S.; Wang, Z.T.; Hu, S.B. Thermal and maturation history for Carboniferous source rocks in the Junggar Basin, Northwest China: Implications for hydrocarbon exploration. *Pet. Sci.* **2020**, *17*, 36–50. [CrossRef]

17. Feng, M.Y.; Liu, T.; Lin, T.; Liu, X.H.; Li, N.X.; Xi, A.H. Fracture Fillings and Implication of Fluid Activities in Volcanic Rocks: Dixi Area in Kelameili Gas Field, Junggar Basin, Northwestern China. *Minerals* **2019**, *9*, 154. [CrossRef]
18. Liu, B.B.; Tan, C.P.; Yu, X.H.; Qu, J.H.; Zhao, X.M.; Zhang, L. Sedimentary characteristics and controls of a retreating, coarse-grained fan-delta system in the Lower Triassic, Mahu Depression, northwestern China. *Geol. J.* **2019**, *54*, 1141–1159. [CrossRef]
19. Zhang, M.M.; Liu, Z.J.; Qiu, H.J.; Xu, Y.B. Characteristics of organic matter of oil shale in the sequence stratigraphic framework at the northern foot of Bogda Mountain, China. *Oil Shale* **2016**, *33*, 31–44. [CrossRef]
20. Jiao, Y.Q.; Yan, J.X.; Li, S.T.; Yang, R.Q.; Lang, F.J.; Yang, S.K. Architectural units and heterogeneity of channel reservoirs in the Karamay formation, outcrop area of Karamay oil field, Junggar basin, northwest China. *Aapg Bull.* **2005**, *89*, 529–545. [CrossRef]
21. Wilhem, C.; Windley, B.F.; Stampfli, G.M. The Altaids of Central Asia: A tectonic and evolutionary innovative review. *Earth-Sci. Rev.* **2012**, *113*, 303–341. [CrossRef]
22. Allen, M.B.; Windley, B.F.; Chi, Z. Palaeozoic collisional tectonics and magmatism of the Chinese Tien Shan, central Asia. *Tectonophysics* **1993**, *220*, 89–115. [CrossRef]
23. Liu, X.J.; Xiao, W.J.; Xu, J.F.; Castillo, P.R.; Shi, Y. Geochemical signature and rock associations of ocean ridge-subduction: Evidence from the Karamaili Paleo-Asian ophiolite in east Junggar, NW China. *Gondwana Res.* **2017**, *48*, 34–49. [CrossRef]
24. Bai, J.K.; Chen, J.L.; Yan, Z.; Tang, Z.; Xu, X.Y.; Li, J.L. The timing of opening and closure of the Mayile oceanic basin: Evidence from the angular unconformity between the Middle Devonian and its underlying geological body in the southern West Junggar. *Acta Petrol. Sin.* **2015**, *31*, 133–142.
25. Sengor, A.M.C.; Natal'In, B.A.; Sunal, G.; van der Voo, R. A new look at the altaids: A superorogenic complex in northern and central asia as a factory of continental crust. part I: Geological data compilation (exclusive of palaeomagnetic observations). *Austrian J. Earth Sci.* **2014**, *107*, 169–232.
26. Weng, K.; Xu, X.Y.; Ma, Z.P.; Chen, J.L.; Sun, J.M.; Zhang, X. The geochemistry and chronology characteristics and the geological significance of ultramafic rock in Mayile ophiolite, West Junggar, Xinjiang. *Acta Petrol. Sin.* **2016**, *32*, 1420–1436.
27. Chen, S.; Guo, Z.J. Time constraints, tectonic setting of Dalabute ophiolitic complex and its significance for Late Paleozoic tectonic evolution in West Junggar. *Acta Petrol. Sin.* **2010**, *26*, 2336–2344.
28. Xu, X.; Zhou, K.F.; Wang, Y. Study on extinction of the remnant oceanic basin and tectonic setting of West Junggar during Late Paleozoic. *Acta Petrol. Sin.* **2010**, *26*, 3206–3214.
29. Zhang, M.; Wang, G.C.; Zhang, X.H.; Liao, Q.A.; Wang, W.; Guo, R.L.; Zhang, P. Reconstruction of the Silurian to Devonian stratigraphic succession along the northeastern margin of the Junggar block, Xinjiang, NW China, and its tectono-paleogeographic implications for the southwestern Central Asian Orogenic Belt. *Sediment. Geol.* **2021**, *411*, 23. [CrossRef]
30. Novikov, I.S. Reconstructing the stages of orogeny around the Junggar basin from the lithostratigraphy of Late Paleozoic, Mesozoic, and Cenozoic sediments. *Russ. Geol. Geophys.* **2013**, *54*, 138–152. [CrossRef]
31. He, D.F.; Li, D.; Fan, C.; Yang, X.F. Geochronology, geochemistry and tectonostratigraphy of Carboniferous strata of the deepest Well Moshen-1 in the Junggar Basin, northwest China: Insights into the continental growth of Central Asia. *Gondwana Res.* **2013**, *24*, 560–577. [CrossRef]
32. Ji, Y.L.; Zhou, Y.; Kuang, J.; Wan, L.; Zhang, R.; Lu, C.H. The formation and evolution of Chepaizi-Mosuowan paleo-uplift and its control on the distributions of sedimentary facies in the Junggar Basin. *Sci. China-Earth Sci.* **2010**, *53*, 818–831. [CrossRef]
33. Shi, Y.Q.; Ji, H.C.; Yu, J.W.; Xiang, P.F.; Yang, Z.B.; Liu, D.D. Provenance and sedimentary evolution from the Middle Permian to Early Triassic around the Bogda Mountain, NW China: A tectonic inversion responding to the consolidation of Pangea. *Mar. Pet. Geol.* **2020**, *114*, 104–169. [CrossRef]
34. Zhang, Z.H.; Qin, L.M.; Qiu, N.S.; Zhong, N.N.; Zhang, Z.Y.; Li, W. Combination and superimposition of source kitchens and their effects on hydrocarbon accumulation in the hinterland of the Junggar Basin, west China. *Pet. Sci.* **2010**, *7*, 59–72. [CrossRef]
35. Qiu, N.S.; Ming, Z.; Wang, X.L.; Yang, H.B. Tectono-thermal evolution of the Junggar Basin, NW China: Constraints from R-o and apatite fission track modelling. *Pet. Geosci.* **2005**, *11*, 361–372.
36. Peters, K.E. Guidelines for evaluating petroleum source rock using programmed pyrolysis. *Aapg Bull.-Am. Assoc. Pet. Geol.* **1986**, *70*, 318–329.
37. Bray, E.; Evans, E. Distribution of n-paraffins as a clue to recognition of source beds. *Geochim. Et Cosmochim. Acta* **1961**, *22*, 2–15. [CrossRef]
38. Cheng, B.; Xu, J.B.; Lu, Z.Q.; Li, Y.H.; Wang, W.C.; Yang, S.; Liu, H.; Wang, T.; Liao, Z.W. Hydrocarbon source for oil and gas indication associated with gas hydrate and its significance in the Qilian Mountain permafrost, Qinghai, Northwest China. *Mar. Pet. Geol.* **2018**, *89*, 202–215. [CrossRef]
39. Herrera-Herrera, A.V.; Mallol, C. Quantification of lipid biomarkers in sedimentary contexts: Comparing different calibration methods. *Org. Geochem.* **2018**, *125*, 152–160. [CrossRef]
40. Silliman, J.E.; Schelske, C.L. Saturated hydrocarbons in the sediments of Lake Apopka, Florida. *Org. Geochem.* **2003**, *34*, 253–260. [CrossRef]
41. Holtvoeth, J.; Whiteside, J.H.; Engels, S.; Freitas, F.S.; Grice, K.; Greenwood, P.; Johnson, S.; Kendall, I.; Lengger, S.K.; Lucke, A.; et al. The paleolimnologist's guide to compound-specific stable isotope analysis-An introduction to principles and applications of CSIA for Quaternary lake sediments. *Quat. Sci. Rev.* **2019**, *207*, 101–133. [CrossRef]
42. Schinteie, R.; Brocks, J.J. Paleoecology of Neoproterozoic hypersaline environments: Biomarker evidence for haloarchaea, methanogens, and cyanobacteria. *Geobiology* **2017**, *15*, 641–663. [CrossRef]

43. Tulipani, S.; Grice, K.; Greenwood, P.F.; Haines, P.W.; Sauer, P.E.; Schimmelmann, A.; Summons, R.E.; Foster, C.B.; Bottcher, M.E.; Playton, T.; et al. Changes of palaeoenvironmental conditions recorded in Late Devonian reef systems from the Canning Basin, Western Australia: A biomarker and stable isotope approach. *Gondwana Res.* **2015**, *28*, 1500–1515. [CrossRef]
44. Greenwood, P.F.; Summons, R.E. GC-MS detection and significance of crocetane and pentamethylicosane in sediments and crude oils. *Org. Geochem.* **2003**, *34*, 1211–1222. [CrossRef]
45. Farzadnia, S.; Nimmagadda, R.D.; McRae, C. A comparative structural study of nitrogen-rich fulvic acids from various Antarctic lakes. *Environ. Chem.* **2018**, *14*, 502–514. [CrossRef]
46. Walters, C.C.; Moldowan, J.M. *The Biomarker Guide: Biomarkers and Isotopes In the Environment and Human History*; Cambridge University Press: Cambridge, UK, 2005.
47. Darnet, S.; Blary, A.; Chevalier, Q.; Schaller, H. Phytosterol Profiles, Genomes and Enzymes-An Overview. *Front. Plant Sci.* **2021**, *12*, 18. [CrossRef]
48. Tomazic, M.L.; Poklepovich, T.J.; Nudel, C.B.; Nusblat, A.D. Incomplete sterols and hopanoids pathways in ciliates: Gene loss and acquisition during evolution as a source of biosynthetic genes. *Mol. Phylogenetics Evol.* **2014**, *74*, 122–134. [CrossRef]
49. Fang, J.S.; Chan, O.; Joeckel, R.M.; Huang, Y.S.; Wang, Y.; Bazylinski, D.A.; Moorman, T.B.; Clement, B.J.A. Biomarker analysis of microbial diversity in sediments of a saline groundwater seep of Salt Basin, Nebraska. *Org. Geochem.* **2006**, *37*, 912–931. [CrossRef]
50. Hakimi, M.H.; Abdullah, W.H. Geochemical characteristics of some crude oils from Alif Field in the Marib-Shabowah Basin, and source-related types. *Mar. Pet. Geol.* **2013**, *45*, 304–314. [CrossRef]
51. Skret, U.; Fabianska, M.J. Geochemical characteristics of organic matter in the Lower Palaeozoic rocks of the Peribaltic Syneclise (Poland). *Geochem. J.* **2009**, *43*, 343–369. [CrossRef]
52. Peters, K.E.; Moldowan, J.M. *The Biomarker Guide: Interpreting Molecular Fossils in Petroleum and Ancient Sediments*; Prentice Hall: Hoboken, NJ, USA, 1993.
53. Irwin, H.; Meyer, T. Lacustrine organic facies—A biomarker study using multivariate statistical-analysis. *Org. Geochem.* **1990**, *16*, 197–210. [CrossRef]
54. Ma, S.H.; Zhang, S.C.; Su, J.; Wang, X.M.; He, K.; Feng, Y.; Mi, J.K. The Biomarkers in the Mesoproterozoic Organic-rich Rocks of North China Craton: Implication for the Precursor and Preservation of Organism in the Prokaryotic Realm. *Acta Geol. Sin.-Engl. Ed.* **2022**, *96*, 293–308. [CrossRef]
55. Fabianska, M.J.; Cmiel, S.R.; Misz-Kennan, M. Biomarkers and aromatic hydrocarbons in bituminous coals of Upper Silesian Coal Basin: Example from 405 coal seam of the Zaleskie Beds (Poland). *Int. J. Coal Geol.* **2013**, *107*, 96–111. [CrossRef]
56. Zhang, S.; Wu, T.; Zhang, S.; Cao, C.; Ma, W.; Shi, J.a.; Sun, G. Organofacies and paleoenvironment of lower Carboniferous mudstones (Dishuiquan Formation) in Eastern Junggar, NW China. *Int. J. Coal Geol.* **2015**, *150–151*, 7–18. [CrossRef]
57. Li, D.; He, D.; Santosh, M.; Ma, D.; Tang, J. Tectonic framework of the northern Junggar Basin part I: The eastern Luliang Uplift and its link with the East Junggar terrane. *Gondwana Res.* **2015**, *27*, 1089–1109. [CrossRef]
58. Li, D.; He, D.; Santosh, M.; Ma, D. Tectonic framework of the northern Junggar Basin Part II: The island arc basin system of the western Luliang Uplift and its link with the West Junggar terrane. *Gondwana Res.* **2015**, *27*, 1110–1130. [CrossRef]
59. He, D.; Chen, X.; Kuang, J.; Zhou, L.; Tang, Y.; Liu, D. Development and Genetic Mechanism of Chepaizi-Mosuowan Uplift in Junggar Basin, China. *Earth Sci. Front.* **2008**, *15*, 42–55. [CrossRef]
60. Pardo, J.D.; Small, B.J.; Milner, A.R.; Huttenlocker, A.K. Carboniferous–Permian climate change constrained early land vertebrate radiations. *Nat. Ecol. Evol.* **2019**, *3*, 200–206. [CrossRef] [PubMed]
61. Zhang, Y.X.; Wen, H.J.; Zhu, C.W.; Fan, H.F.; Cloquet, C. Cadmium isotopic evidence for the evolution of marine primary productivity and the biological extinction event during the Permian-Triassic crisis from the Meishan section, South China. *Chem. Geol.* **2018**, *481*, 110–118. [CrossRef]
62. Walliser, O.H. Global Events in the Devonian and Carboniferous. In *Global Events and Event Stratigraphy in the Phanerozoic: Results of the International Interdisciplinary Cooperation in the IGCP-Project 216 "Global Biological Events in Earth History"*; Walliser, O.H., Ed.; Springer: Berlin/Heidelberg, Germany, 1996; pp. 225–250.
63. Song, H.J.; Tong, J.N.; Xiong, Y.L.; Sun, D.Y.; Tian, L.; Song, H.Y. The large increase of delta C-13(carb)-depth gradient and the end-Permian mass extinction. *Sci. China-Earth Sci.* **2012**, *55*, 1101–1109. [CrossRef]
64. Bush, A.M.; Bambach, R.K. Paleoecologic Megatrends in Marine Metazoa. In *Annual Review of Earth and Planetary Sciences*; Jeanloz, R., Freeman, K.H., Eds.; Annual Review: San Mateo, CA, USA, 2011; Volume 39, pp. 241–269.
65. Kaiho, K.; Chen, Z.Q.; Ohashi, T.; Arinobu, T.; Sawada, K.; Cramer, B.S. A negative carbon isotope anomaly associated with the earliest Lopingian (Late Permian) mass extinction. *Palaeogeogr. Palaeoclimatol. Palaeoecol.* **2005**, *223*, 172–180. [CrossRef]
66. Kaiho, K.; Aftabuzzaman, M.; Jones, D.S.; Tian, L. Pulsed volcanic combustion events coincident with the end-Permian terrestrial disturbance and the following global crisis. *Geology* **2021**, *49*, 289–293. [CrossRef]
67. Lo, C.H.; Chung, S.L.; Lee, T.Y.; Wu, G.Y. Age of the Emeishan flood magmatism and relations to Permian-Triassic boundary events. *Earth Planet. Sci. Lett.* **2002**, *198*, 449–458. [CrossRef]
68. Abrajevitch, A.; Van der Voo, R.; Bazhenov, M.L.; Levashova, N.M.; McCausland, P.J.A. The role of the Kazakhstan orocline in the late Paleozoic amalgamation of Eurasia. *Tectonophysics* **2008**, *455*, 61–76. [CrossRef]
69. Xiao, W.; Santosh, M. The western Central Asian Orogenic Belt: A window to accretionary orogenesis and continental growth. *Gondwana Res.* **2014**, *25*, 1429–1444. [CrossRef]

minerals

MDPI

Article

Metals in Lake Sediments as Indicators of Human Activities in Prehistory: Case Study of the Southeastern Baltic, Kamyshovoe Lake

Olga Druzhinina [1,*], Laura Gedminienė [2] and Kasper van den Berghe [3]

[1] Faculty of Geography Herzen State Pedagogical, University of Russia, 191186 Saint-Petersburg, Russia
[2] Nature Research Centre, Institute of Geology and Geography, 08412 Vilnius, Lithuania
[3] FindX Research Center, 8031 VK Zwolle, The Netherlands
* Correspondence: olga alex.druzhinina@gmail.com

Abstract: This paper presents the results of geochemical research on the Kamyshovoe Lake sediments (Kaliningrad oblast, Russian Federation). The study of Pb, Ni, Zn, As, Co and Cu concentration and enrichment factors (EF) combined with the results of the lithological, geochronological, magnetic susceptibility and microcharcoal studies revealed possible anthropogenic sources of metals in southeastern Baltic lake sediments from the Neolithic to the Medieval period. Increasing Co EF value and peaks of the Pb EF in Kamyshovoe Lake sediments, starting from ~6000 cal yr BP in the Neolithic, probably show the growing role and usage of metals as dyes and fixatives. Since ~3100 cal yr BP, in the end of the Bronze Age, a simultaneous increase in the content of indicators of metallurgical production Pb, Ni, Zn and As, coinciding with growth of the microcharcoal curve, can show a growing demand for metal objects in the southeastern Baltic region and the input of the local or regional ancient metallurgy into the metal pollution of the lake sediments.

Keywords: metals; lake sediments; prehistory; ancient dyes and fixatives; ancient metallurgy; pollution history; lead; geochemistry; archaeology; southeastern Baltic

Citation: Druzhinina, O.; Gedminienė, L.; van den Berghe, K. Metals in Lake Sediments as Indicators of Human Activities in Prehistory: Case Study of the Southeastern Baltic, Kamyshovoe Lake. *Minerals* **2022**, *12*, 1216. https://doi.org/10.3390/min12101216

Academic Editor: Hermann Kudrass

Received: 5 August 2022
Accepted: 21 September 2022
Published: 27 September 2022

1. Introduction

Presence of metals in human life dates back to Palaeolithic times, when metals were used for ritual purposes and as mineral colourants. And since then, up until modern times, metals have been present in human life in the form of dyes and fixatives (Cu, Fe, Pb, Co, As) and metal objects, etc. [1,2]. The appearance of metallurgy in the Neolithic–Bronze Age period brought crucial changes in prehistoric life. Simultaneously, ancient metallurgy as a new essential activity started to play a role in transforming the landscapes, causing metal pollution due to mining, smelting and working, and deforestation due to demand for wood and charcoal [3].

While studies seem to focus on the objects made of metal, their aesthetics and social function, the level of knowledge about the environmental impact of ancient metal mining and metalworking is very limited [3]. The palaeoecological aspects about pre-metallurgical use of metals seems to be unknown, though from the early days of extensive metal smelting onwards, evidence of the impact of metallurgy on atmospheric chemistry is present, as it is reflected in the Greenland ice cores, European lake sediments and in peatbogs [4,5]. Despite that, the scale and directions of the atmospheric metal dispersal remain disputable as does the impact of ancient metallurgy on surface runoff and groundwater [3,6].

Numerous studies testify that lake sediments are very sensitive indicators of the geochemical processes occurring in the lake and in the catchment area [3,7–10]. Of particular value are lake deposits, which have a continuous succession throughout the Holocene and allow for tracing possible anthropogenic influence on the landscapes at all stages of society development. One of such lakes is Kamyshovoe Lake considered in this study. To date, lake sediment research has provided numerous and diverse data that are shedding light on the various aspects of environmental dynamics, including the ancient anthropogenic impact. The previous research of Kamyshovoe Lake revealed both environmental and anthropogenic processes that affected the geochemical composition of the sediment sequence [8–10]. The study showed an increase in the concentration of heavy metals (Zn, V, Cr, Pb) from depths corresponding to the end of the Neolithic [9]. The simultaneous increase in the microcharcoal value and rate of erosion reflected the intensification of human activities surrounding the lake [9]. The aim of the present paper is to consider the elevated values of heavy metals in the Kamyshovoe Lake sediments in the context of possible regional and local anthropogenic activities, which could be sources of the metal input in the lake during the Neolithic–Medieval period.

2. Methods

2.1. Fieldwork and Sampling, Geographical Background

Kamyshovoe Lake (N 54°22′; E 22°42′) occupies an area of 0.32 km^2, is 850 m long and 570 m wide, and is shallow with a maximum depth of 4.5 m. It is located on the terminal moraine ridge formed during the retreat of the ice-sheet associated with the Baltija stage of Weichselian (Vistulian) glaciation. The Quaternary deposits forming the relief are mosaic, with a thickness of around 200 m: glaciofluvial sand, gravel interlayers, boulder loam (till).

At the present time, the area surrounding the lake is forested and not used for agricultural purposes. Archaeological data, however, provides evidence that the landscapes surrounding the lake were settled and exploited by humans during all historical periods (Figure 1), starting from the late Palaeolithic, while the anthropogenic impact on the environment reflected in the lake sediments becomes visible from the end of Mesolithic [9].

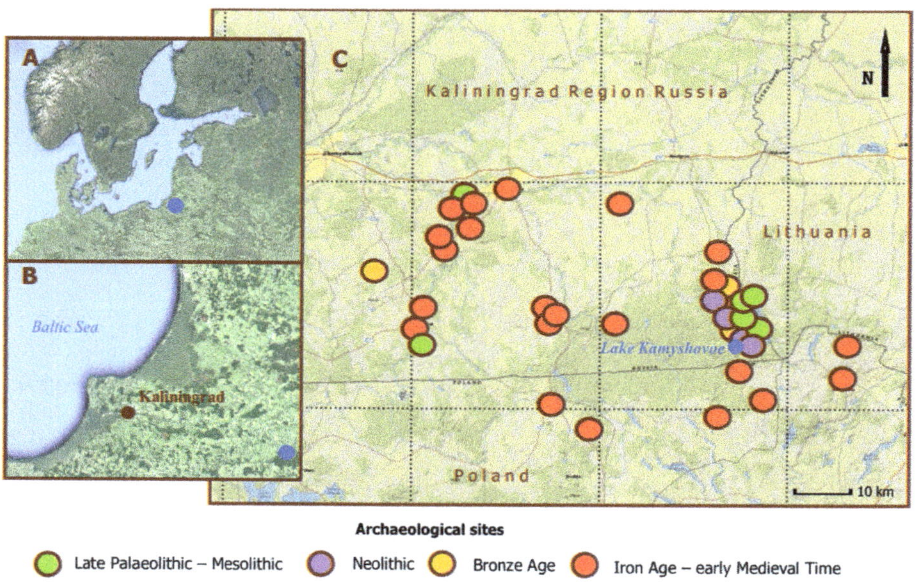

Figure 1. (**A**)—Location of the studied site on the map of Europe. (**B**)—Location of Kamyshovoe Lake in the Kaliningrad region. (**C**)—Archaeological sites of different periods in the vicinity of the lake.

A 9.8 m long core of bottom sediments was obtained from the ice surface using a Russian peat corer (100 cm long chamber; diameter 7 cm). Detailed results for the entire sediment section (1199–230 cm) and its palaeoenvironmental and possible anthropogenic impact interpretation were presented in [8,9]. The recent study is limited to the interval

730–245 cm, coinciding with Neolithic–Medieval period, the time when human activities related to use of metals would be expected to have intensified.

2.2. Geochronological and Lithological Analysis

Sediments were described based on the visual and physical description of the composition and colour of bottom sediments [10]. Thirteen radiocarbon analyses were performed for bulk gyttja for the section analysed in this paper (730–245 cm). From the data obtained, an age–depth relationship using linear interpolation was calculated. Using radioactive carbon (^{14}C), the absolute age of the sediments was determined at the Laboratory of Geochronology, St. Petersburg State University, Russia (laboratory index LU). OxCal v 4.2.4 was used for calibrating the dates [11].

2.3. XRF Geochemical Analysis, Magnetic Susceptibility (MS) Measurements and Microcharcoal Concentration

The geochemical analysis was performed for 1 cm intervals. The samples were dried at 105 °C until the moisture evaporated completely and were then powdered in a mortar. Tablets for geochemical analysis were prepared using boric acid as a base and a laboratory press (pressure 110 bar). A SPECTROSCAN MAKC-GV X-ray fluorescence spectrometer at Herzen State Pedagogical University of Russia, St. Petersburg, was used to perform the analysis. Accuracy of the spectrometer was tested by analysing over 50 certified international standard reference samples (CRMs). The mass fractions of chemical elements and oxides, such as Ba, La, Rb, Nb, Zr, Cr, Sr, As, Pb, Zn, Cu, Ni, Co, V, TiO_2, MnO, Fe_2O_3, CaO, Al_2O_3, SiO_2, P_2O_5, K_2O, MgO and Na_2O, were measured and expressed in ppm or % [10]. In the recent study, only the data on Cu, Ni, Pb, Zn, Co, As and TiO_2 were used for interpretation. TiO_2/Al_2O_3 and MnO/Fe_2O_3, as indicative of the mafic rocks and redox conditions, were also considered in the present study. The total number of samples used for the XRF geochemical analysis was 250: n = 250, Medieval period n = 26, Iron Age n = 44, Bronze Age n = 77, Neolithic n = 87, Mesolithic n = 16.

Magnetic susceptibility analysis was applied for 2 cm intervals. The analysis was performed at the Palaeomagnetic Laboratory, Nature Research Centre, Vilnius, Lithuania, using the standard MFK1-B kappa bridge (AGICO) static specimen method with a manual holder. The data obtained were analysed using SAFYR software (Safyr Metadata Discovery Software for Enterprise Application Packages. Silwood Technology Ltd., Ascot Berkshire, UK), and MS values were expressed in SI units (10^{-9} m^3 kg^{-1}) [8]. The number of samples used for the MS analysis totalled 83: Medieval period n = 8, Iron Age n = 15, Bronze Age n = 26, Neolithic n = 28, Mesolithic n = 6.

Microscopic charred particles with a longer axis of 10–25 µm [12] were counted on the prepared slides in during pollen analysis [9]. The percentage of charred particles was calculated in relation to the total pollen sum (ΣSUM = ΣP + ΣC). Microscopic charcoal particles were counted within a 230–700 cm sediment range. The total number of samples used for the microcharcoal concentration analysis was 137: Medieval period n = 9, Iron Age n = 22, Bronze Age n = 51, Neolithic n = 53, Mesolithic n = 2.

2.4. Enrichment Factors (EF)

Element concentrations can be influenced by the amount and intensity of mineral matter transported from the catchment into the lake. This is clearly observed in the late Pleistocene part of the Kamyshovoe sediment sequence. Here, the concentrations of all metals are generally higher due to more intensive weathering and erosion processes in the vicinity of the lake with unstabilised soils [8] and due to lower organic matter (either allochthonous or autochthonous) concentrations.

To minimise a possible influence of erosional input, metal concentrations were normalised to Ti, which is a conservative element and a good tracer of minerogenic inputs in lakes [13]. Counting was done according to the formula proposed by [13]:

$$XEF = (X/Ti)sample/(X/Ti)bkg$$

where X is element, EF—enrichment factor and bkg—mean relative background values calculated for the depth of 730 cm. This depth is chosen considering lithology: while downwards in the sequence we observe the alternation of sediment types formed by terrigenous and, to a lesser extent, chemogenic processes, here starts the deposition of organogenic gyttja comprising the upper 5 m of the core, nearly to the top, formed in more stable environmental conditions of the Middle and Late Holocene. The depth of 730 cm also coincides with the end of the Mesolithic, a period when the anthropogenic impact related to metal pollution is still expected to be minimal, while it can be increasing upwards in the sediment sequence during the later historical periods.

2.5. Statistical Analysis

Spearman rank R correlation coefficients between the content of trace metal, TiO_2/Al_2O_3 and MnO/Fe_2O_3, MS and microcharcoal were calculated by Statistica software (StatSoft. Inc., Hamburg, Germany (2007). STATISTICA (Data Analysis Software System), version 8.0.). Calculation was done for the entire section using 250 samples (valid n) for all elements and 25 samples (valid n) for microcharcoal data. Significance levels (*p*-values) for the correlations were calculated as well. Significant ($p < 0.05$) Spearman correlation coefficients are marked.

3. Results

3.1. Lithology and Geochronology

Analysis of lithological composition and geochronological modelling revealed the following lithological units: dark brown gyttja (730–270 cm), light brown gyttja (270–247 cm), and gray gyttja (247–240 cm). The analysed interval comprised four historical periods from the Neolithic to the Medieval period.

3.2. Geochemistry, Magnetic Susceptibility and Microcharcoal Concentration

In this study, the focus was on a selected group of elements (Cu, Pb, Zn, Co, As, Ni), which could be tracers of various human activities related to metal use in daily life or tracers of ancient metal smelting. Meanwhile, the detailed description of the full geochemical composition and its palaeoenvironmental and possible anthropogenic impact interpretation was presented in previous articles [8–10].

The average and maximum values of Cu, Zn, Pb, As, Co and Ni in ppm, compared with average values of MS and microcharcoal percentage against the historical periods, are presented in Figures 2 and 3 (Table S1).

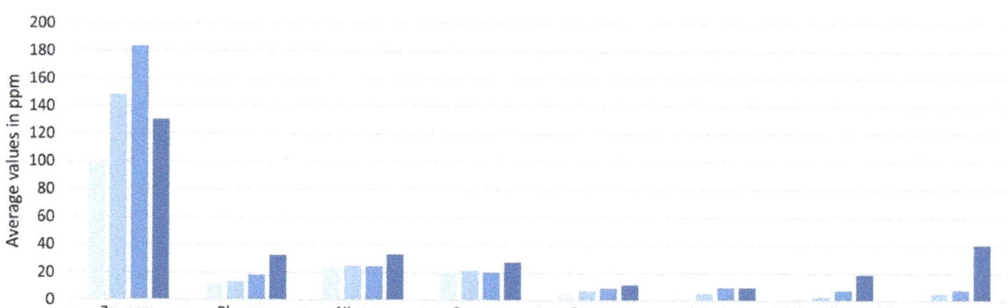

	Zn, ppm	Pb, ppm	Ni, ppm	Cu, ppm	Co, ppm	As, ppm	Microcharcoal	MS
Av Neolithic	98.69 (n = 87)	12.085 (n = 87)	24.46 (n = 87)	20.76 (n = 87)	4.86 (n = 87)	2.19 (n = 87)	1.5 (n = 53)	0.91 (n = 28)
Av Bronze age	148.55 (n = 77)	13.63 (n = 77)	25.18 (n = 77)	21.97 (n = 77)	7.48 (n = 77)	5.31 (n = 77)	2.6 (n = 51)	5.11 (n = 26)
Av Iron age	183.76 (n = 44)	18.22 (n = 44)	24.68 (n = 44)	20.49 (n = 44)	9.14 (n = 44)	9.59 (n = 44)	7.6 (n = 22)	7.71 (n = 15)
Av Medieval	130.43 (n = 26)	32.46 (n = 26)	33.13 (n = 26)	27.78 (n = 26)	11.63 (n = 26)	9.39 (n = 26)	18.98 (n = 9)	40.25 (n = 8)

Figure 2. The average values of element concentrations in the Kamyshovoe Lake sediments according to historical period.

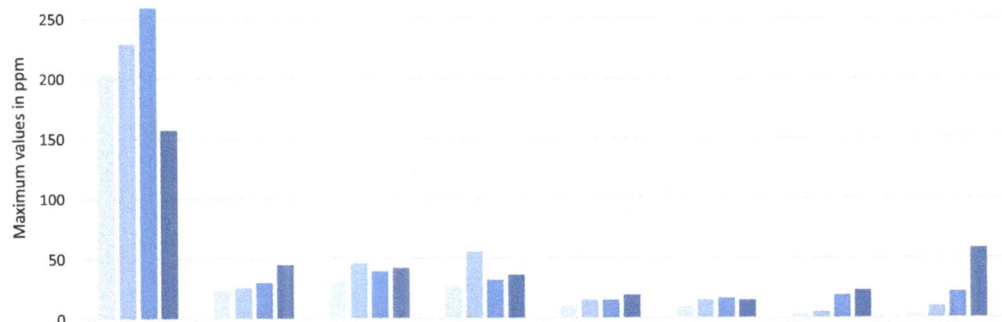

	Zn, ppm	Pb, ppm	Ni, ppm	Cu, ppm	Co, ppm	As, ppm	Microcharcoal	MS
Max Neolithic	203.23 (n = 87)	23.18 (n = 87)	30.80 (n = 87)	26.46 (n = 87)	10.48 (n = 87)	9.88 (n = 87)	3.1 (n = 53)	3.08 (n = 28)
Max Bronze age	229.11 (n = 77)	25.78 (n = 77)	46.22 (n = 77)	55.57 (n = 77)	15.51 (n = 77)	15.31 (n = 77)	5 (n = 51)	9.95 (n = 26)
Max Iron age	259.52 (n = 44)	29.94 (n = 44)	39.2 (n = 44)	31.96 (n = 44)	15.3 (n = 44)	16.7 (n = 44)	19.2 (n = 22)	21.78 (n = 15)
Max Medieval	157.16 (n = 26)	44.84 (n = 26)	42.06 (n = 26)	36.23 (n = 26)	19.38 (n = 26)	14.76 (n = 26)	23.2 (n = 9)	58.24 (n = 8)

Figure 3. The maximum values of element concentrations in the Kamyshovoe Lake sediments according to historical period.

The results showed that the geochemical elements under consideration do not follow a uniform trend from the Neolithic to the early Medieval period, though the concentrations of almost all the elements show increased values towards the upper part of the core. The average values of most of the elements together with microcharcoal and MS reached a maximum during the Medieval period. The exception was Zn, which had a clear peak during the Iron Age, and As with nearly the same concentration during the Iron Age and the Middle Ages. The values of such elements as Pb, Co and As in the Iron Age and the Medieval period became 2–3 times higher than in Neolithic, while concentration of Cu and Ni grew smoothly.

Regarding the highest concentrations throughout the historical periods, the pattern was slightly different. While maximum value trends of Pb, Zn, Co and As coincided with those of average values, Ni and Cu reached the maximum concentration not in the Medieval period but during the Bronze Age.

TiO_2/Al_2O_3 ratio reflected the mild intensity of weathering during the period studied (Figure S1). Compared to the lower part of the sediment profile [9,10], increased TiO_2/Al_2O_3 ratio showed that weathering indicated at the depth 730–245 cm was minimal; however, it increased a few times at about 3640–3150 cal yr BP and from 1950 cal yr BP with short, less intensive weathering events at the end of the Iron Age and beginning of the Medieval period. TiO_2/Al_2O_3 value varied between about 0.06 (more intense weathering) and 0.1 (minimal weathering).

The recorded low MnO/Fe_2O_3 ratio allows speculation about anoxic conditions throughout the interval. The decreased MnO/Fe_2O_3 ratio point to the hypothesis that the internal conditions of the lake, especially at the end of sedimentation, gradually changed to a reducing environment, which most probably led to increased precipitation of some trace metal such as Co and Pb and, especially, As [10]. MnO/Fe_2O_3 ratio varied between about 0.01 to 0.05 (Figure S1).

The highest MS and microcharcoal concentrations were observed during the Medieval period and had several high peaks during the Iron Age. On average, it increased almost 12 times for microcharcoal and almost 40 times for MS in comparison to the values observed in the Neolithic and Medieval period.

3.3. EF and Background Values

The results demonstrated different trends of average EF values for elements under consideration (Figures 4 and 5; Tables S2 and S3). Co, As and Zn showed the tendency to increase from the Neolithic (1.53, 3.40 and 0.93) to the Iron Age, when they reached their highest values (2.53, 10.93 and 1.41, respectively). Cu and Ni demonstrated a similar fluctuating trend: their values decreased from the Neolithic to the Bronze Age (from 0.67 and 0.68 to 0.46 for both elements) with some rise during the Iron Age (0.52 and 0.54). The average EF of Pb generally increased from the Neolithic to the Medieval period (from 1.01 to 1.24) with some drop during the Bronze Age (0.75). Average EF of all elements except Pb decreased during the Medieval period.

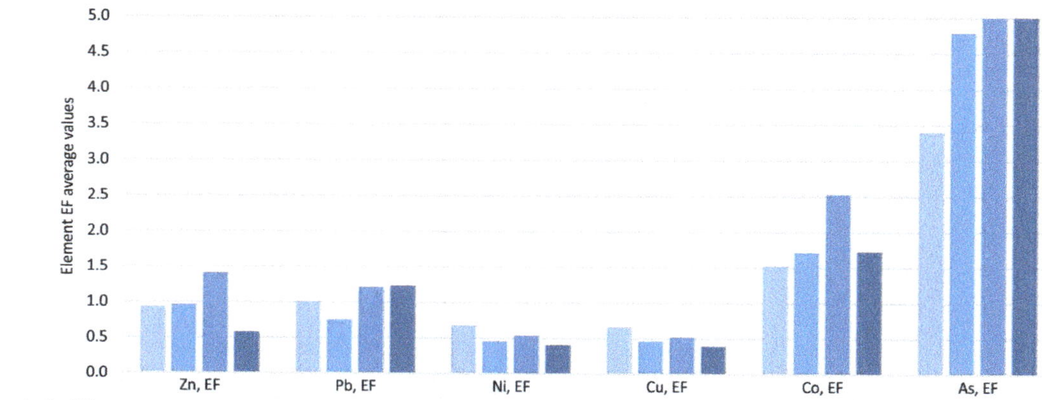

	Zn, EF	Pb, EF	Ni, EF	Cu, EF	Co, EF	As, EF
Av Neolithic	0.93 (n = 87)	1.01 (n = 87)	0.68 (n = 87)	0.67 (n = 87)	1.53 (n = 87)	3.40 (n = 87)
Av Bronze age	0.96 (n = 77)	0.75 (n = 77)	0.46 (n = 77)	0.46 (n = 77)	1.72 (n = 77)	4.79 (n = 77)
Av Iron Age	1.41 (n = 44)	1.21 (n = 44)	0.54 (n = 44)	0.52 (n = 44)	2.53 (n = 44)	10.93 (n = 44)
Av Medieval	0.58 (n = 26)	1.24 (n = 26)	0.41 (n = 26)	0.40 (n = 26)	1.73 (n = 26)	6.17 (n = 26)

Figure 4. The average values of element enrichment factors (EF) in the Kamyshovoe Lake sediments according to historical period.

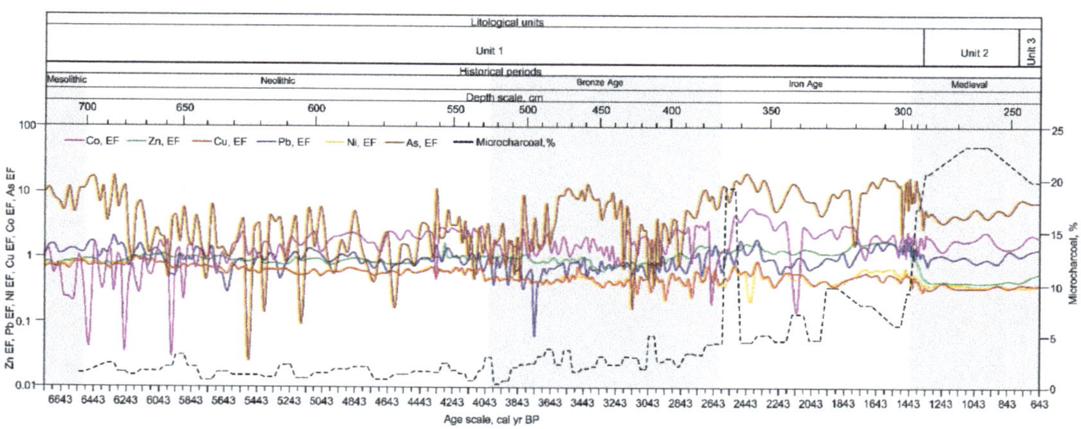

Figure 5. Co, Ni, Zn, Pb, As and Cu enrichment factors (EF) and the microcharcoal concentration in the Kamyshovoe Lake sediments. Modelled age (linear interpolation), cal yr BP.

The study showed that the EF of Co (5.88), As (19.91), Zn (1.81) and Fe (1.33) had a clearly expressed maximum during the Iron Age, while Cu, Ni and Pb did not display similar sharp fluctuations during the entire study period.

3.4. Statistical Analysis

Statistical results are presented in Table 1 and Figures S2 and S3. The highest positive correlation is seen between microcharcoal and TiO2, Co, Ni, MS; TiO_2 and Co, Ni; Co and MS; Ni and Cu; Cu and As, etc. The highest negative correlation coefficients are between Zn, MS and microcharcoal. We observe no correlation between, e.g., microcharcoal and As; Co and Pb. TiO_2 and Zn. TiO_2/Al_2O_3 ratio has the highest positive correlation with Pb (also Ni, Cu and Zn), showing that it precipitates more when weathering is less intensive; however, Co precipitation is higher when weathering is more intense. The same tendency is shown by the high negative correlation with MS, proving that the higher MS values can predict weathering intensity. MnO/Fe_2O_3 ratio has a positive correlation only with Zn and Pb. These two elements most probably precipitate in more oxic conditions and other elements, e.g., Co, Ni and As, precipitate in redox conditions.

Table 1. Spearman rank order correlations (MD pairwise deleted) of trace metal concentrations, TiO_2/Al_2O_3 and MnO/Fe_2O_3, MS and charcoal. Marked in **red** are correlations significant at $p < 0.05$.

	Microcharcoal,%	TiO_2,%	Co, ppm	Ni, ppm	Cu, ppm	Zn, ppm	Pb, ppm	As, ppm	MS, 10^{-9} m^3/kg	TiO_2/Al_2O_3	MnO/Fe_2O_3
Microcharcoal, %	1.00	0.76	0.79	0.66	0.39	−0.84	−0.53	−0.09	0.89	0.04	−0.95
TiO_2,%	0.76	1.00	0.65	0.60	0.45	−0.07	0.26	0.24	0.66	0.19	−0.6
Co, ppm	0.79	0.65	1.00	0.41	0.09	−0.28	−0.05	0.10	0.82	−0.35	−0.53
Ni, ppm	0.66	0.60	0.41	1.00	0.84	0.50	0.61	0.32	0.73	0.56	−0.18
Cu, ppm	0.39	0.45	0.09	0.84	1.00	0.56	0.53	0.60	0.56	0.67	−0.12
Zn, ppm	−0.84	−0.07	−0.28	0.50	0.56	1.00	0.84	−0.05	−0.98	0.68	0.63
Pb, ppm	−0.53	0.26	−0.05	0.61	0.53	0.84	1.00	−0.23	−0.18	0.72	0.52
As, ppm	−0.09	0.24	0.10	0.32	0.60	−0.05	−0.23	1.00	−0.20	0.23	−0.43
MS, 10^{-9} m^3/kg	0.89	0.66	0.82	0.73	0.56	−0.98	−0.18	−0.20	1.00	−0.69	−0.82

4. Discussion

The previous study shows that human activity in the Kamyshovoe Lake surroundings grew, starting from ~6500 cal BP [9]. This process is reflected not only in archaeological data, but in the simultaneous start of the microcharcoal curve, increased level of soil erosion and in the pollen record of plants-indicators of human activity [9]. The increased anthropogenic input is also confirmed by geochemical indicators. From the Neolithic onwards, the concentration of elements such as Co, Zn and As increased. The same trend can be seen in the dynamics of enrichment factors against the background of minimised contribution of erosion. As the present study shows, the EF of Co, Zn and As increase from the Neolithic and have maximum values in the Iron Age. It should be emphasised that the Iron Age is the period when all the elements in consideration have the highest values of the EF (such as Co, As and Zn) or increased values compared to the previous period (Pb, Cu and Ni). Apparently, the decrease in values observed later, in the Middle Ages, is related to local historical events that took place in the Vyshtynets Upland. After the conquest and devastation of this region by the Teutonic Order, the landscapes surrounding the lake were transformed for several centuries into 'wild lands' not involved in any economic activities.

The possible sources of the anthropogenic metal input into Kamyshovoe Lake over the last six millennia are considered further.

4.1. Use of Metals in Prehistory

Long before the mass use of metals for the production of metal objects and the emergence of 'classical' metallurgy, the use of metals in the form of minerals has its roots in the Palaeolithic, where burials accompanied with finds of mineral dyes are known. Probably just as ancient is the use of mineral dyes in 'cosmetics' for ancient rituals and ceremonies [2].

Before the emergence of metallurgy, metals could also be used by people in two other important spheres of life, such as fixatives for organic dyes in the production of textiles (and possibly as dyes themselves) and in colouring and decorating ceramics. The most commonly used dyes in prehistory were Cu, Fe, Pb, Co and As [1,2]. The data regarding the oldest regional textile production are provided by Polish archaeologists [14]. The first evidence for fibre and textile production appeared around 7450 cal yr BP (5500 BC) in southeastern Poland, and during the next two millennia the agricultural societies became well established and increasingly sophisticated. The Bronocice settlement (southern Poland) dated to 5650–4850 cal yr BP (3700–2900 BC) is an example of economically and technologically well-developed centre of fibre and cloth production, increasing in scale during the life of the settlement [14].

Archaeological data testify that new ways of settling and an economy with stationary settlements, which were also established in the Neolithic in the northern Poland [15] and Kaliningrad region [9], led to a growth of population in this part of the southeastern Baltic. It can be assumed that the gradual increase in population, accompanied by an increase in the production of daily-use items such as clothing, utensils and other household articles, could result in growing demand for metals as mineral dyes and colourants or fixatives for woodwork, leather production, textiles and pottery. Mineral dyes were certainly the subject of the oldest trade and exchange systems that already existed in prehistory [16]. In this context, it seems plausible that increasing from the Neolithic onwards Co EF and peaks of the Pb EF at about 6400–5700, 5100 and 4300–4200 cal yr BP in Kamyshovoe Lake sediments probably reflect the growing role and usage of metals in Neolithic society on local or regional scale (Figures 4 and 5).

At the same time, the idea of tracing the anthropogenic input of the most commonly and widely used mineral dyes such as Cu and Fe seems to be a challenging task. Owing to the strong affinity of Cu for complexation with organic ligands, the Cu value is strongly dependent on the influx of organic matter into the lake and autochthonous geochemical processes [13], while iron ores ($nFe_2O_3 + nH_2O$) are formed in the lakes and bogs themselves. Therefore, archaeological but not palaeolimnological data remain so far the only rigid evidence regarding usage of Cu and Fe in the life of prehistoric people. High values of As EF in the beginning of the Neolithic, before the expected growing usage of metals, shows that the dynamic of this element should be interpreted with caution as well, because As is a highly redox-sensitive element and its dynamics can also be related to autochtonic oxic/redox condition changes in the lake. However, the elevated values of As EF in later periods, when correlated with other metals, give more basis to connect it with the start of metal smelting in the region.

4.2. Metallurgy as a Source of Anthropogenic Input of Metals in the Lakes

Following the emergence of metallurgy—metal ore mining, metal smelting and production of metal objects—a new source of anthropogenic input of metals in the lakes occurred. While individual archaeological bronze artefacts found in Serbia, dating back to ~7300–6550 cal yr BP (5350–4600 BC), revealed the earliest evidence for copper smelting in Europe [17], since the late Neolithic–early Bronze Age numerous copper and polymetallic copper-bearing deposits in Europe had already been exploited [18,19]. The earliest impact of metal production on lake sediments is recorded in layers dated to 5750 cal yr BP (3800 BC) [13].

It is hypothesised that the emergence of metallurgy is closely related to the development of the production of ceramics [20,21]. The production of ceramics in the pottery kiln and usage of metals as colourants and glazes could lead to the inadvertent discovery of new qualities of metals and their melting properties. Though recent studies in the Balkans show that probably these two processes appear as related technologies and not as one being the precursor to the other [21], it is still in line with the view that people obtained their experience of metal working far before the Bronze Age. In the Eneolithic, which is interpreted by many researchers as a transitional stage from the Neolithic to the Bronze Age, metal products played an increasing role in the life of people, and a transition from a ritual meaning and an attribute of power to a more utilitarian use of objects took place [20,22].

According to studies [16,23], during the Bronze Age in Europe there was already a dynamic exchange and trade in raw metals, semifinished and finished products made from bronze and other alloys. Ling et al. [23] came to the conclusion that there were two routes for metal supply: the 'Atlantic' (western maritime route) and the continental route, 'repeating' the one for amber. During this period, the southeastern Baltic was situated at the crossroads between different centres of trade and exchange, but mainly between those located in Scandinavia and Central Europe [16].

The significance and scale of the production of bronze products during this period is evidenced by the excavation of the oldest among the currently known settlements of metallurgists in Central Europe, Szczepidło (Poland) [24]. The settlement dates back to 3300–3100 cal yr BP (1350–1150 BC). As semifinished and finished products, production waste, fragments of crucibles and casting ladles with traces of usage, and tools were found, the archaeological finds represent markers of the whole production cycle [25].

Besides metallurgy, the settlers of Szczepidło had a wide range of activities. Evidence for the production of wood for fuel and for building purposes were found, so too for pottery, clothing and agricultural practices and cattle breeding [24]. Approximately from this time, ~3100 cal yr BP, a major simultaneous increase in the content of metals—indicators of metallurgical production (Pb, Ni, Zn, As)—starts to emerge in the Kamyshovoe Lake sediments as well. This probably reflects the general picture of growing demand for metals in the southeastern Baltic. The observed trend continues during the entire Iron Age until a drop during the Migration Period.

Despite the fact that no metal production locations in the lake surroundings have been found, archaeological sites dated from the Bronze Age to the Middle Ages, as well as archaeological finds of bronze and iron items are numerous in this area and in the broader Kaliningrad region. Archaeometallurgical analysis has been applied on two of those finds, a Bronze Age hoard from the Svetly site (six items) and an early Roman neck ring from Kalinovo site.

The bronze hoard from the Svetly site was studied by means of energy dispersive X-ray fluorescence analysis. The studied samples contain significant concentrations of Ag (up to 0.61%), Ni (up to 0.74%), As (up to 1.4%), Sn (up to 15.6%) and Pb (up to 7.1%), classifying the objects as dilute fahlore copper with nickel, which is common in Central Europe in the Middle and Late Bronze Age [26].

The neck ring from the Kalinovo site, studied by electron probe microanalysis of the enamel decoration, revealed a high concentration of PbO (34.8%) and CuO (5.62%) in its chemical composition [27]. Enamel with high Cu and Pb concentration is known in Europe from ~2450 cal yr BP (500 BC), representing one more application of these metals in metalworking activities [28].

Further archaeometallurgical research of finds and expanding the database will probably help to sharpen ideas regarding the origin of the metal ores used for production of items, though this issue will remain complicated due to the long, complex and sometimes untraceable chain of transformations, which metals pass through on the way from ore to final object [29].

At the same time, the history of local and regional metallurgy—existence of metal smelting workshops, metalworking traditions and trade routes—can be reconstructed only if archaeological excavations are supplemented by targeted geochemical research on the culture-bearing layers. Geochemical analysis at archaeological sites has already demonstrated its high potential regarding the indicators of various human activities [30,31]. Another important point is the proper location, collection and analysis of metal slags, including microslags, which until now have received undeservedly much less attention from archaeologists than, for example, pottery sherds [18,32]. Microslags, if found in lake sediments probably could be considered as a reliable indicator of local metal smelting in the surrounding of the lakes, as well as on excavation sites.

4.3. Long-Range Atmospheric Transport or Local Input?

Available regional data show that Kamyshovoe Lake displays an elevated concentration of Co (Table 2) compared to Latvian lakes [7]. The same observation applies to values of Ni in the Kamyshovoe Lake sediments (max = 46 ppm) compared to both Latvian (max = 35.7 ppm) and Polish lakes (max = 25 ppm) from the Suwalki region [7,33]. Kamyshovoe Lake shows a higher concentration of Pb than Latvian lakes but falls into range with medium values compared to Suwalkian lakes, where values of Pb in some lakes exceed 50 ppm and even 75 ppm in the sediments of Mulaczysko Lake [33].

Table 2. Concentration (ppm) of metals in the sediments of Kamyshovoe Lake (interval 730–245 cm; min and max values), Suwalki Lakeland (Preindustrial interval 52–50 cm; median value and standard deviations) and Pilvelis Lake (interval 400–0 cm; min and max values) [7,33].

Element	Kamyshovoe Lake	Suwalki Lakeland	Pilvelis Lake
Cu	16–34	16.7 ± 7.3	7.6–19.2
Ni	18–46	15.4 ± 9.1	3.1–17.8
Pb	5–45	27.6 ± 28.6	0.6–30.9
Zn	54–260	112.2 ± 60.7	46–198
Co	0.1–19	–	0.73–6.08

Concerning lead pollution identified in the lake sediments of northern Europe, it is interesting to compare the Kamyshovoe Lake Pb values during certain intervals correlating with particular historical time periods (Table 3).

Table 3. Pb concentration (ppm) in the sediments of Kamyshovoe Lake (min and max values within interval), Swedish lakes (mean value and range within different lakes) and Pilvelis Lake (Latvia) during different historical times [5,7].

Historical Time	Kamyshovoe Lake	Swedish Lakes	Pilvelis Lake
Before 2000 BC	6–18	7 (range = 2–17)	0.6–7
2000–500 BC	6–30	7	0.6–7
500 BC–400 AD	10–28	11–20 (range = 2–41)	7–15
400 AD–900 AD	12–36	-	15–30.9
900 AD–1400 AD	24–45	10–50 (range = 3–110)	20

In Central and Southern Europe, increased Hg and Pb enrichment values in the lake sediments, compared with crustal values and Pb isotopes, provide evidence for the impact of human activities during the Chalcolithic (~3800 BC) [13], while studies of the northern lakes show different results. According to Brännvall et al. [5], the first traces of atmospheric lead pollution in Swedish lake sediments are recorded ~3500–3000 cal BP (1550–1050 BC) with a clearly expressed peak of pollution at around 2550 cal BP (600 BC), while Pb values of earlier periods should be considered rather as natural background. The next pollution peak in the Iron Age–Roman Times (500 BC–400 AD) is followed by a decline of pollution in Early Medieval period (400 AD–900 AD); then (from about 1000 AD), a significant and permanent increase in lead pollution starts with clear peaks at 1200 AD and 1530 AD.

The most clearly expressed difference between the Kamyshovoe Lake Pb concentrations and other lakes (Table 3) concerns the Bronze Age–beginning of the Iron Age interval (2000–500 BC). Here, pronounced elevated values of Pb (up to 30 ppm) are recorded compared to other studied lakes, where the concentrations remain at the natural background values, despite pollution revealed by isotope analyses in some Swedish lakes [5]. The results show that peaks of Pb EF in the Kamyshovoe sediments are expressed at 850–350 BC (end of the Bronze Age–first half of the Iron Age), 150–400 AD (Roman Period) and after 1050 AD.

Considering the historical metal pollution in the Kamyshovoe Lake, the following question deserves further deeper study: Does the observed phenomenon reflect only the long-range atmospheric transport associated with metal mining in the main centres of Southern Europe, or is there a contribution of a more regional or even local component to the increased content of metals?

The hypothesis, that peaks of heavy metals (Pb in particular) found in the sediments of various water bodies in Northern Europe are the result of long-range atmospheric transport from centres of active metallurgy (Iberia, the Alps, the British Isles, etc.), is based mainly on results of isotopic analysis of metals, suggesting nonlocal origin [5]. However, when attempting to explain the peaks of metals with 'nonlocal signature' in the sediments of northern Europe by long-range atmospheric transport, two points must be taken into account. First, as Elbaz-Poulichet et al. [13] point out, different metals have different residence time in the atmosphere. This, along with the windrose for a particular time period and area, must be taken into account when reconstructing long-range transport of metals. Respective residence time in the atmosphere for different metals ranges from approximately 6 months for Hg to a few days or weeks for Pb, Sb and Cu, which can result in their atmospheric dispersal. An example from the Falu mine in Sweden, provided by [5], shows that lead pollution from the copper mines can be limited to a very local scale, not ranging beyond the first tens of kilometres. Secondly, as Ling et al. [23] and Melheim et al. [16] show, the sources of raw materials for local metalworking workshops in Northern Europe during the Bronze Age were mostly nonlocal but imported from the Mediterranean, the Alps, the British Isles and other mining and smelting centres. Since processing of this imported metal took place in local smelting workshops, it is possible that the recorded elevated content of heavy metals with a 'nonlocal geochemical signature' in lakes is rather the result of regional or local atmospheric and surface runoff or groundwater transport and pollution. Probably, some difference in the metal concentration and in the timing of the major Pb elevated values revealed in Kamyshovoe Lake, compared to Polish, Latvian and Swedish lakes, correlates better with an assumption about the local or regional, rather than long-range atmospheric metal input. This version is also supported by the fact that the reason for the observed decline of elemental EF in the Middle Ages has a local origin as well. Undoubtedly, the concentration and dynamics of metals in bottom sediments largely depend on the intensity of erosion, the input of organic matter, changes in the oxygen regime, etc.; however, it should be emphasised that anthropogenic impact is precisely one of the most important factors affecting the listed natural parameters. Still, to approach further understanding of local vs. long-range metal pollution, joint palaeolimnological and archaeological studies are required, for example, the study of lake sediments in the vicinity of ancient metallurgical centres such as Szczepidło archaeological site in Poland.

5. Conclusions

The results show that Cu, Pb, Zn, Co, As and Ni do not follow a uniform trend from the Neolithic to the early Middle Ages, though the concentrations of almost all of them show an increasing trend. The values of elements such as Pb, Co and As in the Iron Age and the Medieval period become 2–3 times higher than in the Neolithic, while the concentration of Cu and Ni grows insignificantly. Probably, the increasing Co EF value and the peaks of Pb EF in the Kamyshovoe Lake sediments, starting from ~6000 cal yr BP and during the entire Neolithic, show an increasing role and use of metals as dyes and fixatives. A major simultaneous increase in content of the Pb, Ni, Zn and As indicators of metallurgical production since ~3100 cal yr BP may show a growing demand for metal objects in the southeastern Baltic area, and input of local or regional metal smelting workshop pollution into lakes. The observed trend continues during the entire Iron Age until a decrease during the Migration Period. To give greater foundation to assumptions regarding metal use in prehistoric society, further combined palaeolimnological and archaeological research is necessary, with a focus on the study of microslags and detailed geochemical investigation of the culture-bearing layers of archaeological sites.

Supplementary Materials: The following supporting information can be downloaded at: https://www.mdpi.com/article/10.3390/min12101216/s1, Table S1: Basic statistics for the entire sediment section under consideration: values of the selected elements, microcharcoal and MS; Table S2: Natural background values; Table S3: Basic statistics for the EF of selected elements on the background of historical periods; Figure S1: TiO_2/Al_2O_3 and MnO/Fe_2O_3 as indicatives of the weathering and redox conditions; Figure S2: Box and whisker plot, mean values and the 25th and 75th percentiles; Figure S3: Box and whisker plot, mean values, SE and SD.

Author Contributions: Conceptualisation, O.D.; methodology, O.D. and L.G.; formal analysis, L.G.; data curation, L.G.; writing—original draft preparation, O.D. and K.v.d.B.; writing—review and editing, O.D., L.G. and K.v.d.B.; visualisation, L.G.; supervision, O.D.; project administration, O.D.; funding acquisition, O.D. All authors have read and agreed to the published version of the manuscript.

Funding: This research was funded by The Russian Science Foundation, project 22-17-00113.

Data Availability Statement: Raw data are available upon request; please contact the corresponding author.

Acknowledgments: The authors are grateful to S. Ratkai for editing the English version of this article.

Conflicts of Interest: The authors declare no conflict of interest.

References

1. Banck-Burgess, J.; Nübold, C.; Mitschke, L. (Eds.) Textile Faseranalytik. In *North European Symposium for Archaeological Textiles 11 (NESAT XI), Rahden, Germany, 10–13 May 2011*; Marie Leidorf GmbH Rahden/Westf: Esslingen, Germany, 2013; pp. 45–56.
2. Siddall, R. Mineral pigments in Archaeology: Their analysis and the range of available materials. *Minerals* **2018**, *8*, 201. [CrossRef]
3. Dubois, N.; Saulnier-Talbot, É.; Mills, K.; Gell, P.; Battarbee, R.; Bennion, H.; Chawchai, S.; Dong, X.; Francus, P.; Flower, R.; et al. First human impacts and responses of aquatic systems: A review of palaeolimnological records from around the world. *Anthrop. Rev.* **2018**, *5*, 28–68. [CrossRef]
4. Nriagu, J.O. A history of global metal pollution. *Science* **1996**, *272*, 223. [CrossRef]
5. Brännvall, M.-L.; Bindler, R.; Emteryd, O.; Renberg, I. Four thousand years of atmospheric lead pollution in northern Europe: A summary from Swedish lake sediments. *J. Paleolimn.* **2001**, *25*, 421–435. [CrossRef]
6. Camarero, L.; Masqué, P.; Devos, W.; Ani-Ragolta, I.; Catalan, J.; Moor, H.C.; Sanchez-Cabeza, J.A. Historical variations in lead fluxes in the Pyrenees (Northeast Spain) from a dated lake sediment core. *Water Air Soil Pollut.* **1998**, *105*, 439–449. [CrossRef]
7. Stankevica, K.; Vincevica-Gaile, Z.; Klavins, M.; Kalnina, L.; Stivrins, N.; Grudzinska, I.; Kaup, E. Accumulation of metals and changes in composition of freshwater lake organic sediments during the Holocene. *Chem. Geol.* **2020**, *539*, 1–14. [CrossRef]
8. Druzhinina, O.; Kublitskiy, Y.; Stancikaite, M.; Nazarova, L.; Syrykh, L.; Gedminiene, L.; Uogintas, D.; Skipityte, R.; Arslanov, K.; Vaikutiene, G.; et al. The Late Pleistocene–Early Holocene palaeoenvironmental evolution in the SE Baltic region: A new approach based on chironomid, geochemical and isotopic data from Kamyshovoye Lake, Russia. *Boreas* **2020**, *49*, 544–561. [CrossRef]
9. Druzhinina, O.; Stančikaitė, M.; Gedminienė, L.; Vaikutiene, G.; Lavrova, N.; Kublitsky, J.; Subetto, D. Anthropogenic impact on the landscape of the Vishtynets Upland (Kaliningrad region, SE Baltic) in prehistory and Middle Ages: A multi-proxy palaeoenvironmental study. *Quat. Int.* **2022**. [CrossRef]
10. Kublitskiy, Y.; Kulkova, M.; Druzhinina, O.; Subetto, D.; Stančikaitė, M.; Gedminienė, L.; Arslanov, K. Geochemical Approach to the Reconstruction of Sedimentation Processes in Kamyshovoye Lake (SE Baltic, Russia) during the Late Glacial and Holocene. *Minerals* **2020**, *10*, 764. [CrossRef]
11. Bronk Ramsey, C. *OxCal Project*; Research Lab for Archaeology, University of Oxford: Oxford, UK, 2020; Available online: https://c14.arch.ox.ac.uk (accessed on 1 February 2022).
12. Tolonen, K. Charred particle analysis. In *Handbook of Holocene Palaeoecology and Palaeohydrology*; Berglund, B.E., Ed.; John Wiley: Chichester, UK, 1986; pp. 485–496.
13. Elbaz-Poulichet, F.; Guedron, S.; Develle, A.-L.; Freydier, R.; Perrot, V.; Rossi, M.; Piot, C.; Delpoux, S.; Sabatier, P. A 10,000-year record of trace metal and metalloid (Cu, Hg, Sb, Pb) deposition in a western Alpine lake (Lake Robert, France): Deciphering local and regional mining contamination. *Quat. Sci. Rev.* **2020**, *228*, 106076. [CrossRef]
14. Pipes, M.L.; Kruk, J.; Milisauskas, S. Neolithic Textile Production Artifacts at Bronocice. *Spraw. Archeol.* **2018**, *70*, 87–118. [CrossRef]
15. Wačnik, A.; Gumiński, W.; Cywa, K.; Bugajska, K. Forests and foragers: Exploitation of wood resources by Mesolithic and para–Neolithic societies in north–eastern Poland. *Veg. Hist. Archaeobot.* **2020**, *29*, 717–736. [CrossRef]
16. Melheim, L.; Grandin, L.; Persson, P.; Billström, K.; Stos-Gale, Z.; Ling, J.; Williams, A.; Angelini, I.; Canovaro, C.; Hjärthner-Holdar, E.; et al. Moving metals III: Possible origins for copper in Bronze Age Denmark based on lead isotopes and geochemistry. *JAS* **2018**, *96*, 85–105. [CrossRef]
17. Radivojevič, M.; Rehren, T.; Kuzmanović-Cvetković, J.; Jovanović, M.; Northover, J.P. Tainted ores and the rise of tin bronze metallurgy, c. 6500 years ago. *Antiquity* **2013**, *87*, 1030–1045. [CrossRef]

18. Artioli, G. *Archaeometallurgy: The Contribution of Mineralogy, Archaeometry and Cultural Heritage: The Contribution of Mineralogy Seminar*; Herrero, J.M., Vendrell, M., Eds.; Spanish Mineralogical Society (SEM): Bilbao, Spain, 2012.
19. Cooke, C.A.; Bindler, R. Lake sediment records of preindustrial metal pollution. In *Environmental Contaminants*; Springer: Berlin/Heidelberg, Germany, 2015; pp. 101–119.
20. Radivojevič, M.; Roberts, B.W. Early Balkan Metallurgy: Origin, Evolution and Society, 6200–3700 BC. *J. World Prehistory* **2021**, *34*, 195–278. [CrossRef]
21. Amicone, S.; Radivojevič, M.; Quinn, P.; Berthold, C.; Rehren, T. Pyrotechnological connections? Re-investigating the link between pottery firing technology and the origins of metallurgy in the Vinca Culture, Serbia. *JAS* **2020**, *118*, 105–123. [CrossRef]
22. Johnson, P.; Turner, S. (Eds.) *The Archaeology of Northern Europe, 1*; Brepols Publishers n.v.: Turnhout, Belgium, 2020; p. 282.
23. Ling, J.; Stos-Gale, Z.; Grandin, L.; Billström, K.; Hjärthner-Holdar, E.; Persson, P. Moving metals II: Provenancing Scandinavian Bronze Age artefacts by lead isotope and elemental analyses. *JAS* **2014**, *41*, 106–132. [CrossRef]
24. Makarowicz, P. *Szczepidło. Osada Metalurgów Kultury Mogiłowej Nad Warta*; Archeologia Bimaris: Poznań, Poland, 2016; p. 554.
25. Garbacz-Klempka, A.; Makarowicz, P.; Tokarski, T. The Analysis of Foundry Engineering of Copper Alloys Based on the Research of a Metallurgist Settlement in Szczepidło. *Arch. Foundry Eng.* **2017**, *17*, 45–50. [CrossRef]
26. Čivilytė, A.; Duberow, E.; Pernicka, E.; Skvortzov, K. The new Late Bronze Age hoard find from Kobbelbude (former Eastern Prussia, district Fischhausen) and the first results of its archaeometallurgical investigations. *Archaeol. Anthropol. Sci.* **2017**, *9*, 755–761. [CrossRef]
27. Skvortsov, K.; Rumyantseva, O.; Khanin, D. Grivna rannerimskogo vremeni iz Kalinovo (Kaliningradskaya oblast RF): Sostav emali i dannye o khronologii i proiskhozhdenii. *Kratk. Soobsheniya Inst. Arkheologii RAN* **2021**, *264*, 422–437.
28. Freestone, I.C.; Stapleton, C.P.; Rigby, V. The Production of Red Glass and Enamel in the Late Iron Age, Roman and Byzantine Periods. In *Through a Glass Brightly: Studies in Byzantine and Medieval Art and Archaeology Presented to David Buckton*; Entwistle, E.C., Ed.; Oxbow Books: Oxford, UK, 2003; pp. 142–154.
29. Radivojevič, M.; Roberts, B.W.; Pernicka, E.; Stos-Gale, Z.; Martinón-Torres, M.; Rehren, T.; Bray, P.; Brandherm, D.; Ling, J.; Mei, J.; et al. The provenance, use, and circulation of metals in the European Bronze Age: The state of debate. *J. Arch. Res.* **2019**, *27*, 131–185. [CrossRef]
30. Wilson, C.A.; Davidson, D.A.; Cresser, M.S. Multi-element soil analysis: An assessment of its potential as an aid to archaeological interpretation. *JAS* **2008**, *35*, 412–424. [CrossRef]
31. Druzhinina, O.; Gedminienė, L.; van den Berghe, K.J. Geochemical Study of the Iron Age Settlement Occupational Layer and the Early Roman Time Agricultural Layer at Voorthuizen, The Netherlands. *Minerals* **2022**, *12*, 373. [CrossRef]
32. Vodyasov, E.; Zaytseva, O. What can iron slag tell an archaeologist? *Vestn. Tomsk. Gos. Univ. Istor. -Tomsk. State Univ. J. Hist.* **2017**, *47*, 107–115. [CrossRef]
33. Tylmann, W.; Łysek, K.; Kinder, M.; Pempkowiak, J. Regional pattern of heavy metal content in lake sediments in Northeastern Poland. *Water Air Soil Pollut.* **2011**, *216*, 217–228. [CrossRef]

minerals MDPI

Article

Major Factors Controlling the Elemental Composition of Al-Kharrar Lagoon Bottom Sediments, Rabigh, Saudi Arabia

Ibrahim M. Ghandour [1,2,*], Ammar Mannaa [1], Omar Alharbi [3], Alaa A. Masoud [2] and Mohammed H. Aljahdali [1]

[1] Marine Geology Department, Faculty of Marine Science, King Abdulaziz University, P.O. Box 80200, Jeddah 21589, Saudi Arabia
[2] Geology Department, Faculty of Science, Tanta University, Tanta 31527, Egypt
[3] Geography Department, College of Social Sciences, Umm Al-Qura University, P.O. Box 5555, Makkah 21955, Saudi Arabia
* Correspondence: author: ighandour@kau.edu.sa

Abstract: Bottom sediments of Al-Kharrar Lagoon, Rabigh area, Saudi Arabia were analyzed for mineralogical (26 samples) and major oxides, minor, and trace elements (46 samples). The prime objective is to document the controlling factors of the mineralogical and chemical composition of the lagoon bottom sediments. Hierarchical cluster (HCA) and the principal component (PCA) analyses are used to disclose the degree of similarities among elements to distinguish them into statistically significant groups. Results clarified the interplay of terrestrial sediment influx through the temporarily active Wadi Rabigh, hydrological regime and the autochthonous biogenic sedimentation, and to a lesser extent rare anthropogenic influence that impacted the lagoon sediments. The spatial distribution of minerals shows a southward increase in the siliciclastic-related minerals (quartz, clay minerals, k-feldspars, and plagioclase along with traces of amphiboles), whereas carbonate minerals (high Mg-calcite and aragonite) dominate the northern sector of the lagoon in areas far from the influence of detrital influx. The concentrations levels of oxides, minor, and trace elements display spatial variability. Three main distinctive elemental groups were delineated appraising the analysis of the elemental interrelationships and associated statistical analysis. The first group includes the positively correlated SiO_2, TiO_2, Al_2O_3, Fe_2O_3, MnO, MgO, K_2O, Na_2O, V, Cr, Ni, Zn, Rb, and Ba, which are concentrated in the southern sector of the lagoon. The second group is the carbonate-related elements (CaO and Sr) that dominate the northern sector. The distribution patterns of P_2O_5 and Cu varied highly across the lagoon. Enrichment factors revealed moderate levels of Cu in some sites supporting the anthropogenic source. The results showed the hospitable bottom ecological status of the lagoon despite local anthropogenic stressors such as an influx of flood water that contain a mixture of lithogenic and dissolved Cu from local farming.

Keywords: Al-Kharrar Lagoon; sediment geochemistry; lithogenic elements; anthropogenic effect; Red Sea coastal lagoons; carbonate-related elements; mineralogical composition

Citation: Ghandour, I.M.; Mannaa, A.; Alharbi, O.; Masoud, A.A.; Aljahdali, M.H. Major Factors Controlling the Elemental Composition of Al-Kharrar Lagoon Bottom Sediments, Rabigh, Saudi Arabia. *Minerals* **2022**, *12*, 1096. https://doi.org/10.3390/min12091096

Academic Editors: Marianna Kulkova and Dmitry Subetto

Received: 21 July 2022
Accepted: 24 August 2022
Published: 29 August 2022

1. Introduction

Coastal lagoons are environmentally and socio-economically significant. They are highly productive and important sites for the ecological preservation of biodiversity, spawning, and nursery of many habitats as well as important fisheries and aquaculture sites [1–3]. In addition, they are the most desirable human living and recreation areas. Coastal lagoons are dynamic and environmentally complex affected by the interplay of terrestrial and marine influences and display rapid fluctuation in physical, chemical, and biological properties [4,5]. On the other hand, they are a highly sensitive and vulnerable ecosystem, and are negatively impacted by increased urbanization and anthropogenic activities such as fisheries, tourism, demographic expansion, and massive discharge of untreated domestic and industrial wastewaters as well as urban drainage [6]. Monitoring of the environmental quality of these ecosystems is rather difficult because of the simultaneous interaction of several parameters such as hydrodynamics, grain size, mineralogical and chemical composition, and the contribution from anthropogenic activities. The study of the chemical composition of lagoon bottom sediments is one of the most applicable strategies that have been used to evaluate the environmental changes in response to coastal zone development and to monitor the quality of these highly fluctuating environments [6–10].

The economic plan of the Saudi Arabian Vision 2030 has paid great attention to the development of the Red Sea coastal zone. Intensive urbanization, recreation, and industrial projects have significantly increased in the last decade along the Red Sea coast of major cities such as Jazan, Jeddah, Rabigh, Yanbu, Umluj, Al-Wajh, and Duba. Recent interest in Red Sea coastal development has motivated geochemical research on the sediments of coastal lagoon ecosystems. Numerous coastal lagoons of different morphologies, surface areas, habitats, and depths extend along the Saudi Red Sea coast and connect to the Red Sea through single or multiple inlets [11]. The composition of their bottom sediments is largely dominated by autochthonous biogenic sediments with very rare lithogenous constituents of limited spatial distribution reflecting the extreme arid climate and the absence of perennial riverine input [12]. These lagoons are important fishing sites holding important living resources of commercial and ecological interest. Several geochemical studies have been conducted on the Red Sea lagoon sediments to determine the controlling factors on elemental distribution [12–14], to establish a geochemical background for the Red Sea coastal sediments [15], and to assess the pollution with heavy metals [16–18].

Al-Kharrar Lagoon, which is located about 10 km to the north of Rabigh City, receives allochthonous sediments from the drainage of some temporarily active wadis such as Wadi Rabigh to the south and W. Al-Habbak to the east, and autochthonous sediments from reef debris as well as indogenic skeletal remains either from the Red Sea or that inhabit the lagoon. Little is known about the compositional heterogeneity of lagoon bottom sediments. The discharge of untreated industrial and domestic wastewaters is absent or very negligible. It is expected to be impacted by rapid urbanization and increased industrial activities in the Rabigh area in the coming years. With the increase of human activities along the Red Sea coast, geochemical studies of lagoon bottom sediments are important in evaluating environmental degradation related to coastal zone development. The study of the chemical composition of bottom sediments in the Al-Kharrar Lagoon is important not only from the geochemical point of view but also from an environmental perspective. Several studies on the recent sediments of Al-Kharrar Lagoon have been conducted, but with dissimilar subjects including hydrographic, sedimentological, mineralogical, micropaleontological, and geochemical [13,15,19–24].

Characterizing the mineralogical and chemical composition of surface sediments in the Al-Kharrar Lagoon is important not only from the geochemical point of view but also from an environmental perspective. It is essential to determine the quality and to quantify the environmental stress that may arise with future development and urbanization. The aim of the present work is to identify the sources of sediments in the lagoon and the controlling factors of the minerals and elemental spatial distribution. The study provides a database for future biogeochemical and environmental research in the area of study and similar ecosystems along the Red Sea. It provides furthermore a monitoring tool needed to detect changes and to help in early warning regarding qualitative and quantitative risks.

2. Area of Study

Al-Kharrar Lagoon is a relatively shallow (5 ± 2.8 m deep) and elongated (20 km long and 5 km maximum width) coastal basin that connects to the Red Sea through a shallow and narrow inlet through which the water of the lagoon and Red Sea exchange. It extends between latitudes 22°50′ to 22°59′ N and longitudes 38°55′ to 38°57′ E, about 10 km north of Rabigh City (Figure 1a). The southern and eastern shorelines of the lagoon are bounded by extensive tidal flat (sabkha), whereas the western shoreline is separated from the Red Sea by carbonate terraces trending parallel to the Red Sea shoreline. It includes two small islands; Al-Ultah and Um Dinar [13,20]. Al-Kharrar Lagoon is characterized by diverse ecosystems including coral reefs, mangroves, sea grasses, and macro-algae. The climate in the region is hot and arid with a very low precipitation rate (6 cm/y) and a relatively high evaporation rate (205 cm/y). The high evaporation and low precipitation rates promote a relatively high salinity of the water body varying between 38.8 and 41.5‰ (average 40.5‰). The surface water temperature varies between 23 °C and 34 °C in winter, and from 29.8 °C to 31.5 °C in summer. Like the other Red Sea lagoons, Al-Kharrar Lagoon is the terminal wetland of some inactive wadis that drain the Arabian-Nubian Shield (ANS) to the east. These wadis include Rabigh, Rehab, Murayykh, and Al-Khariq, which are intermittently and sporadically active during major flood seasons and supply the lagoon with freshwater and terrigenous sediments. The freshwater and sediment supply from land are very rare because wadis draining into the lagoons are almost inactive and dry most of the year. Wind stress and tidal currents are the major controls on water circulation in the lagoon [22,25]. The speed of tidal currents varies from 50 cm/s at the entrance to 5 and 20 cm/s inside the lagoon depending on the spring-neap cycle and sea level variations [22,25]. The tide is semi-diurnal with a small range varying from 0.20 to 0.30 m during a spring-neap cycle [22,26].

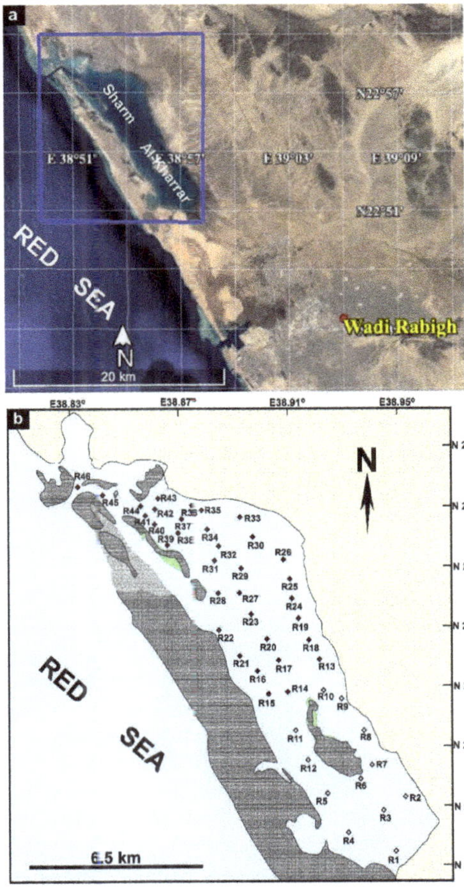

Figure 1. Google Earth map showing the location map of the Sharm Al-Kharrar (**a**) and a map showing sampling sites (**b**).

Based on grain size and mineralogical variations, Al-Washmi [20] differentiated the bottom sediments of Al-Kharrar Lagoon into two sedimentary facies; siliciclastic sediments to the south and calcareous dominated facies with abundant biogenic remains to the north and near the inlet. The chemical composition of Al-Kharrar Lagoon bottom sediments was a focus for many geochemical investigations in order to display the lateral variations within different grain size fractions [13] and to elucidate possible anthropogenic contaminations [17]. Basaham et al. [15] tried to establish a geochemical background for the Red Sea coastal sediments by selecting three pristine lagoons; Al-Kharrar Lagoon was an example of the lagoons at the terminal of Wadi. The modern environmental conditions in the Al-Kharrar Lagoon including pH, salinity, temperature, dissolved oxygen, and bathymetry have recently been the focus of some studies [11,24,27].

3. Materials and Methods

A total of 46 surface sediment samples were collected by the Van Veen Grab sampler to cover Al-Kharrar Lagoon (Figure 1b). The upper 2 cm was sampled and stored refrigerated. Samples were washed with deionized water to remove salts and subsequently wet sieved by 0.5 mm sieve and a fraction less than medium sands were used in this study. Samples were dried, finely ground with an agate pestle and mortar, and then analyzed for their mineralogical and chemical composition. The mineralogical composition of 26 samples was determined by Shimadzu X-ray diffraction equipped with Ni-filtered Cu Kα radiation at 15 kV to 40 mA at the Faculty of Marine Science, King Abdulaziz University. A gram of dry sediments was finely powdered in an agate mortar and packed into a cavity-bearing slide that was scanned from 2° to 60° 2θ at a speed of 1°/min. The relative abundances of minerals were estimated based on the height of the major peaks [28,29]. The concentrations of major oxides, and minor (1% > conc. > 0.1%) and trace (conc. < 0.1%) elements in 46 sediments samples were determined by a RIGAKU RIX 2100 X-ray fluorescence spectrometer (XRF), equipped with a Rh/W dual-anode X-ray tube at the Department of Geosciences, Osaka City University, Japan. The analysis was carried out under 50 kV and 50 mA accelerating voltage and tube current, respectively. About 1.8 g of dry powdered sediments were mixed with 3.6 g of Spectroflux ($Li_2B_4O_7$ 20 %, $LiBO_2$ 80 %, dried at 450 °C for 4 h), 0.54 g of oxidant $LiNO_3$ (dried at 110 °C for 4 h) and traces of LiI to form fused glass discs [30].

The degree of elemental enrichment was determined for the environmentally significant trace elements (V, Cr, Cu, and Zn) using the enrichment factor (EF). The EF of element X was determined by comparing the ratios of elemental concentrations to Al (X/Al) in sediments and the average chemical composition in shallow subsurface lagoon facies [31] as a background. The EF f is computed following the method described by Covelli and Fontolan [32] as:

$$EFx = Conc.\ (X/Al)_{sample} / Conc.\ (X/Al)_{Back} \quad (1)$$

EF values less than 2 suggest a depletion with no or minimal pollution, while values in the range level of 2–5 suggest moderate enrichment/pollution, range level of 5–20 indicates significant enrichment/pollution, level of 20–40 clarifies very high enrichment/pollution, and EF values greater than 40 disclose extreme enrichment/pollution [33].

Data distribution and correlation among the variables were analyzed appraising the use of the SPSS Statistical package for Windows version 20. Pearson's correlation analysis was conducted to describe the strength and direction of the linear relationship between the studied elements at a significance level of 95%. Hierarchical cluster (HCA) and the principal component (PCA) analyses were applied to clarify the degree of similarity between elements. Variogram modeling using krigging implemented in ArcGIS 9.3 was appraised for rendering the spatial distribution of the variables where the spherical model was of major use. Surfaces produced were evaluated from the original sampling point data values applying cross-validations in a trial and error step, where maps with the least errors were further interpreted.

4. Results

The results of mineralogical and chemical analyses of the bottom sediments of Al-Kharrar Lagoon are summarized in Tables 1 and 2. Mineralogical and geochemical mapping showing the spatial distribution of the most common minerals (Figure 2), major oxides, minor, and trace elements (Figures 3 and 4, respectively) were used to evaluate the sediment quality in the lagoon.

Table 1. The mineral content in the bottom sediments of the Sharm Al-Kharrar as revealed by XRD.

Sample	Clay Minerals	Mica	Amphiboles	Quartz	K-Feldspars	Plagioclase	Aragonite	HMC	LMC	Dolom.
R1	10	-	7	44	20	-	4	4	11	2
R2	18	4	3	38	-	13	6	7	7	3
R3	9	-	5	42	5	18	6	7	5	2
R4	13	3	3	32	-	11	18	6	12	2
R5	12	-	4	36	-	13	14	7	11	2
R6	9	-	4	31	-	10	18	9	17	2
R8	14	-	4	35	4	13	10	8	10	3
R9	7	-	9	29	23	-	-	9	21	2
R10	11	-	3	31	-	11	17	9	17	2
R11	12	-	3	29	3	11	18	8	15	2
R12	11	-	5	33	4	10	14	7	13	2
R15	7	2	2	20	-	6	33	8	20	-
R17	10	-	2	23	4	9	23	10	16	3
R18	8	-	5	21	3	15	23	8	15	1
R19	6	-	2	19	3	13	30	8	19	-
R20	11	-	1	23	-	9	28	9	18	-
R22	5	-	2	16	-	5	38	9	22	2
R23	12	-	3	26	-	9	25	9	17	-
R24	5	-	5	22	-	24	22	7	14	1
R28	5	-	2	15	-	7	26	15	28	2
R29	5	3	3	23	-	8	25	11	22	-
R35	2	-	2	17	-	5	39	14	20	2
R36	10	3	-	17	-	7	34	12	21	-
R38	7	-	-	19	-	6	29	15	23	1
R39	5	-	2	14	-	9	37	11	20	2
R43	4	-	3	12	-	7	44	11	19	1

HMC; high Mg-calcite, LMC; low Mg-calcite, (-) undetected.

Table 2. Summary statistics showing the concentration ranges and average of oxides (%) and minor and trace elements (μg/g), elemental ratios, and enrichment factor (EF) of the environmentally significant trace elements in the bottom sediments of Sharm Al-Kharrar.

Oxides/elements	Minimum	Maximum	Mean
SiO_2	12.85	51.62	31.25
TiO_2	0.26	1.12	0.62
Al_2O_3	4.23	16.33	10.32
T-Fe_2O_3	2.06	9.12	5.35
MnO	0.04	0.15	0.09
MgO	3.39	6.26	5.16
CaO	8.12	61.27	35.20
Na_2O	0.69	2.8	1.65
K_2O	0.22	1.62	0.67
P_2O_5	0.13	0.51	0.24
V	48	168	104.74
Cr	34	155	82.26

Table 2. *Cont.*

Oxides/elements	Minimum	Maximum	Mean
Ni	14	93	51.24
Cu	30	64	47.72
Zn	25	99	61.57
Rb	3	45	15.50
Sr	624	7932	4155
Ba	46	232	134
Mg/Ca	0.05	0.59	0.17
Sr/Ca	107	185	160
EF_V	0.55	0.68	0.61
EF_{Cr}	0.65	0.91	0.73
EF_{Cu}	0.64	3.63	1.52
EF_{Zn}	0.88	1.77	1.42

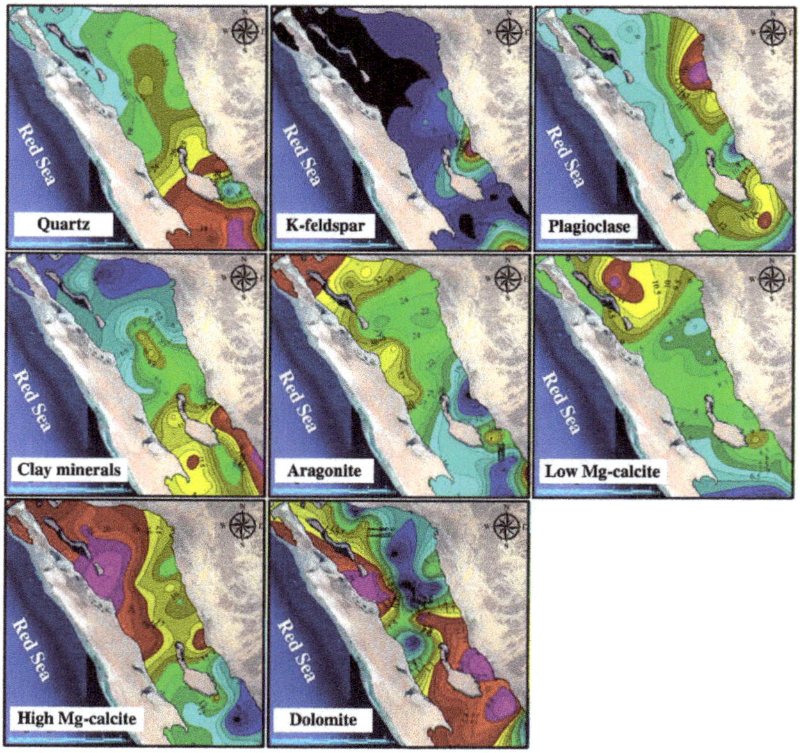

Figure 2. Contour maps showing the spatial distribution of the most common minerals in the bottom sediments of the Sharm Al-Kharrar. Light blue scale bar on maps equals 6 km.

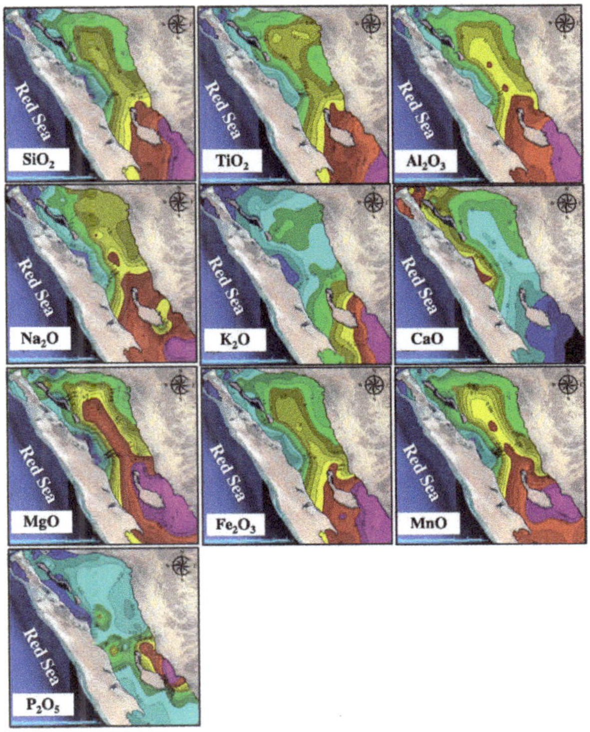

Figure 3. Contour maps showing the spatial distribution of the concentrations of oxides (%) in the bottom sediments of the Sharm Al-Kharrar. Light blue scale bar on maps equals 6 km.

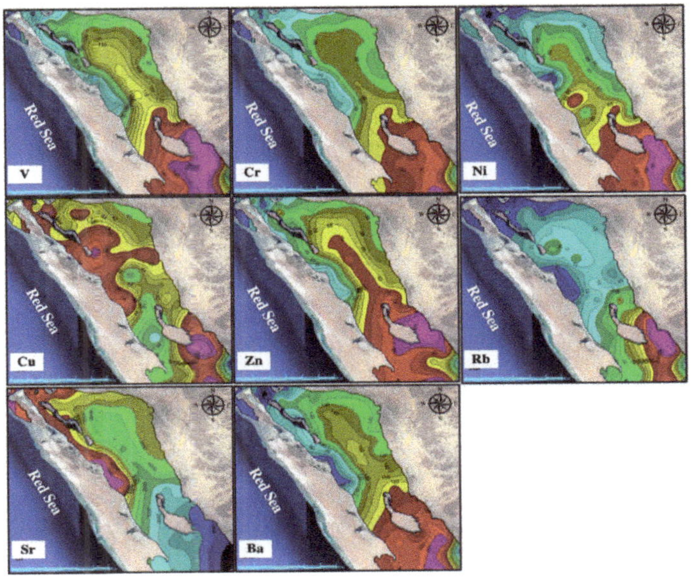

Figure 4. Contour maps showing the spatial distribution of the concentrations of trace elements (µg/g) in the bottom sediments of the Sharm Al-Kharrar. Light blue scale bar on maps equals 6 km.

4.1. Mineralogical Composition

XRD analysis of the sediment from the entire lagoon system revealed the presence of the two mineral groups for all sediments (Figure 2). The first group (clays, quartz, k-feldspars, plagioclase, and traces of mica and amphiboles) probably of detrital origin, derived from the hinterland. On the other hand, the second group (including aragonite, high and low Mg-calcite, and traces of dolomite) probably precipitated within the lagoon mainly as biogenous constituents. The distribution and abundance of the two groups are opposite. The minerals of the first group are more abundant in the southern sector of the lagoon than in the northern part.

4.2. Oxides

Descriptive statistics of oxides and minor and trace element concentrations in the bottom sediments of Al-Kharrar Lagoon are shown in Table 2. The concentrations of oxides in the analyzed sediments are consistent with the mineralogical composition. The chemical data showed a heterogeneous distribution throughout the lagoon. SiO_2, TiO_2, Al_2O_3, Fe_2O_3, MnO, MgO, Na_2O, and K_2O display a similar spatial distribution in that they are highly enriched towards the head of the lagoon near the southern coast. SiO_2 is the most abundant major oxide, ranging from 12.85 to 51.62 %; total concentrations of Al_2O_3 and TiO_2 vary from 4.23 to 16.33% and from 0.26 to 1.12%, respectively. On average, the concentrations of iron oxide (Fe_2O_3) (2.06–9.12%) and MnO (0.04–0.15%) decrease northward. Similarly, the concentrations of MgO, Na_2O, and K_2O decrease northward. The concentrations of Na_2O and K_2O display a similar spatial distribution. They vary between 0.69 and 2.8%, and 0.22 and 1.62%, respectively. The concentration of CaO displays an opposite distribution to other oxides and it is enriched towards the mouth of the lagoon (Figure 3). It varies from 8.12 and 61.27% (mean; 35.2%). Total phosphorus (P_2O_5: 0.13–0.51%) shows a relatively even distribution with a slight enrichment east of the Al-Ultah Island (Figure 3).

4.3. Minor and Trace Elements

The most abundant minor and trace elements are Sr > Ba > V > Cr > Zn> Ni > Cu > Rb. Vanadium (V), Cr, Ni, Zn, and Rb, display a similar spatial distribution to Al_2O_3. They show a southward increase in concentrations, ranging from 48 to 168 μg/g, 34 to 155 μg/g, from 14 to 93 μg/g, from 25 to 99 μg/g, 3 to 45 μg/g, from 3 to 11 μg/g, and from 8 to 30 μg/g, respectively. Copper (Cu) shows a fairly homogeneous spatial distribution over the study area with values ranging between 30 and 47.7 μg/g. The higher concentrations of Cu are recorded from areas near the mangrove. Similar to CaO, the highest Sr values (7932 μg/g) were obtained from the northern sector of the lagoon (Figure 4) coinciding with samples that are enriched in aragonite. Such high Sr levels are common in calcareous sediments containing abundant aragonite [34].

The ratio of Mg/Ca decreases more northward. It varied between 0.05 and 0.59 (mean; 0.17). On the other hand, the Sr/Ca ratio does not change remarkably from north to south. It ranges from 107×10^{-4} to 185×10^{-4} (mean; 160×10^{-4}).

4.4. Elemental Interrelationships

Pearson's correlation coefficients of elements in Al-Kharrar Lagoon are summarized in Table 3, showing various degrees of relationship among elements. The correlation matrix shows that most oxides and trace elements have a strong positive correlation with Al_2O_3 ($r > 0.8$) except for CaO and Sr, which show a strong negative correlation. Barium (Ba) displays a strong positive correlation with K_2O and Rb ($R > 0.9$). Iron oxide (Fe_2O_3) shows a strong positive correlation with other transitional elements (MnO, V, Cr, Co, Ni, and Zn). P_2O_5 and Cu show poor correlation with Al_2O_3 and CaO (Table 3).

Table 3. Correlation matrix of oxides and minor and trace element concentrations in the bottom sediments (n = 46 samples) of Sharm Al-Kharrar.

Oxides/Elements	SiO_2	TiO_2	Al_2O_3	T-Fe_2O_3	MnO	MgO	CaO	Na_2O	K_2O	P_2O_5	V	Cr	Ni	Cu	Zn	Rb	Sr	Ba
SiO_2	1																	
TiO_2	0.99	1																
Al_2O_3	0.99	0.97	1															
T-Fe_2O_3	0.98	0.97	0.99	1														
MnO	0.95	0.93	0.98	0.97	1													
MgO	0.89	0.86	0.93	0.90	0.92	1												
CaO	−0.99	−0.98	−0.99	−0.99	−0.96	−0.91	1											
Na_2O	0.88	0.87	0.86	0.83	0.82	0.75	−0.88	1										
K_2O	0.92	0.93	0.91	0.93	0.84	0.76	−0.92	0.73	1									
P_2O_5	0.50	0.48	0.50	0.50	0.47	0.54	−0.47	0.29	0.53	1								
V	0.98	0.97	0.99	0.99	0.97	0.90	−0.99	0.85	0.92	0.48	1							
Cr	0.98	0.98	0.98	0.99	0.95	0.87	−0.98	0.84	0.93	0.49	0.98	1						
Ni	0.84	0.83	0.87	0.88	0.83	0.78	−0.84	0.69	0.84	0.50	0.87	0.88	1					
Cu	0.07	0.08	0.09	0.12	0.03	0.07	−0.07	−0.29	0.26	0.23	0.09	0.13	0.20	1				
Zn	0.93	0.91	0.95	0.94	0.95	0.92	−0.93	0.74	0.81	0.54	0.93	0.92	0.82	0.02	1			
Rb	0.89	0.89	0.89	0.91	0.82	0.75	−0.88	0.64	0.98	0.56	0.90	0.91	0.84	0.35	0.82	1		
Sr	−0.99	−0.98	−0.99	−0.98	−0.96	−0.92	0.99	−0.87	−0.91	−0.48	−0.98	−0.98	−0.83	−0.06	−0.94	−0.88	1	
Ba	0.99	0.97	0.98	0.97	0.94	0.88	−0.99	0.90	0.90	0.44	0.97	0.97	0.83	0.05	0.91	0.86	−0.98	1

4.5. Enrichment Factor (EF)

The enrichment factors of V, Cr, Cu, and Zn were calculated (Table 2) and according to the classification of Southerland (2000), the bottom sediments of Al-Kharrar Lagoon are significantly depleted in V (0.55–0.68), Cr (0.65–0.91) and Zn (0.88–1.7). On the other hand, 8 sites out of 46 show moderate enrichment in Cu (2–3.6), whereas other sites are significantly depleted (EF < 2).

4.6. Statistical Analysis

Cluster analysis using correlation coefficient between all pairs of samples enabled discrimination of two groups of samples; A and B, representing samples of the southern and northern portions, respectively (Figure 5a). The samples within the two groups display a relatively similar geochemical composition. In addition, cluster analysis enabled classifying variables (oxides, minor, and trace elements) into two geochemical associations; C and D (Figure 5b). The association C displays a great similarity among lithogenic silicate and heavy mineral-related (SiO_2, Al_2O_3, Fe_2O_3, TiO_2, MnO, V, Cr, Ni, Zn, Rb, and Ba), and feldspars-related (K_2O and Rb) elements. These elements display a similar spatial distribution in the lagoon, with higher concentrations recognized from the southern portion of the lagoon. On the other hand, association D represents carbonate-related elements (CaO and Sr) that display a dissimilar distribution pattern to the elements of group C. Copper (Cu) and P_2O_5 are not associated with the two geochemical associations.

Factor analysis allowed the identification of two factors (eigenvalues > 1) that explain 91.52% of the total variance (Table 4). Each factor represents compositional variables that identify a distinct group of elements. Factor 1 accounts for 84.85% of the total variance with a marked bipartition: on one hand, elements with significant positive factor loadings (SiO_2, Al_2O_3, Fe_2O_3, MgO, K_2O, TiO_2, Ni, Rb, V, and Ba) are related to silicates and clay minerals; on the other hand, elements with negative factor loadings (CaO and Sr) can be related to carbonates. Factor 2 accounts for 6.672% of the total variance. This factor is characterized by high positive loadings of Cu and P_2O_5. The source and spatial distribution of Cu and P_2O_5 are different from other elements. They display poor correlation either with Al_2O_3 or CaO suggesting a different source, possibly anthropogenesis.

Table 4. Factor loadings of geochemical variables in the bottom sediments of Sharm Al-Kharrar.

Variable	F1	F2
SiO_2	0.99	−0.055
TiO_2	0.98	−0.038
Al_2O_3	0.99	−0.032
Fe_2O_3	0.99	0.003
MnO	0.97	−0.087
MgO	0.91	−0.029
CaO	-0.99	0.062
Na_2O	0.85	−0.438
K_2O	0.93	0.195
P_2O_5	0.53	0.414
V	0.99	−0.027
Cr	0.99	0.007
Ni	0.88	0.138
Cu	0.11	0.925
Zn	0.94	0.002
Rb	0.91	0.3
Sr	−0.99	0.068
Ba	0.98	−0.10
Eigen value	17.18	1.4
% of variance	84.85	6.67

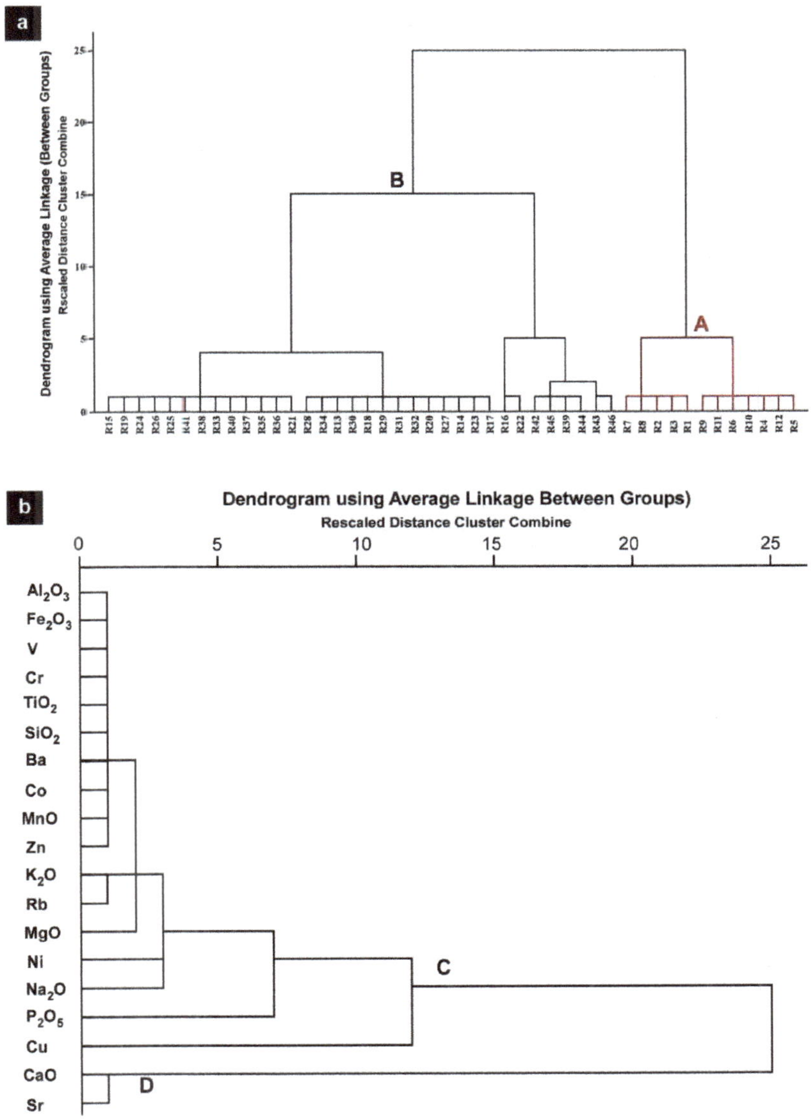

Figure 5. Dendrogram showing the clustering of samples (**a**) into two groups; A and B, representing the southern and northern sectors of the lagoon, respectively. (**b**) Clustering of elements into lithogenous-related (C) and carbonate-related elements (D).

5. Discussion

The spatial variation in the mineralogical and chemical composition of Al-Kharrar Lagoon bottom sediments reflects multiple sources. Statistical analysis and elemental interrelationships allowed the discrimination of three geochemical associations of dissimilar sources. The correlation of elements with Al_2O_3 (an indicator of the terrigenous influx; [35]) provides a useful tool in determining the source and behavior of elements. Oxides (SiO_2, TiO_2, Fe_2O_3, K_2O_3, MgO, and Na_2O) and trace elements (V, Cr, Co, Ni, Zn, Rb, and Ba) of the first geochemical association displayed significant positive correlations with

Al_2O_3 suggesting their association with the insoluble terrigenous aluminosilicates. The distribution patterns of SiO_2, TiO_2, Al_2O_3, Fe_2O_3, MnO, V, Cr, Co, Ni, Zn, Rb, and Ba mirror each other, suggesting a similar detrital source. A considerable part of these geogenic elements is preferentially contributed from the igneous and metamorphic rocks of the Arabian Shield which are abundant in the Wadi Rabigh catchment basins. The elevated concentrations of Cr and Ni are mostly related to the occurrence of heavy minerals such as pyroxenes and amphiboles [36]. The positive correlation of K_2O, Na_2O, and MgO with Al_2O_3, suggests their detrital origin, mainly associated with K- and plagioclase feldspar, as well as clay minerals [37]. Barium (Ba) displayed a strong positive correlation with K_2O and Rb ($r > 0.9$) imposing a possible similar terrestrial source; possibly feldspars and mica. Transitional elements (MnO, V, Cr, Ni, and Zn) displayed a strong positive correlation with Fe_2O_3 suggesting their association with Fe-oxyhydroxides [38]. The second geochemical association (CaO and Sr) displayed a negative correlation with Al_2O_3 a group that relates to autochthonous biogenic carbonates. Carbonate content exhibits great variability, with the highest values observed in the northern sector of the lagoon, which is far from the dilution effect of terrigenous influx and dominated by calcareous skeletal remains. The absence of perennial riverine discharge and the low flow conditions are responsible for the short distance of terrigenous influx into the lagoon and therefore, the decrease in the concentration of lithogenic elements basin-ward.

The northward decrease in the Mg/Ca ratio suggests that the two elements have different origins, where a larger portion of Mg is of lithogenic origin, whereas Ca is mostly related to biogenic carbonate origin. Slight Mg enrichment in the sediment of the northern sector is attributed to the abundant high Mg-calcite. On the other hand, the Sr/Ca ratio notably does not change from north to south suggesting that Sr appears to be mainly controlled by the carbonate content particularly the presence of biogenic calcite and aragonite that constitute the tests of foraminifera, calcareous algae and coral remains and to a less and negligible contribution from feldspars [37]. The carbonate fraction is mainly of biogenic origin including shells and shell fragments of foraminifera, mollusks, coral remains, and echinoids.

The source of the Red Sea coastal lagoon bottom sediments includes the allochthonous detritus derived from the weathered parent rocks in the hinterland, the autochthonous biogenic calcareous remains, and the authigenic (hydrogenous) material precipitated at or near the sediment-water interface [12,18]. The content of detrital minerals (quartz, feldspars, and clay minerals) and lithogenic elements decrease from the shallow nearshore sites in the lagoon, whereas the abundance of carbonate minerals and carbonate-related elements (CaO and Sr) increases in the deeper sites and areas adjacent to the mouth of many Red Sea coastal lagoons [12,16]. This is attributed to the reduction of terrigenous materials mouth-ward and/or the dilution effect by biogenic carbonates [12]. Youssef and El-Sorogy [17] attributed the elevated concentrations of metals in the bottom sediments of Al-Kharrar Lagoon to the discharge of siliciclastic sediments by Wadi Rabigh during major floods and to a lesser extent the urbanization in the nearby areas such as King Abdullah Economic City.

Similar to cluster analysis, a ternary plot of major constituents of sediments; silica (SiO_2) vs. iron and aluminum oxides ($Al_2O_3 + Fe_2O_3$) vs. ($CaO + Na_2O + K_2O + MgO$) differentiates the bottom sediments of Al-Kharrar Lagoon into two groups; A and B displaying spatial variability based on their chemical composition he ternary diagram (Figure 6). The sediments near the head (group A; southern sector) are dominated by terrigenous minerals and elements supplied mainly by Wadi Rabigh during major flood seasons in addition to the contribution from the aeolian supply. Towards the mouth of the lagoon (group B; northern sector), the carbonate-related minerals and elements dominate (Figure 6).

The spatial distribution of Cu and P_2O_5 (third geochemical association) does not follow any of the two geochemical associations. Copper (Cu) and P_2O_5 are poorly correlated with Al_2O_3 and CaO suggesting a non-lithogenic and non-carbonate origin of the two elements.

Elevated concentrations and the local moderate enrichment of Cu indicate that a part of Cu in lagoon bottom sediments may be of an anthropogenic origin possibly anti-fouling boat paints and influxes from agrochemicals in farms bordering Wadi Rabigh, and may produce toxic and irreversible effects on living organisms in the future [39]. Copper (Cu) has been historically used as seed disinfectants, fungicides, herbicides, and added to fertilizers to meet the demand of plant growth [40,41]. Therefore, it may be contributed by the agricultural activities surrounding Wadi Rabigh. The source of Cu pollution in the bottom sediments of the Aetoliko (Etoliko) Lagoon Western Greece, resulted from the use of olive orchards Cu-based fungicide [42]. The elevated concentration of P_2O_5 adjacent to the Al-Ulta Island may reflect an increase of P input from the bird and mammal guano.

The quality of the sediments of Al-Kharrar Lagoon was assessed using enrichment factor (EF) of the environmentally significant trace elements (V, Cr, Cu, and Zn). The bottom sediments are depleted in V, Cr, and Zn (EF < 2), and depleted to moderately enriched for Cu (EF; 0.63–3.63). The EF values for V and Cr are below the regional geochemical background level and these elements are considered to be of a geogenic origin. The slight local enrichment of Cu may be attributed to urban pollution such as chemical fertilizers.

The chemical composition of the sediments in this study shows that the lagoon is not under anthropogenic stresses. However, it is important to take into account the negative impact and the potential accumulation of heavy metals in sediments even if they are trapped in minerals. With future changes in the pH, redox potential, salinity, and dissolved organic carbon of the lagoon, heavy metals trapped in sediments can be released into the water column and become bioavailable and posing a negative effect on biota. If the pH of bottom water drops to 5 and 5.5, Cu, and Zn can be rapidly bio-available in the aquatic environment, respectively [43,44]. In addition, the warm and dry climate and the construction of dams reduced the freshwater supply to the lagoon allowing increasing water salinity. The increase in salinity can indirectly increase metal bioavailability through the reduction of SO_4^{2-} and the decomposition of organic matter in sediment [45–47].

The local enrichment of Zn and Cu near the edges of the lagoon may be related to the presence of macrophytes (mangroves, seagrasses, and macro-algae) which provide the organic matter on which the two metals scavenge [48,49]. The elevated concentrations of Zn and Cu in the northern sector were attributed to biosorption by mucus algae that proliferate on the coarse shell fragments with the inlet channel [23].

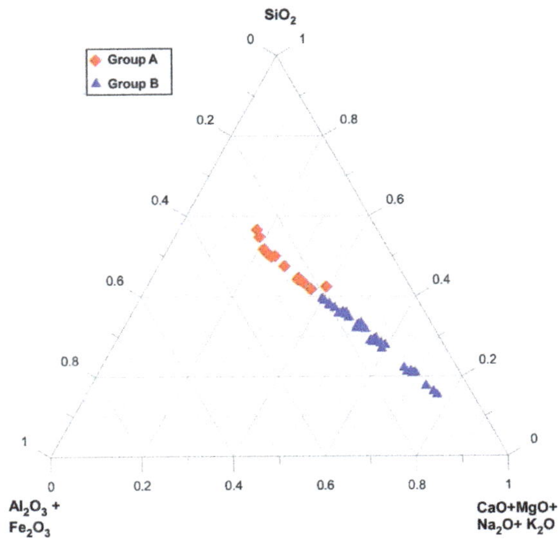

Figure 6. Ternary diagram of the oxide in the bottom sediments of Sharm Al-Kharrar.

6. Conclusions

Mineralogical and chemical analyses of Al-Kharrar bottom sediments provided the base for better understanding the factors interplayed to govern their variability summarized as follows:

1. The mineralogical and chemical composition of Al-Kharrar Lagoon bottom sediments reflects the interplay of terrestrial sediment influx through the temporarily active Wadi Rabigh, hydrological regime, and the autochthonous biogenic sedimentation, and to a lesser extent rare anthropogenic influence.
2. Minerals of detrital origin (quartz, k-feldspars, plagioclase, and amphiboles) and highly enriched in the nearshore shallow areas of the southern sector of the lagoon at the mouth of Wadi Rabigh, whereas carbonate minerals (aragonite, high and low Mg-calcite) dominate the marine-influenced northern sector.
3. Factor analysis distinguished two factors and allowed the subdivision of Al-Kharrar Lagoon bottom sediments geochemically into three geochemical groups. The first factor is characterized by positive loadings of lithogenic elements and negative values of variables linked to carbonate minerals. This factor records high concentrations of siliciclastic-related oxides and trace elements as well as high concentrations of carbonate-related elements. The second factor is characterized by positive loadings of Cu and P_2O_5 which have no relationship with the groups of the first factor. Statistical analysis has allowed the identification of two main groups of sediments.
4. The lithogenic-related oxides (SiO_2, TiO_2, Al_2O_3, Fe_2O_3, MgO, Na_2O, and K_2O) and trace elements (V, Cr, Ni, Zn, Rb, and Ba) are enriched in the southern sector near the head of the lagoon. The spatial distribution patterns reveal that lithogenous sediments settled rapidly mostly near the Wadi mouth, and the sediments are restricted to the shallow area and not transported far into the deeper parts of the lagoon due to the low flow conditions.
5. The carbonate-related oxides (CaO) and trace element (Sr) are significantly enriched in the northern sector where autochthonous shell fragments are the main source of carbonates.
6. The levels of heavy metals are generally below the recommended allowable limits and the lagoon is in good ecological condition. The EF values of Cu show a local slight enrichment supporting a possible anthropogenic source. However, there has currently no clear evidence of heavy metal contamination by anthropogenic sources in the Al-Kharrar Lagoon.
7. The obtained data are of particular significance providing a base to monitor any future environmental deterioration associated with development and urbanization.

Author Contributions: I.M.G., A.M., O.A., A.A.M. and M.H.A. proposal of the idea, conceptualization, sample collection and coordination for analysis, data analysis, and writing the first draft of the manuscript. I.M.G. and A.A.M. data analysis, preparing tables and figures, reviewing the manuscript and participation in the discussion. A.A.M. preparation of the GIS layers and reviewing the article. All authors have read and agreed to the published version of the manuscript.

Funding: This project was funded by the Deanship of Scientific Research (DSR) at King Abdulaziz University, Jeddah, under grant No. G: 320-150-1442.

Data Availability Statement: The data are available upon request at KAU and can be requested by contacting the corresponding author.

Acknowledgments: The authors thank the DSR for this technical and financial support. The authors are grateful to Aaid Al-Zubairi (KAU) for his assistance in the fieldwork. We are very grateful to the editor and the reviewers for their constructive comments and editorial handling.

Conflicts of Interest: The authors declare no conflict of interest

References

1. Morton, R.A.; Ward, G.H.; White, W.A. Rates of sediment supply and sea-level rise in a large coastal lagoon. *Mar. Geol.* **2000**, *167*, 261–284. [CrossRef]
2. Forrest, B.M.; Creese, R.G. Benthic impacts of intertidal oyster culture, with consideration of taxonomic sufficiency. *Environ. Monit. Assess.* **2006**, *112*, 159–176. [CrossRef]
3. Abidi, M.; Amor, R.; Gueddari, M. Assessment of the Trophic Status of the South Lagoon of Tunis (Tunisia, Mediterranean Sea): Geochemical and Statistical Approaches. *J. Chem.* **2018**, *2018*, 1–17. [CrossRef]
4. Benoit, J.M.; Gilmour, C.C.; Mason, R.P.; Heyes, A. Sulfide controls on mercury speciation and bioavailability to methylating bacteria in sediment pore waters. *Environ. Sci. Techol.* **1999**, *33*, 1780. [CrossRef]
5. Pérez-Ruzafa, A.; Marcos, C.; Pérez-Ruzafa, I.M.; Pérez-Marcos, M. Are coastal lagoons physically or biologically controlled ecosystems? Revisiting r vs. K strategies in coastal lagoons and estuaries. *Estua. Coast. Shelf Sci.* **2013**, *132*, 17–33. [CrossRef]
6. Arcega-Cabrera, F.; Garza-Pérez, R.; Noreña-Barroso, E.; Oceguera-Vargas, I. Impacts of geochemical and environmental factors on seasonal variation of heavy metals in a coastal lagoon Yucatan, Mexico. *Bull. Environ. Contam. Toxico.* **2015**, *94*, 58–65. [CrossRef]
7. Förstner, U.; Ahlf, W.; Calmano, W. Studies on the transfer of heavy metals between sedimentary phases with a multi-chamber device: Combined effects of salinity and redox variation. *Mar. Chem.* **1989**, *28*, 145–158. [CrossRef]
8. Yu, K.C.; Tsai, L.J.; Chen, S.H.; Ho, S.T. Correlation analyses on binding behavior of heavy metals with sediment matrices. *Water Res.* **2001**, *35*, 2417–2428. [CrossRef]
9. Karageorgis, A.P.; Anagnostou, C.L.; Kaberi, H. Geochemistry and mineralogy of the NW Aegean Sea surface sediments: Implications for river runoff and anthropogenic impact. *Appl. Geochem.* **2005**, *20*, 69–88. [CrossRef]
10. Ahmed, F.; Bibi, M.H.; Seto, K.; Ishiga, H.; Fukushima, T.; Roser, B.P. Abundances, distribution, and sources of trace metals in Nakaumi–Honjo coastal lagoon sediments, Japan. *Environ. Monit. Assess.* **2010**, *167*, 473–491. [CrossRef]
11. Rasul, N.M. Lagoon sediments of the Eastern Red Sea: Distribution processes, pathways and patterns. In *The Red Sea*; Springer: Berlin/Heidelberg, Germany, 2015; pp. 281–316.
12. Basaham, A.S.; Ghandour, I.M.; Haredy, R. Controlling factors on the geochemistry of Al-Shuaiba and Al-Mejarma coastal lagoons, Red Sea, Saudi Arabia. *Open Geosci.* **2019**, *11*, 426–439. [CrossRef]
13. Basaham, A.S. Mineralogical and chemical composition of the mud fraction from the surface sediments of Al-Kharrar, a Red Sea coastal lagoon. *Oceanologia* **2008**, *50*, 557–575.
14. Abu-Zied, R.H.; Al-Mur, B.A.; Orif, M.I.; Al Otaibi, A.; Ghandourah, M.A. Concentration distribution, enrichment and controlling factors of metals in Al-Shuaiba Lagoon sediments, Eastern Red Sea, Saudi Arabia. *Environ. Earth Sci.* **2021**, *80*, 1–20. [CrossRef]
15. Basaham, A.S.; El Sayed, M.A.; Ghandour, I.M.; Masuda, H. Geochemical background for the Saudi Red Sea coastal systems and its implication for future environmental monitoring and assessment. *Environ. Earth Sci.* **2015**, *74*, 4561–4570. [CrossRef]
16. Ghandour, I.M.; Basaham, S.; Al-Washmi, A.; Masuda, H. Natural and anthropogenic controls on sediment composition of an arid coastal environment: Sharm Obhur, Red Sea, Saudi Arabia. *Environ. Monit. Assess.* **2014**, *3*, 1465–1484. [CrossRef]
17. Youssef, M.; El-Sorogy, A. Environmental assessment of heavy metal contamination in bottom sediments of Al-Kharrar lagoon, Rabigh, Red Sea, Saudi Arabia. *Arab. J. Geosci.* **2016**, *9*, 474. [CrossRef]
18. Ghandour, I.M.; Aljahdali, M.H. Elemental Enrichment in Shallow Subsurface Red Sea Coastal Sediments, Al-Shuaiba, Saudi Arabia: Natural vs. anthropogenic Controls. *Minerals* **2021**, *11*, 898. [CrossRef]
19. Abou-Ouf, M.A. Variation of benthic foraminiferal assemblages in different microenvironments along the shore zone north of the Rabigh Coast, eastern Red Sea, Saudi Arabia. *Neues Jahrbuch für Geologie und Paläontologie-Monatshefte* **1996**, *9*, 129–139. [CrossRef]
20. Al-Washmi, H.A. Sedimentological aspects and environmental conditions recognized from the bottom sediments of Al-Kharrar Lagoon, eastern Red Sea coastal plain, Saudi Arabia. *J. King Abdulaziz Univ. Mar. Sci.* **1999**, *10*, 71–87. [CrossRef]
21. Hariri, M.S. Effect of hydrographic conditions on the ecology of benthic foraminifera in two different hypersaline lagoons, eastern Red Sea coast, Kingdom of Saudi Arabia. *J. King Abdulaziz Univ. Mar. Sci.* **2008**, *19*, 3–13. [CrossRef]
22. Al-Barakati, A.M.A. Application of 2-D tidal model, Shuaiba Lagoon, eastern Red Sea coast. *Can. J. Comput. Math. Nat. Sci. Med.* **2010**, *1*, 9–20.
23. Hariri, M.S.; Abu-Zied, R.H. Factors influencing heavy metal concentrations in the bottom sediments of the Al-Kharrar Lagoon and Salman Bay, eastern Red Sea coast, Saudi Arabia. *Arab. J. Geosci.* **2018**, *11*, 1–20. [CrossRef]
24. Al-Dubai, T.A.; Bantan, R.A.; Abu-Zied, R.H.; Al-Zubieri, A.G.; Jones, B.G. Distribution of Benthic Foraminifera in Intertidal Sabkha of Al-Kharrar Lagoon, Saudi Arabia: Tools to Study Past Sea-Level Changes. *Front. Mar. Sci.* **2022**, *351*. [CrossRef]
25. Al-Barakati, A.M.; Ahmad, F. Water column conditions in a coastal lagoon near Jeddah, Red Sea. *Oceanologia* **2012**, *54*, 675–685. [CrossRef]
26. Lisitzin, E. *Sea-Level Changes*; Elsevier Sci. Publ. Co.: Amsterdam, The Netherlands; Oxford, UK; New York, NY, USA, 1974.
27. Al-Dubai, T.A.; Abu-Zied, R.H.; Basaham, A.S. Present environmental status of Al-Kharrar Lagoon, central of the eastern Red Sea coast, Saudi Arabia. *Arab. J. Geosci.* **2017**, *10*, 305. [CrossRef]
28. Hardy, R.; Tucker, M.E. *Techniques in Sedimentology*; Blackwell Scientific Publications: Hoboken, NJ, USA, 1988; pp. 191–228.
29. Moore, D.M.; Reynolds, R.C., Jr. *X-ray Diffraction and the Identification and Analysis of Clay Minerals*; Oxford University Press: New York, NY, USA, 1989.

30. Tawfik, H.A.; Ghandour, I.M.; Maejima, W.; Armstrong-Altrin, J.S.; Abdel-Hameed, A.M. Petrography and geochemistry of the siliciclastic Araba Formation (Cambrian), east Sinai, Egypt: Implications for provenance, tectonic setting and source weathering. *Geol. Magaz.* **2017**, *154*, 1–23. [CrossRef]
31. Haredy, R.; Ghandour, I.M. Geochemistry and mineralogy of the shallow subsurface Red Sea coastal sediments, Rabigh, Saudi Arabia: Provenance and paleoenvironmental implications. *Turk. J. Earth Sci.* **2020**, *29*, 257–279. [CrossRef]
32. Covelli, S.; Fontolan, G. Application of a normalization procedure in determining regional geochemical baseline. *Environ. Geol.* **1997**, *30*, 34–35. [CrossRef]
33. Sutherland, R.A. Bed sediment-associated trace metals in an urban stream, Oahu, Hawaii. *Environ. Geol.* **2000**, *39*, 611–627. [CrossRef]
34. Calvert, S.E. The mineralogy and geochemistry of nearshore sediments. In *Chemical Oceanography*; Riley, J.P., Chester, R., Eds.; Academic Press: London, UK, 1976; Volume 6, pp. 187–280.
35. Loring, D.H.; Rantala, R.T. Manual for the geochemical analyses of marine sediments and suspended particulate matter. *Earth-Sci. Rev.* **1992**, *32*, 235–283. [CrossRef]
36. Mil-Homens, M.; Vale, C.; Raimundo, J.; Pereira, P.; Brito, P.; Caetano, M. Major factors influencing the elemental composition of surface estuarine sediments: The case of 15 estuaries in Portugal. *Mar. Poll. Bull.* **2014**, *84*, 135–146. [CrossRef]
37. Calvert, S.E.; Pedersen, T.F.; Thunell, R.C. Geochemistry of the surface sediments of the Sulu and South China Seas. *Mar. Geol.* **1993**, *114*, 207–231. [CrossRef]
38. Taylor, G.; Eggleton, R.A. *Regolith Geology and Geomorphology*, 1st ed.; John Wiley and Sons: Hoboken, NJ, USA, 2001.
39. Lejon, D.P.; Martins, J.M.; Lévêque, J.; Spadini, L.; Pascault, N.; Landry, D.; Milloux, M.J.; Nowak, V.; Chaussod, R.; Ranjard, L. Copper dynamics and impact on microbial communities in soils of variable organic status. *Environ. Sci. Technol.* **2008**, *42*, 2819–2825. [CrossRef]
40. He, Z.L.; Yang, X.E.; Stoffella, P.J. Trace elements in agroecosystems and impacts on the environment. *J. Trace Elements Medic. Biol.* **2005**, *19*, 125–140. [CrossRef]
41. Ruiz-Fernández, A.C.; Frignani, M.; Hillaire-Marcel, C.; Ghaleb, B.; Arvizu, M.D.; Raygoza-Viera, J.R.; Páez-Osuna, F. Trace metals (Cd, Cu, Hg, and Pb) accumulation recorded in the intertidal mudflat sediments of three coastal lagoons in the Gulf of California, Mexico. *Estu.Coast.* **2009**, *32*, 551–564. [CrossRef]
42. Avramidis, P.; Barouchas, P.; Dünwald, T.; Unkel, I.; Panagiotaras, D. The influence of olive orchards copper-based fungicide use, in soils and sediments-The case of Aetoliko (Etoliko) Lagoon Western Greece. *Geosciences* **2019**, *9*, 267. [CrossRef]
43. Adriano, D.C.; Wenzel, W.W.; Vangronsveld, J.; Bolan, N.S. Role of assisted natural remediation in environmental cleanup. *Geoderma* **2004**, *122*, 121–142. [CrossRef]
44. Badassan, T.E.; Avumadi, A.M.; Ouro-Sama, K.; Gnandi, K.; Jean-Dupuy, S.; Probst, J.L. Geochemical Composition of the Lomé Lagoon Sediments, Togo: Seasonal and Spatial Variations of Major, Trace and Rare Earth Element Concentrations. *Water* **2020**, *11*, 3026. [CrossRef]
45. Huo, S.; Xi, B.; Yu, X.; Su, J.; Zan, F.; Zhao, G. Application of equilibrium partitioning approach to drive sediment quality criteria for heavy metals in a shallow eutrophic lake, Lake Chaohu, China. *Environ. Earth Sci.* **2013**, *69*, 2275–2285. [CrossRef]
46. Nizoli, E.C.; Luiz-Silva, W. Seasonal AVS–SEM relationship in sediments and potential bioavailability of metals in industrialized estuary, southeastern Brazil. *Environ. Geochem. Health* **2012**, *34*, 263–272. [CrossRef]
47. Zhang, C.; Yu, Z.G.; Zeng, G.M.; Jiang, M.; Yang, Z.Z.; Cui, F.; Zhu, M.Y.; Shen, L.Q.; Hu, L. Effects of sediment geochemical properties on heavy metal bioavailability. *Environ. Int.* **2014**, *73*, 270–281. [CrossRef]
48. Guigue, J.; Mathieu, O.; Lévêque, J.; Denimal, S.; Steinmann, M.; Milloux, M.J.; Grisey, H. Dynamics of copper and zinc sedimentation in a lagooning system receiving landfill leachate. *Waste Manag.* **2013**, *33*, 2287–2295. [CrossRef]
49. Almahasheer, H.; Serrano, O.; Duarte, C.M.; Irigoien, X. Remobilization of heavy metals by mangrove leaves. *Front. Mar. Sci.* **2018**, *13*, 484. [CrossRef]

minerals

MDPI

Article

Geochemical Indication of Functional Zones at the Archaeological Sites of Eastern Europe

Marianna Kulkova

Department of Geology and Geoecology, Herzen State Pedagogical University, 191186 St. Petersburg, Russia; kulkova@mail.ru

Abstract: The article considers a new approach for determining the functional zones of the prehistoric archaeological sites in Eastern Europe by the method of geochemical indication: the use of mathematical statistics for processing the geochemical data of cultural deposits at archaeological sites, and the identification of groups of interrelated chemical elements and compounds that reflects the processes of natural sedimentation and anthropogenic activity. It makes it possible to separate the lithological and anthropogenic components. This approach is important for the identification of geochemical element groups associated with different functional zones. The reconstructions were conducted at the Neolithic, Early Metal Age, and the Bronze-Early Iron Age sites in Eastern Europe. Abnormal concentrations of the association (P_2O_{5antr}, CaO_{antr} and Sr_{antr}) in sediments are attributed to zones of accumulation of bone remains. Anomalous concentrations of a group of elements (K_2O_{antr}, Rb_{antr}) in deposits are associated with wood ash and fireplaces, ash residues from ritual activities, and fires. The group of elements (Ba, MnO, C_{org}) reflects the accumulation of humus and organic remains, and can characterize areas with food residues, skins, and rotten wood. With the help of the distribution of the main lithological elements (SiO_2, Al_2O_3) in sediments, it is possible to reconstruct the paleorelief at the sites.

Keywords: geochemical indication; functional zones; archaeological sites; Eastern Europe; mathematical statistic; XRF analysis; cultural layers

Citation: Kulkova, M. Geochemical Indication of Functional Zones at the Archaeological Sites of Eastern Europe. *Minerals* **2022**, *12*, 1075. https://doi.org/10.3390/min12091075

Academic Editors: Domenico Miriello and Donatella Barca

Received: 1 July 2022
Accepted: 23 August 2022
Published: 25 August 2022

1. Introduction

The problem of the functional zone determinations on archaeological sites is very important for an understanding of the spatial distribution of different structures connected with ancient human activity. The living and household buildings had not been preserved on most sites of the Stone, Bronze, and Iron Ages and their space reconstructions are impossible through the archaeological methods only. The activities of prehistoric people influenced the variations in the chemical compositions of the soil by enriching it with or depleting it of certain chemical elements that form archaeological soils and cultural layers [1]. The geochemical markers of anthropogenic activity are conserved in the deposits for many years. In this connection, the application of the geochemical methods at the archaeological sites is a useful instrument for the analysis of cultural layers and determination of the ancient anthropogenic activity at the sites and the reconstruction of functional zones. The last time, the method of geochemical indication for the functional zone is actively being developed in different archaeological sites in Eastern Europe.

The spatial distribution of anomalous concentrations of some chemical elements in the places of ancient settlements makes it possible to establish the boundaries of archaeological sites and their locations; obtain information about the landscape features; establish functional zones and features of various structures within the settlements [2,3]. As a rule, elements for determining anthropogenic activity on ancient sites are P, Ca, K, Na, and Mg, as are the trace elements Cd, Cr, Cu, Pb, and Zn [4–10]. Aston et al. [4] identified the main functions of human activity in which the accumulation of anthropogenic chemical elements occurs: the development of ancient settlements, animal breeding in close quarters, the use of fire (fireplaces, slash-and-burn cultivation), ancient metallurgy, or subsistence activities (production of leather, processing of crops). One of the important anthropogenic indicators is phosphorus. Phosphorus is the main component of human and animal bones, a component of living tissues (in the form of nucleic acids, phospholipids, nucleotides, and so on), and a component of everyday products (such as wood, plant, or meat products) [11]. Phosphorus can accumulate in the soils of ancient settlements as a result of food preparation and the utilization of waste products [8]. The concentration of organic materials containing phosphorus, accumulated in the process of human activity, is proportional to the time of human occupation and the growth of the population [12]. Therefore, the phosphorus content in cultural layers is an indicator of the intensity of human occupation in the area. The phosphorus concentration in the soil increases depending on the supply of different organic materials, such as plants and animals, that are used by people. The decomposition of animal and plant organisms in settlement areas, the use of animal dung as fertilizer, and the physiological activity of humans and animals in their habitation areas increase phosphorus concentration in soils [7]. The different phosphorus compounds in soils are stable to oxidation, reduction, leaching, and dissolution [13,14] and newly mineralized inorganic phosphorus, while generally retained in the soil, is subject to some vertical translocation in the soil due to factors directly affecting adsorption, such as pH, cation exchange capacity, and cation availability [15,16]. The distribution of inorganic phosphate is therefore determined by the chemistry and adsorption kinetics of the soil throughout decomposition and precipitation [7].

At archaeological sites, cultural deposits usually consist of household waste, bones, metal slag, ashes, dung, and the remains of burials and cremations, etc. Application of other geochemical elements and indicators derives additional information about site occupation. Such elements as potassium (K) and sodium (Na) can be connected with the presence of fire ash in the areas of fireplaces [17]. However, it needs to be noted that these elements are the main components of feldspar and plagioclase minerals from lithological sediments. Rubidium (Rb) could be taken up by plants as a substitute for potassium (K), which it chemically resembles. These elements can be considered as anthropogenic components at the archaeological site just in comparison with the background or a correlation between high concentrations and any evidence of charcoal or fire ashes at the ancient places.

According to Wells et al. [10], the complex of elements (Fe, Mn, Zn, and Cu) can indicate areas of waste disposal, burials, cesspools, or rubbish after feasts. Increased concentrations of iron (Fe) and mercury (Hg) at settlements can be explained by the use of different natural pigments in rituals. It should also be noted that the accumulation of geochemical components depends also on the different natural factors of sedimentation such as geomorphology, diagenetic transformations of sediments, climatic conditions, and others. For example, if the archaeological sites are located in areas with complex relief, the primary distribution of chemical elements may have changed as a result of erosion after the elements were originally deposited in the soil. The destroyed material accumulates in depressions and at the feet of slopes as colluvium [1]. These cases require the development of a uniform methodological approach, the determination of a precise geochemical background, and the understanding of geochemical processes at the site. Application of indicator ratios of chemical elements reflects the degree of enrichment of anthropogenic elements in comparison with their background concentrations in the soils of the settlement and outside of it [1,18]. The method of "multi-element" analysis [3,8,13], which has been applied relatively recently, allows us to establish a complex of several chemical components and to consider their connection with different functional zones. Scholars established some groups of the chemical elements characterizing certain functional zones connected with anthropogenic activity [1]: abnormal concentrations of P (phosphorous), Cu (copper), Mn (manganese), Ca (calcium) connect with burials/cemeteries [19–21]; zones of fireplaces–P, K, Mg (magnesium), Zn (zinc), Rb [22,23]; waste piles-P, K [10,24,25]; farm areas, inner space of living buildings-P (phosphorus), Ca, Mg, Fe, K, Th (thorium), Rb, Cs (cesium), Pb (lead), Zn (zinc), Sr (strontium), Ba (barium) [3,10,24–29]; paints–the heavy metals [10]; ore mines, metallurgy-Cu (copper), Pb (lead), Mn (manganese) [30–34]; places of archaeological sites-B (boron), Cu (copper), Mg, Mn, Ni (nickel), P, Se (selenium), Zn (zinc), K (potassium), Ba (barium), Ca (calcium), and Na (sodium) [13,19,21,34].

At the archaeological sites of Eastern Europe, the geochemical approach using multi-component analysis based on the mathematical statistics was applied [18,35–38].

2. Materials and Methods

The samples of cultural deposits were collected at sites of different ages (Figure 1). The main characteristics of sites are presented in Table 1. The sediment samples were taken at intervals of 0.6 m on the coordinate grid on the surface of cultural layer at each archaeological site under consideration. The coordinates of the primary point of sampling were registered by GPS.

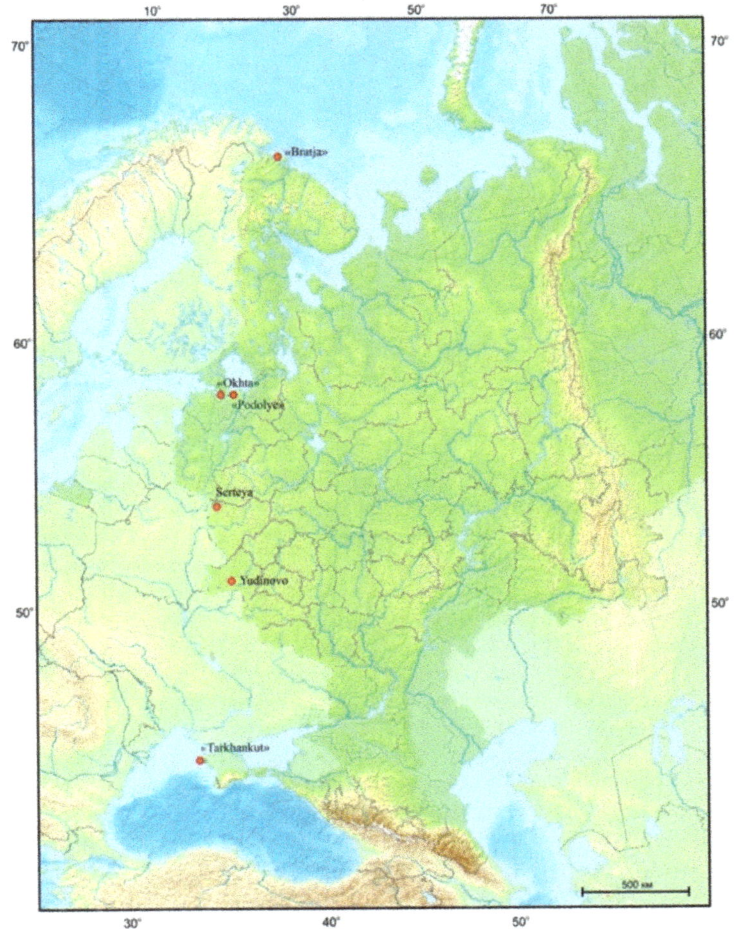

Figure 1. Map of studied archaeological sites of Eastern Europe.

Table 1. The main characteristics of places of sampling.

Archaeological Site	Geology and Soils	Climatic Zone	Age of Cultural Layer	Number of Analyzed Samples	Archaeological Excavations
The Serteya site	fluvioglacial sediments; sod-podzolic	temperate zone; average temperature in July +17 °C; average temperature in January −9 °C; precipitation 615–751 mm per year	Early Neolithic	35 from the cultural 1st layer and 3 control samples; 28 from the cultural 1st layer and 3 control samples; square of 20 × 6 m	The State Hermitage Museum, St. Petersburg, A.N. Mazurkevich [36,37].
The cult object "Bratja"	magmatic rocks; tundra soils	Northern coastline, tundra;average temperature in July +8 °C; average temperature in January −14 °C; precipitation 400 mm per year	Middle Ages	40 from the cultural layer; 8 control samples; square of 15 × 6 m	Expedition of the National Museum republic of Karelia (Petrozavodsk), Dr. M.M. Shakhnovich [39]
Okhta 1 and Podolye 1 sites	morane and lake-glacial deposits; podzolic-gley soils	tundra; average temperature in July +17 °C; average temperature in January +3.3 °C; precipitation 635 mm per year	Neolithic and Early Metal Ages	25 from the cultural layer at Okhta site and 3control samples; square of 500 × 700 cm 32 from cultural layer and 3 control samples at Podolye site; square of 8 × 6 m	The excavations of Dr. T.M.Gusentsova and P. Sorokin [40–42]
Tarkhankut-18 site	the aeolian deposits on limestones; light-chestnut soils	coastline, steppe zone; average temperature in July +24 °C; average temperature in January −8 °C; precipitation 120–700 mm per year	Late Bronze-Early Iron Age	47 from the cultural layer 600 × 180 cm square; and 18 control samples from square of 90 × 90 cm	The excavations of Dr. M.T.Kashuba (IIMC RSA, St. Petersburg) and Prof. T.N. Smekalova (Crimea Federal University, Simferopol) [43]
Yudinovo site	loess deposits; loess soils	Temperate continental; average temperature in July +19 °C; average temperature in January −8 °C; precipitation 550–600 mm per year	Upper Paleolithic	from cultural layer control samples	The excavations of Dr. G.A.Khlopachev (MAE RSA, Kunstkamera Museum, St. Petersburg) [44]

The Serteya archaeological microregion (Figure 1) is located in the basin of the Dvina-Lovat' basin, in the southern part of the Pskov and northwestern parts of the Smolensk region. Ancient lakes in the area of the Serteya archaeological microregion were formed after the retreat of the glacier and were located in a chain, separated by narrow isthmuses, which were eroded during periods of rising water levels. Along their shores, as well as in the central parts of the lake basins, there are unique archaeological sites that were dated from the 8th millennium BC up to 10 c. AD. At present, the lake basins are swamped and inherited by the narrow channel of the Serteya River, which flows into the Western Dvina River. The archaeological excavations are conducted by the northwestern expedition of the State Hermitage Museum (St.Petersburg, the head of the expedition is A. Mazurkevich) [36,37]. Thirty-five samples from the 1st cultural layer, three control samples beyond the archaeological area and 28 samples from the 2nd cultural layer, and three control samples were taken for analysis at the Serteya 3-3 site.

Rocky remains "Bratja"—an object of the post-glacial geological history of Fennoscandia, located directly on the seashore, in the northwestern part of the Sredny Peninsula (Figures 1 and 2). Two picturesque stone pillars of a bizarre shape stand out in the surrounding landscape, which is similar to natural "statues" with a mythological meaning [39]. The pillars are located on a flat, first sea terrace-an ancient pebble beach, at an altitude of about 15 m above sea level. Sediments of medium grain size sand were sampled under the turf layer for determination of cultural layer features. Forty samples from the cultural layer at the Bratja object and eight control (background) samples were taken outside of the place. The investigations were provided together by the archaeologist M. Shakhnovich (National Museum Republic of Karelia, Petrozavodsk, Russia).

Figure 2. Rocky remains "Bratja"—an object of the post-glacial geological history of Fennoscandia. The photo is by M. Shakhnovich [39].

Sites of the Neolithic-Early Metal Age (Okhta 1 and Podolye 1) are located in the Ladoga Lake basin (Figure 1). The excavations were conducted under the head of T. Gusentsova and P. Sorokin (Institute of the Cultural Heritage, St. Petersburg, Russia). The Okhta 1 site is located on the border of the Russian Platform and the Baltic-Scandinavian crystalline shield, on a sandy cape formed after the Littorina transgression [18,36,40,41]. On the Okhtinsky Cape, at the confluence of the Okhta and Neva rivers, a unique multi-layered monument was discovered containing cultural layers of the Neolithic and Early Metal Ages, the Early Iron Age, and Roman times. The site Podolye 1 is located on the southern coast of Lake Ladoga, about 4 km from the coast [41,42]. At the Podolye 1 site, as well as at Okhta 1, wooden fishing structures of different ages were found. The lower cultural layer is light yellow silt with thin interlayers of peat. The upper cultural horizon consists of sandy-silty sediments with thin black organic layers. Dense grey loam overlaps these layers. Twenty-five samples from the cultural layer at the Okhta 1 site, three control samples and 32 samples from the upper cultural layer at the Podolye 1 site and five control samples were sampled.

At the Tarkhankut-18 site (Figure 1), located on the Tarkhankut Peninsula in Crimea, there are the Late Bronze-Early Iron Age complexes. The excavations were provided by M.T. Kashuba (IIMC RSA, St. Petersburg, Russia) and T.N. Smekalova (Crimea Federal University. Simferopol, Autonomous Republic of Crimea) [43]. The cultural layer has been formed on the surface of the aeolian deposits. The samples have been taken from the surface of the cultural layer and from the aeolian deposits (background pit) to establish the background contents of geochemical components. The background pit was laid outside the archaeological site for determination of the composition of geologically similar deposits that are not subject to anthropogenic influence. Forty-seven samples at the Tarkhankhut_18 excavation were taken, and 16 probes were sampled from the background pit.

The Yudinovo site is one of the most studied sites of the Upper Paleolithic (15–12 thousand years ago), located in the Desna river basin, in the Pogarsky district of the Bryansk region of Russia (Figure 1). The excavations were carried out by G. Khlopachev (Kunstkamera Museum, St. Petersburg, Russia). The household objects of different types have been discovered on the site: the remains of five dwellings of the Anosovo-Mezin type, six large household pits, ten large "ash pans"—the accumulations of bone coal and ash, two small hearths, two hearth pits, several "clad" pits, numerous production sites associated with the primary treatment of flint and tusk, and cutting carcasses of Arctic foxes [44]. The strategy of sample preparation is the following: the total weight of the sample was about 100 g, taking into account possible duplicates. The deposit samples were dried in an oven for 24 h at a temperature of 105 °C. For geochemical analysis, after each crushing cycle, control sieving was carried out, the residue is crushed to the desired grain size. Then, all crushed material is combined. Manual reduction is carried out by the method of multiple (no more than 3 with one crushing) quartering and combining the material of two opposite quadrants to continue processing according to the accepted scheme. Due to the heterogeneity of the distribution of components in large and small fractions, before reducing the samples, they are mixed and then quartered. The sample is grinded up in powder to a particle size of 200 mesh, 0.074 mm (to the state of powder). Next, the loss on ignition (LOI) at a temperature of about 550 °C for 40 min was determined for the powder samples. For X-Ray Spectral fluorescence analysis using the Spectroscan Max-GV spectrometer, Spectron, St.Petersburg (Lab of Geochemistry of the Environment of Herzen State University, St.Petersburg, Russia), 300 mg samples were pressed into a tablet using a boric acid substrate. The measuring system of the spectrometer is in the vacuum chamber, while the samples are at the ambient pressure, so no He is required for the sample chamber and all samples (including liquid and powder) may be studied with no special measures taken. XRF-WD with an energy resolution of 9 eV (Si Kα) and 60 eV (Fe Kα). Spectrum scanning and high resolution of crystal analyzers exclude close lines of different elements overlapping (therefore, there is no need in their mathematical differentiation) and provides correct background accounting. The Specroscan determines the elements from ^{11}Na to ^{92}U. A vacuum line along the optical way allows us to determine the light elements including (11 Na). All 15 chemical elements have been determined. The range of determined concentrations from 0.0001% to 100%. Limits of detection, L for light matrix (exposure-100 sec): Na–1 $\times$ 10^{-1}%, Mg–1 $\times$ 10^{-2}%, Al–1 $\times$ 10^{-3}%, Si–5 $\times$ 10^{-4}%, P–5 $\times$ 10^{-4}%, Cd, Pb–5 $\times$ 10^{-4}%; S, Ti, V, Cr–1 $\times$ 10^{-4}%, Co, Ni–5 $\times$ 10^{-5}%. Each sample was measured for 2 h once. The main elements are transformed into oxides by a spectrometer automatically. Several samples were measured twice for control of stability.

The Surfer Mapping Software, Golden Software, Colorado, USA (Versions 10.0, 13.0) was applied to map the distribution of geochemical indicators of anthropogenic activity in the studied area. According to geochemical indicators and groups of indicators, the functional zones were reconstructed.

After measuring, the results of geochemical elemental composition were processed by the method of multivariate statistical analysis (correlation analysis and factor analysis on the base of PCA) to determine the main factors influencing sedimentogenesis. Correlation analysis allows the grouping of the geochemical elements and their compounds depending on their genesis and certain conditions of sedimentogenesis basing the mineralogical and geochemical composition of sediments. A method of multicomponent analysis based on the identification of groups of geochemical components of anthropogenic genesis on the base of correlation and method PCA of factor analyses was developed by the author. This gives the possibility to characterize various functional zones at ancient settlements [18,35,36,38]. The data on the chemical composition of deposits from two sites (the Serteya and the Bratja) were separately calculated by means of the factor analysis based on the principle component method (PCA) for determination of environmental and ancient anthropogenic factors that influenced the sedimentation. The key concept of factor analysis is that multiple observed variables have similar patterns of responses because they are all associated with a latent variable. The number of principal components was determined according to how complex our model will be. As a rule, the factor corresponding to the largest eigenvalue (7.14, 7.8 for the Serteya 3-3 for the 1st and 2nd cultural layers separately and 7.1 for the Bratja site, respectively) accounts for approx. 27%, 30%, and 26% of the total variance. The second factor corresponding to the second eigenvalue (4.9, 5.2, and 4.7) accounts for approx. 19%, 21%, and 18% of the total variance, and so on. When analyzing correlation matrixes, the sum of the eigenvalues is equal to the number of (active) variables from which the factors were extracted (computed). Four main factors were taken into account for determination of sedimentation characteristics for sites. For other sites, the geochemical modules were applied without PCA calculation because the results can be compared with archaeological information directly.

The ratios of geochemical components and geochemical modules for individual elements and their oxides, for example, potassium, calcium, strontium, and phosphorus, which can form compounds associated both with the mineral composition of deposits and the activity of ancient people, were applied. This makes it possible to separate the components associated with the lithological component of the deposits from the anthropogenic ones [18,36,38]. Comparison with samples taken outside the anthropogenic activity from the same lithological layer. Other factors that affect the behavior of individual chemical elements are pH (alkaline acid), Eh (reductive-oxidative) potentials of the soil environment, the content of organic matter in soils, etc., which should also be taken into account when assessing the formation of a cultural layer at a settlement. To assess the overall anthropogenic impact on the settlements, the author proposed the following ratios of components for calculation: $P_2O_{5antr} = P_2O_5/(P_2O_5 + Na_2O)$ (%), $Ca_{antr} = CaO_{tot}/(CaO_{tot} + Na_2O)$ (%), $K_2O_{antr} = K_2O/K_2O + Na_2O)$ (%) and $Rb_{antr} = Rb/(Rb + Na_2O)$ (%). Elevated values of indicators (P_2O_{5antr}, CaO_{antr}) correlate well with the areas of sediments in which bone remains were found. In this case, such ratios make it possible to establish and separate anthropogenic calcium and phosphorus from lithogenic ones, which are associated with the lithology of the deposits. Indicators (K_2O_{antr}, Rb_{antr}) were used to identify focal clusters [18,35,38].

3. Results

3.1. Serteya Archaeological Microregion

The reconstruction of the functional zones was carried out at the multilayer Neolithic site Serteya 3-3 [37]. The remains of the material culture were found in sandy-silty deposits. Mineral composition of deposits: quartz, mica, clay minerals, feldspars, accessory minerals: zircon, garnet, titanomagnetite, hydrogoethite, and ilmenite. According to the correlation matrix basing on the results of the geochemical composition of deposits (n = 35), several geochemical groups of components can be distinguished. These groups will be associated with different minerals, organo-mineral complexes of the anthropogenic component of deposits at the site:

- Group I: SiO_2-quartz;
- Group II: Al_2O_3, Fe_2O_3, MgO, TiO_2, Na_2O-feldspars, mica, clay minerals; hydrogoethite; ilmenite;
- Group III: K_2O, Ba-the main components of wood ash according to [17];
- Group IV: P2O5, CaO, MnO, C_{tot}-components of carbonate-apatite of bone tissues and teeth, with organic residues according to [8,11,12].

An analysis of the stratigraphy and paleorelief on the site allowed the conclusion that the place for the settlement was chosen taking into account the features of the ancient relief and composite from sandy dune deposits. According to the factor analysis, it is possible to identify the factors associated with anthropogenic transformations of sediments at the site. For the lower (1st) cultural horizon, the first factor (FI) characterizes the features of the relief and reflects the antagonism of the components of the sand (SiO_2, Zr) and the components of the clay-mica minerals (Al_2O_3, TiO_2, Fe_2O_3, MgO, Na_2O). It is based on positive and negative correlation between components. Accumulation of the clay component enriched in iron oxides and hydroxides is typical for depressions and pits, including those left from plant roots. The second, third, and fourth factors can be interpreted as related to various anthropogenic activities at the site, which led to a change in the composition of sediments and the appearance of different types of components:

FII (CaO, Na_2O, SiO_2/K_2O, Ba, TiO_2);
FIII (K_2O,Al_2O_3, Ba, CaO/Fe_2O_3, MnO, C_{tot});
FIV (MgO, MnO, Fe_2O_3,TiO_2/P_2O_5, C_{tot}).

The second factor (FII) shows the antagonism of components associated with the fireplaces (K_2O, Ba, TiO_2) to lithogenic components (CaO, Na_2O, SiO_2). In places of increased concentrations of elements associated with wood ash, residues of charcoal and ash stains from hearths were found. The third factor (FIII) shows the antagonism between the complex of lithogenic components (K_2O, Al_2O_3, Ba, CaO) and components (Fe_2O_3, MnO, C_{tot}), which are associated with the accumulation of humus and coincide with traces of habitation and internal areas of the dwellings. The fourth factor FIV (MgO, MnO, Fe_2O_3, TiO_2/P_2O_5, and C_{tot}) is associated with areas of accumulation of organic matter and bone residues (P_2O_5, C_{tot}), which could be concentrated in areas of animal cutting and food preparation. Sherds from pottery were also found in these zones. Figure 3 shows geochemical maps on the base of indicators-factor load data, giving an idea of the position of various sites associated with anthropogenic activity and reflecting the spatial arrangement of various functional zones in the area of the settlement.

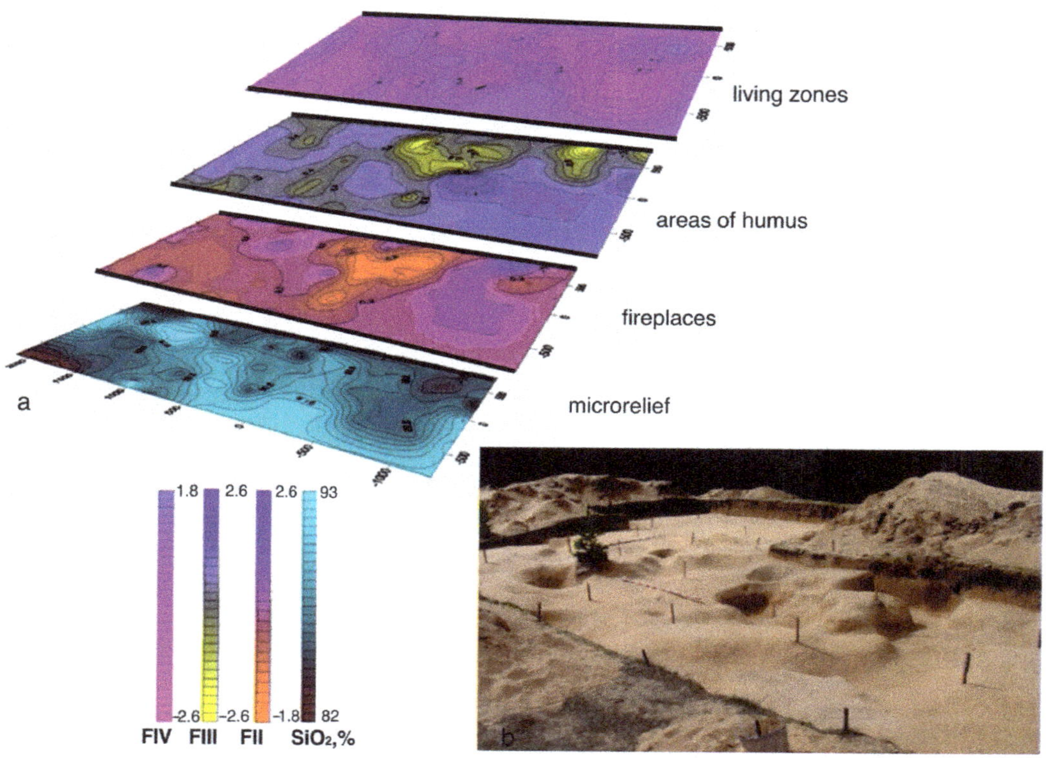

Figure 3. (**a**) Spatial distribution of functional zones on the surface of the 1st cultural horizon of the Serteya 3-3 site according to geochemical mapping adapted from [37]; (**b**) surface of the 1st cultural layer [37], designed by M.Kulkova.

In deposits of the 2nd upper cultural horizon at the Serteya 3-3 site, the remains of temporary structures are recorded. According to the geochemical data of deposits from the 2nd cultural horizon, the factors influencing the formation of these deposits were also established (n = 28). FI (SiO_2/Al_2O_3, Fe_2O_3, MgO, CaO) reflects the features of the microrelief. In depressions in this case, sand deposits were accumulated, which may indicate small ancient river channels. This matches well with the 3D reconstruction of the ancient surface. The areas of the fireplaces were determined by the indicator ratio $K_2O_{antr} = K_2O/(K_2O + Al_2O_3)$. Comparison of the anomalous K_2O_{antr} values with charcoal accumulations in these zones shows a good correlation. It allows us to trace the change in the location of the hearths in space. Areas of concentration of bone residues were determined by the CaO/SiO_2 ratio. The abnormal values match the locations where calcified bones were found. The distribution map of the positive values of the second factor shows the zones of humus content (Ba, MnO, TiO_2). Negative values of the third factor FIII (MgO, K_2O/P_2O_5, C_{tot}) indicate habitat areas (P_2O_5, C_{tot}). The integrated application of the geochemical method and the geomagnetic method [37], as well as the use of more accurate methods for measuring magnetic susceptibility with a kappameter, makes it possible to detect traces of unreadable structures during excavations and obtain data for determination of zones associated with human activity. Such sites have been recorded in the cultural layer at the multilayer site of the Neolithic period Serteya XIV [18,37]. In addition to the main indicators of anthropogenic activity, the comparison of geochemical data and geophysical anomalies also made it possible to obtain additional information on other indicator elements that may be associated with biogenic processes and human activities. Elevated concentrations of iron oxides Fe_2O_3 and FeO appear around plant roots and may characterize the remains of decayed stumps [37]. High anomalous concentrations of iron oxides (Fe_2O_3 and FeO) also correspond to the localization of long-term fires. Accumulation of lead (Pb) may be associated with hearths, or with deposits of humus under reducing conditions [24–29]. In this archaeological context, the distribution of Pb correlates with the accumulation of K_2O_{antr} in wood ash. The accumulation of copper (Cu) can also be associated with reducing conditions and concentrates in the areas of the fireplaces. According to geochemical indicators, high values of such indicators as K_2O_{antr}, Cu, and Pb were established, which can be explained by the accumulation of ash or the localization of a small hearth that was not visually detected during excavations.

3.2. *Sredny Peninsula. the Cult Place "Bratja"*

The method of geochemical indication makes it possible to establish several main factors influencing the process of sedimentation at the archaeological sites and to identify anthropogenic components, if they were present in the deposits (n = 40). The formula of the first factor (FI): CaO, P_2O_5 (Sr)/Al_2O_3, SiO_2, TiO_2 (Zr) shows antagonism based on negative and positive correlation between the groups of components associated with anthropogenic activity; the association of components (CaO, P_2O_5 (Sr)), which are characterized by high correlations and rock-forming and accessory components (Al_2O_3, SiO_2; TiO_2; (Zr)), was associated with the minerals of the sandy deposits [39]. The contents of anthropogenic components CaO, P_2O_5; (Sr; Rb) were also determined for sediments at two sites. In a comparison of the concentration of these elements with the background samples taken outside the site, anomalous concentrations in the sediments at the site stand out. Two anomalous zones are distinguished on both sites of the monument investigated by the complex of anthropogenic components. These zones can be considered places of ritual actions. In such zones, anomalous contents of anthropogenic potassium and rubidium have been recorded. Elevated concentrations of K_2O_{antr} are associated with wood ash in the cultural layer and are a marker of the fireplace zone. This is also noted for the Rb_{antr} values, which correlate well with K_2O_{antr}. In the areas of fireplaces, which are enriched with ash, increased concentrations of zinc (Zn) and copper (Cu) are also registered. Thus, basic data obtained on the zones of fireplaces, as well as areas of ash emissions from them, located nearby, were reconstructed. It can be assumed that zones with elevated contents of a complex of elements such as P, Ca, K, Sr, Rb, Zn, and Cu mark areas of ancient ritual practice where sacred offerings could be left, animal meat was butchered, and there were fireplaces. Thus, according to geochemical studies, it can be concluded that, in the space between the two stone columns of the "Bratja" ceremonial, ritual actions were performed, which can be considered pagan sacrifices associated with hunting and household magic. These rocks could be perceived as expressive natural objects, endowed with sacred properties, and the areas near them were used for ritual practices. The ethnographical evidence of the ritual activity is known at this place [39].

3.3. Ladoga Lake Basin (Sites Okhta 1 and Podolye 1)

At the sites of Okhta 1 and Podolye 1, various types of functional zones were considered on the surface of cultural layers, which formed around 3300 calBC [35,38,40–42]. They are presented by soils and medium grain sands. At the Okhta 1 site, geochemical mapping data and the distribution of artifacts show the burial site (Figures 4–6). The burial deposits are characterized by anomalous values of a complex of chemical anthropogenic components, such as P_2O_{5antr}, Ca_{ntr} and Fe, Sr_{antr}. The geochemical composition of the deposits at the Podolye 1 site shows a different character of the distribution of geochemical elements (Figures 7–10) [38]. On the terrace of the small river channel, there were seasonal fishing camps with fireplaces and places for cutting and cooking animals and fish. This is supported by the archaeological evidence [35,38]. A combination of such anthropogenic elements as K_2O_{anhtr}, Rb_{antr}, CaO_{antr}, and Sr_{antr} stands out in the cultural layer of the sediments of this site. On the shore of the channel, it is possible to distinguish areas for the location of fireplaces (K_2O_{antr} and Rb_{antr}—elements that are accumulated in charcoal and wood ash) and areas for cutting and cooking animals and fish (CaO_{antr} and Sr_{antr}—main elements of bone tissue). In the pit located on the terrace of the channel, another geochemical complex with anomalous values was characterized: K_2O_{antr}, Rb_{antr}, CaO_{antr}, Sr_{antr}, Mn_{antr}, Ba_{antr}, Fe, and P_2O_{5anthr}, which may indicate the accumulation of waste in the channel pit.: K_2O_{antr} and Rb_{anhtr}-components, which are part of ash and charcoal, $P_2O_{5\ antr}$, CaO_{antr}, Sr_{antr} are components of bone tissues, Mn_{antr}, Ba_{antr}, and Fe are components that are part of decomposed organic matter. The distribution geochemistry of major sediment-forming elements such as alumina (Al_2O_3) and silica (SiO_2) is an important factor in determining the ancient microrelief at this site (Figure 11). Associations of geochemical indicators from the same lithological context at the sites of Okhta 1 and Podolye 1 have different anthropogenic loads. The deposits from the burial are characterized by anomalous values of such bioindicators as P_2O_{5anhtr} and CaO_{antr}, which are the main components of bones and tissues. Iron in the form of hematite (Fe_2O_3) is part of the ocher, which was used in the ritual ceremony.

P_2O_{5antr} (%)

Figure 4. Distribution of P_2O_{5antr} (%) on the surface of the burial layer at the Okhta 1 site [38], designed by M.Kulkova, a, b, c, d—burial area.

Caantr (ppm)

Figure 5. Distribution of Ca_{antr} (ppm) on the surface of the burial layer at the Okhta 1 site [38], designed by M.Kulkova.

Fe (%)

Figure 6. Distribution of Fe (%) on the surface of the burial layer at the Okhta 1 site [38], designed by M.Kulkova.

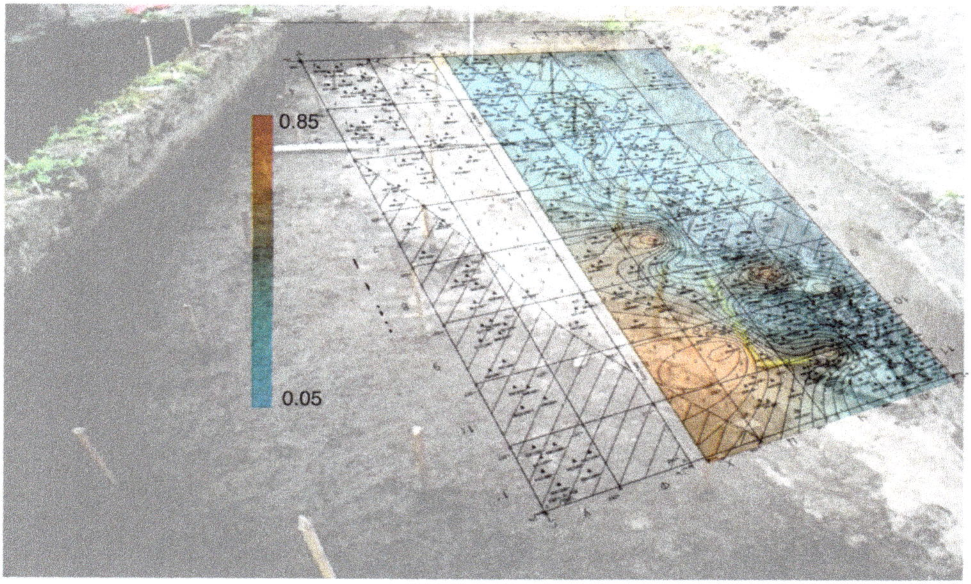

P_2O_{5antr} (%)

Figure 7. Geochemical maps of P_2O_{5anthr} distributions in sediments on the surface of the cultural layer at the Podolye 1 site [38], designed by M.Kulkova.

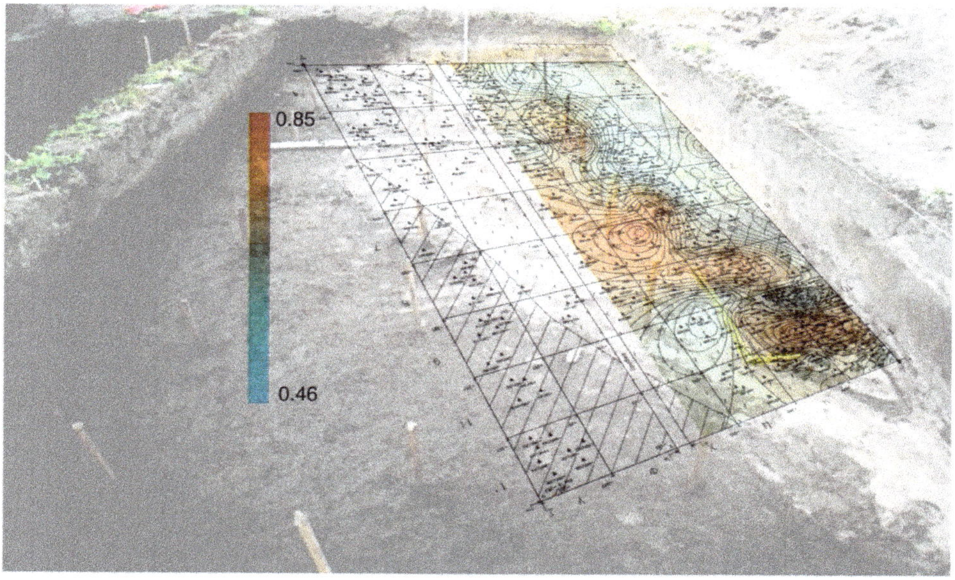

CaO_{antr} (%)

Figure 8. Geochemical maps of CaO_{anthr} distributions in sediments on the surface of the cultural layer at the Podolye 1 site [38], designed by M.Kulkova.

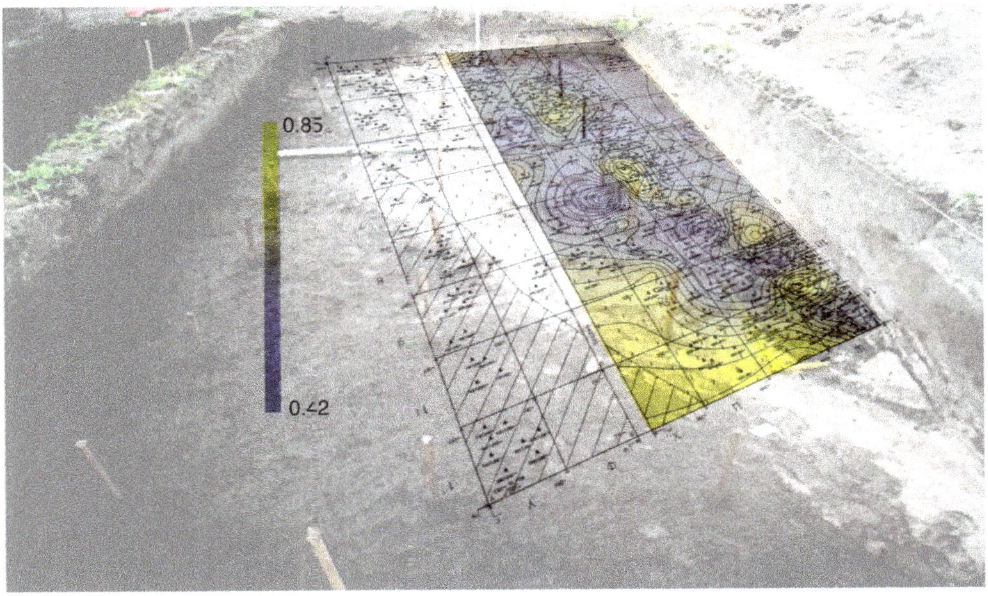

Figure 9. Geochemical map of K_2O_{antr} distribution in sediments on the surface of the cultural layer at the Podolye 1 site [38], designed by M.Kulkova.

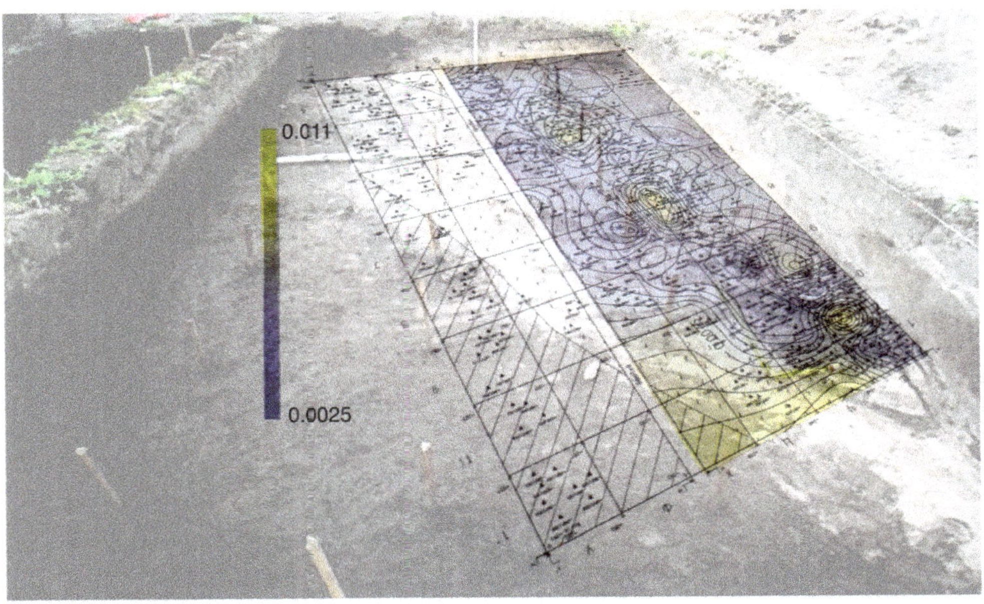

Figure 10. Geochemical map of Rb_{antr} distributions in sediments on the surface of the cultural layer at the Podolye 1 site [33], designed by M.Kulkova.

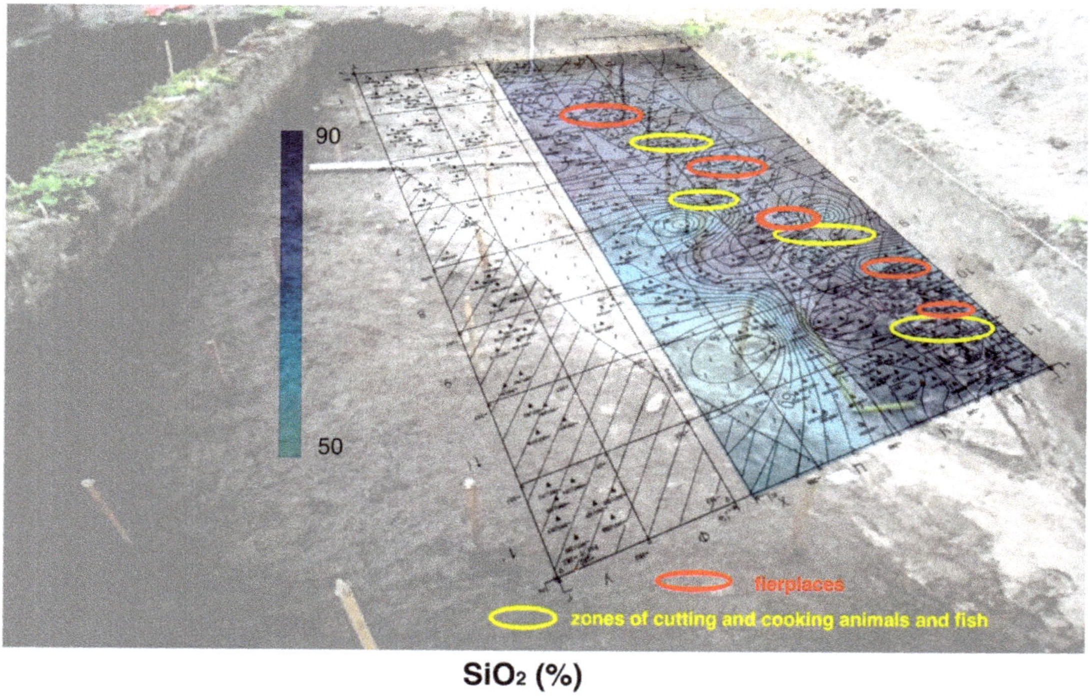

Figure 11. Features of relief on the base of SiO_2 distribution at the site surface and location of different types of functional zones based on geochemical reconstructions at the Podolye site [38], designed by M.Kulkova.

3.4. Tarkhankut Peninsula (Crimea), Bronze Age Cattle Pens

Maps of the distribution of components (SiO_2, Al_2O_3) associated with the lithological composition of the deposits show the features of the microrelief in the excavation area (Figure 12) [43]. Elevated areas are clearly distinguished, on which the stone walls of the corral were built, and low areas, in which cattle were kept inside stone structures. The sediments in these areas are heavily "trampled" and reworked. Elevated values of anthropogenic components (CaO_{antr}, K_2O_{antr}, P_2O_{5antr}) in sediments in the areas of depression inside the paddocks indicate intense anthropogenic activity associated with livestock management (Figure 13). These sites are also characterized by high values of the main anthropogenic components, compared with the background site, outside the paddocks. The aggressive chemical environment, which changed the chemical composition of the sediments and led to the almost complete dissolution of carbonates, was formed in the conditions of the cattle stable, which is also confirmed by archaeological evidence. The content of CaO in the paddocks is much lower than in the background deposits (Figure 13). Anomalies of anthropogenic K_2O_{antr} and Ba, which are associated with the processes of formation of humus, manure, etc. are also noted in paddock areas. Thus, according to the data of geochemical elements, it is possible to determine the main function of this area as a pen-stall for keeping animals.

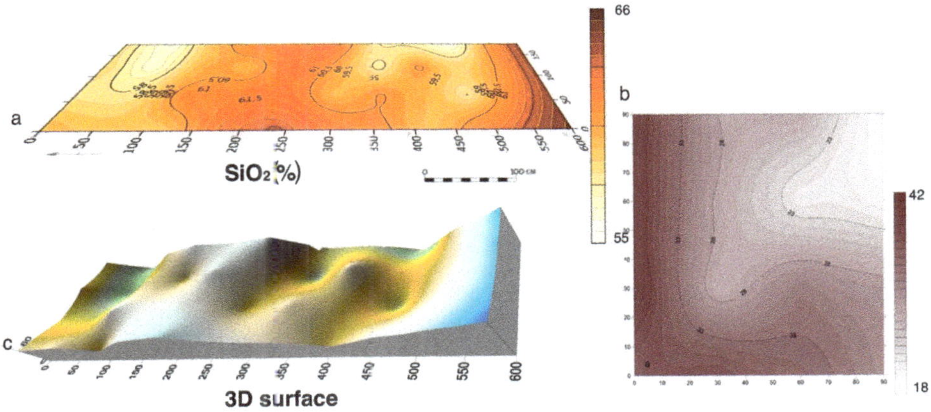

Figure 12. The Tarkhankhut 18 site. Geochemical maps of SiO_2 distribution in the sediments of (**a**) inside cattle pen; (**b**) background pit; (**c**) 3D reconstruction surface based on SiO_2 distribution [43], designed by M.Kulkova.

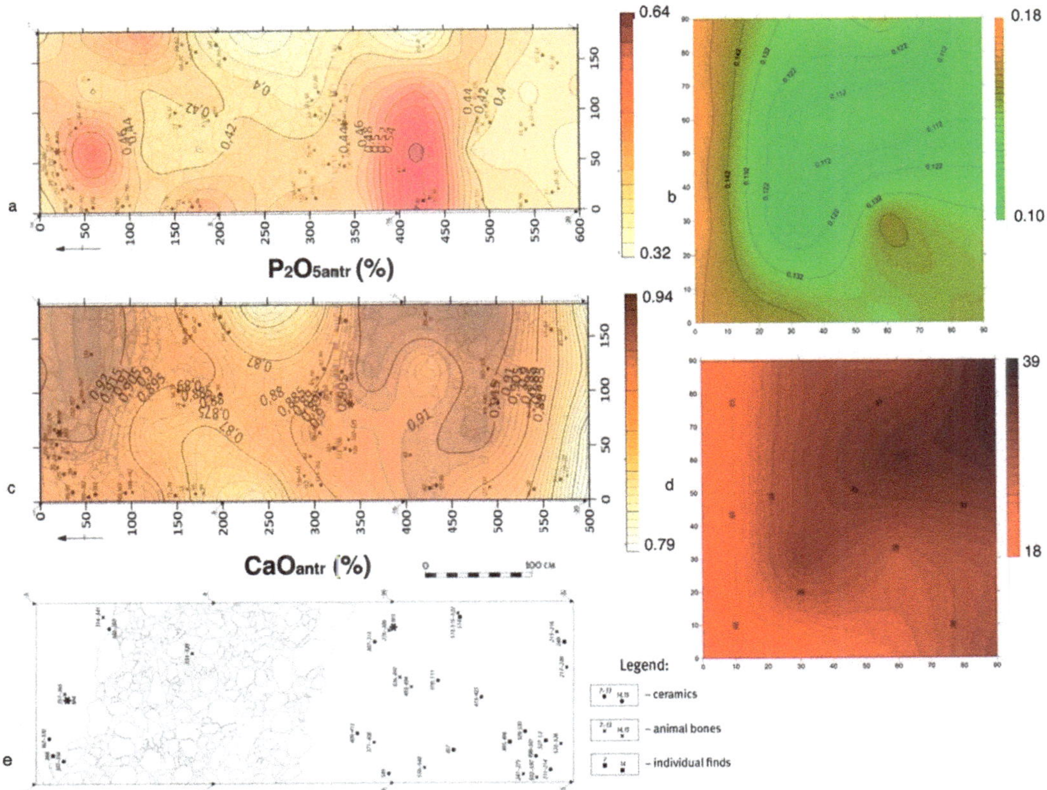

Figure 13. The Tarkhankhut 18 site. Geochemical maps of P_2O_5 distribution in the sediments of (**a**) inside cattle pen; (**b**) background pit; geochemical maps of CaO distribution in the sediments of (**c**) inside cattle pen; (**d**) background pit; (**e**) archaeological plan of cattle pen [43], designed by M.Kulkova.

3.5. The Upper Paleolithic Site Yudinovo

The cultural deposits at the Yudinovo site consist of Pleistocene loess and sandy loam. The geochemical results were considered in the article [44]. Important indicators for assessment of the ancient microrelief are changes in the content of the main rock-forming components (Al_2O_3)-alumina and (SiO_2)-silica. 3D geochemical maps of Al_2O_3 and SiO_2 distributions can be used for reconstruction of ancient microrelief in the area of the excavation (Figure 14). Elevated concentrations of alumina (Al_2O_3) are associated with depressions in the relief, where loam deposits accumulate. Low concentrations of SiO_2 also mark areas of depressions in the relief. Silica, which is associated with the sandy component, is confined to elevated areas and is associated with areas of gray sandy loam in the excavation area. Elevated areas are composed of gray sandy loam enriched in silica and depleted in alumina; they mark the layer in squares t-28 and t-29, and are also noted in the middle part of square t-30, on the border with the pit-depression. It is possible that the sand could have been specially brought by ancient people to this place.

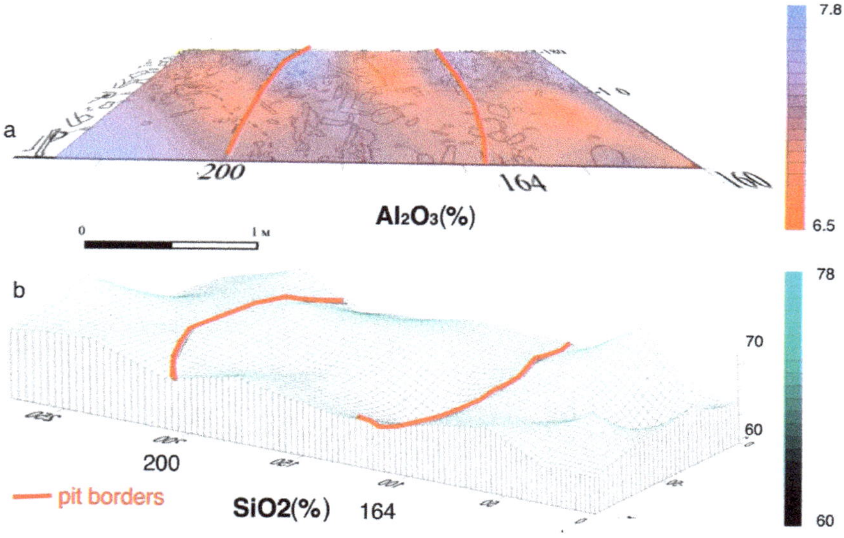

Figure 14. The Yudinovo site. Geochemical maps of (**a**) Al_2O_3 (**b**) 3D distribution of SiO_2 in the sediments [44], designed by M.Kulkova.

Geochemical maps of the distribution of complex of anthropogenic components, such as phosphorus (P_2O_{5antr}), calcium (CaO_{antr}), potassium (K_2O_{antr}), and rubidium (Rb) (Figures 15 and 16) show areas associated with burnt bone remains. The largest accumulations are associated with depressions in the relief. Such zones are observed on the surface of the depression in squares t-28 and t-29 and form ring structures. A zone with high contents of the complex of these anthropogenic elements is also recorded in the T-30 square on the border of the "ash pit" and the pit-depression. The second group of chemical components (Fe_2O_3, MnO, Ba) may be connected with the location of organic residues (skins, wood, etc.), which were subject to decomposition and decay. High concentrations of these elements are noted in the ash pan deposits. The zone of high concentrations of these elements is confined to the border of the squares t-29 and t-30, which are planigraphically connected with the edge of the "ash pan". The layer with high concentrations of these elements "flows" further into the "pit-depression" zone. Anomalous arsenic (As) concentrations are recorded in the area of square t-30, in its middle part. Some contents of arsenic (As), lead (Pb), and zinc (Zn) are also noted in the deposits of the "ash pan".

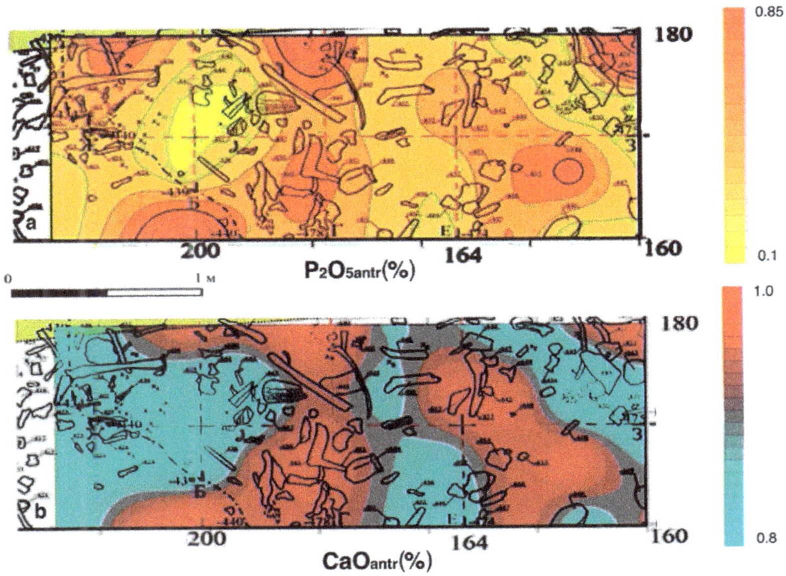

Figure 15. The Yudinovo site. Geochemical maps of (**a**) P_2O_{5anthr} and (**b**) CaO_{antr} distributions in the sediments [44], designed by M.Kulkova.

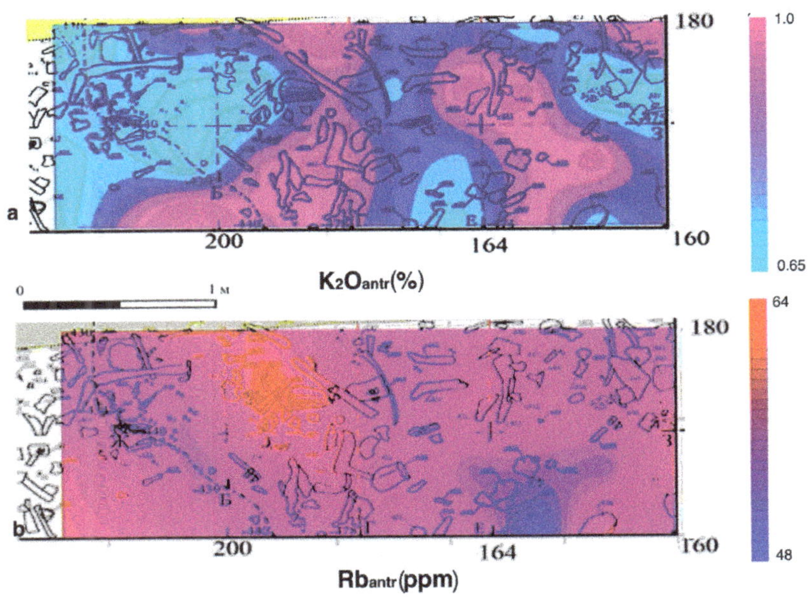

Figure 16. The Yudinovo site. Geochemical maps of (**a**) K_2O_{antr} and (**b**) Rb distributions in the sediments [44], designed by M.Kulkova.

Data on anthropogenic chemical elements make it possible to reconstruct the following structures on the surface of the studied area of the Yudinovo site: firstly, the upper cultural horizon above the pit-depression is characterized by increased contents of a complex of components that are associated with burnt bones: phosphorus (P_2O_{5anthr}), calcium (CaO_{antr}), potassium (K_2O_{antr}), and rubidium (Rb), which allows us to consider it as a separate cultural layer with bone remains; secondly, in the layer of the ash pit, which belongs to the main, lower cultural layer, there are high concentration (Fe_2O_3, MnO, Ba), elements associated with decomposed organic residues, such as skins, wood, etc. The totality of these chemical elements is also concentrated in deposits associated with deposits in the filling of the pit-depression, especially increased concentrations of elements in the deposits on the sides of the pit-depression, into which the ash pan layer falls. Along with this complex of elements, elevated contents of arsenic (As), lead (Pb), and zinc (Zn) are recorded in the ash pan deposits at the edge of the depression. It is known that the concentrations of these elements increase in a reducing environment, during the decomposition of organic matter. In addition, a complex of such elements as arsenic (As), zinc (Zn), and lead (Pb) can be associated with various components that make up the substances, the sources of which were located far from the Yudinovo site. For example, realgar is a red arsenic sulfide mineral, which, according to geological maps, is not found in the region where this archaeological site is located. The nearest sources of that mineral could be regions of Ukraine and Caucasus.

4. Discussion

Multicomponent statistical analysis of the geochemical composition of loose Quaternary deposits at archaeological sites located in Eastern Europe makes it possible to identify geochemical elements associated with both the lithological component of deposits and anthropogenic elements that accumulate in deposits as a result of the activities of prehistoric people. Such an approach allows identification of insoluble mineral and organo-mineral compounds that are formed as a result of the ancient anthropogenic activity at the sites. On the other hand, the distribution of the main rock-forming geochemical elements, such as Si (silicon) and Al (aluminum) in sediments on the area of settlements, allows us to establish the features of the microrelief on the site that existed during the period of occupation of the ancient population. This is important for multi-layer sites. At the Serteya 3-3 site, the upper and lower Neolithic cultural layers were formed as a result of the transformation of dune sand deposits under the influence of various anthropogenic activities. According to geochemical data, it is possible to distinguish zones in the settlement enriched in the sandy component in elevated areas, characterized by an association of elements (SiO_2, Zr) and areas of depressions in the relief enriched in the clay-mica component (Al_2O_3, TiO_2, Fe_2O_3, MgO, Na_2O).

Comparison of the 3D surface of this paleorelief with archaeological data [37] makes it possible to reconstruct the contours of residential structures, pits from tree roots, and natural depressions in the relief that are not associated with an anthropogenic activity. Features of the microrelief reconstructed on the data of lithological geochemical elements were analyzed on the studied sites. At the Paleolithic site Yudinovo on the surface of loess cultural deposits, there are elevated parts with sediments enriched by SiO_2 and lower areas with high Al_2O_3 concentrations of loess. Analysis of the distribution of SiO_2 in loess deposits in different parts of the site and comparison with archaeological data on the distribution of finds and structural features also makes it possible to identify artificial sand beds [44]. Geochemical mapping of the coastal zone of the ancient channel at the Podilye 1 site, composed of sandy-silty deposits, also makes it possible to identify microrelief features according to the data of the distribution of alumina (Al_2O_3) and silica (SiO_2). Comparison with anthropogenic geochemical indicators (K, Rb) determined the lowland zones-pits associated with fireplaces. According to the geochemical data of sand deposits at the Tarkhankut 18 site, the distribution of components (SiO_2, Al_2O_3) in the deposits also fixes the features of the microrelief in the paddock zones and artificial sand bedding, which was used in the construction of the stone fences of the paddocks.

At all sites under studies, despite the deposits having a different genesis, it is possible to reliably identify groups of anthropogenic geochemical elements in cultural layers, which reflect various functional zones of ancient anthropogenic activity. At the Serteya 3-3 site, a group of geochemical elements is distinguished in the cultural deposits of the lower horizon. P_2O_5, CaO, MnO, and C_{tot} are components of carbonate-apatite of bone tissues and teeth, with organic residues. In the sediments of the upper cultural horizon, an association of elements (P_2O_5, C_{tot}) was determined. Anomalous concentrations of these elements in the sediments at the site are associated with animal butchering zones. This is confirmed by archaeological material and the remains of calcined bones and ceramic fragments accumulated inside these areas. At the Okhta 1 site, high concentrations of the group of elements P_2O_{5antr}, CaO_{antr} and Fe_2O_3, Sr_{anthr} are associated with the burial zone. In the sediments, in addition to P_2O_{5antr}, CaO_{antr}, and Sr_{anthr}, a high content of Fe_2O_3 is also recorded, which is associated with the ocher component. Geochemical maps of individual element distributions show the same contours, which coincide with the stonework in which the amber adornments were found. On the shore zone of the Podolye 1 site, a different pattern of distribution of elements P_2O_{5antr}, CaO_{antr}, and Sr_{antr} is observed. They are concentrated in depressions of the relief in separate spots. These zones were probably used for butchering animals. Pottery sherds were found in these zones. At the Yudinovo site, an association of elements (P_2O_{5antr}), calcium (CaO_{antr}) and, additionally, potassium (K_2O_{antr}) and rubidium (Rb) is distinguished, the anomalous values of which are associated with depressions in the relief, in which burnt bones were found. Pits could be both utilitarian and ritual purposes. At the cult site "Bratja", a complex of anthropogenic components CaO, P_2O_5, (Sr, Rb) is distinguished in the sediments, which can be associated with remains of bones and ashes. Along with these elements, increased concentrations of other elements, such as K_2O_{anthr}, as well as microelements Zn, Cu, are recorded. Such an association of elements in sediments may characterize ritual activity. An interesting situation is observed at the Tarkhankut 18 site, which is a cattle pen. An association of anthropogenic components (CaO_{antr}, K_2O_{antr}, P_2O_{5antr}) is distinguished, the anomalous values of which are typical for deposits inside stone pens. At the same time, CaO_{antr} concentrations are an order of magnitude lower than in sediments outside the paddocks. This is due to the dissolution of the carbonate component of sediments as a result of the vital activity of animals.

The complex of elements (K_2O_{antr}, Rb_{antr}) in cultural deposits at all sites is associated with the remains of charcoal and wood ash. At the sites of Serteya 3-3, Okhta 1, and Podolye 1, anomalous concentrations of these elements are connected with fireplaces. The sediments also contained charcoal inclusions. The association of elements (K_2O_{antr}, Ba, MnO, C_{opr}) can characterize zones of decomposed organics (skins, wood, food residues) and humus rich sediment zones.

For the reconstruction of various factors influencing on the cultural deposit formation, an approach has been developed using multivariate factorial and correlation matrix analyses. It possible in this case to identify individual groups of geochemical elements and their associations for characteristics of anthropogenic or natural factors influencing sediment composition [1,18,45]. Another important criterion for detailed interpretation of geochemical data is archaeological information about the site. Using complex research methods, such as geochemical, geophysical, and archaeological analysis in the assessment of an archaeological site, in detail for all the features of life, including household activity, pottery manufacturing, ancient metallurgy, etc. can be reconstructed [18,36,38]. An assessment of the distribution of anomalous concentration of certain elements or associations of elements at archaeological sites has shown that the interpretation of the data obtained cannot always be unambiguous and depends on many factors [46,47]. For example, the technogenic impact on the distribution of the chemical elements at the archaeological site located in the industrial zone has been considered by authors in the Central Greece region [48]. The authors introduce the term "equifinality" for chemical elements which are characterized by multiple sources. An important question in the interpretation of geochemical data is also the geochemical explanation that the accumulation of one or another chemical element is of an anthropogenic nature and is not related to the mineral composition of deposits at a given site and excluded other factors of contamination [49].

5. Conclusions

The assessment of the impact of anthropogenic activity on the transformation of sediments and the formation of cultural layers on the ancient settlements was carried out using the method of geochemical indication of functional zones on settlements. Thus, it is possible to note the main associations of anthropogenic elements in sediments at archaeological sites that characterize certain functional zones. Abnormal concentrations of such complex association of components (P_2O_{5antr}, CaO_{antr} and Sr_{antr}) in sediments are attributed to zones of accumulation of bone remains. A more precise and detailed interpretation of functional zones (animal dressing zone, burial, ritual zone, waste pit) can be possible using other additional geochemical markers and archaeological context. Anomalous concentrations of a group of elements (K_2O_{antr}, Rb_{antr}) at archaeological sites are associated with wood ash residues and fireplaces, ash residues from ritual activities, and fires. The group of components (Ba, MnO, Corg) reflects the accumulation of humus and organic residues, and can characterize areas with food residues, remains of skins, and rotten wood. Combinations of the main groups of elements may also indicate various other types of functional zones, living zones, household areas, and places of ritual purposes. 3D geochemical maps of Al_2O_3 and SiO_2 distributions can be used for the reconstruction of ancient microrelief at the sites. Data from the archaeological context should be studied to carry out more detailed reconstructions. Using a geochemical multicomponent approach, even in complex geo-morphological contexts, anthropogenic geochemical indicators can be used to reconstruct functional zones at ancient settlements.

Funding: This research was funded by RSF, project No. 22-18-00065 "Cultural and historical processes and paleoenvironment in the Late Bronze-Early Iron Age of the North-Western Black Sea region: interdisciplinary approach" and the project with financial support from the Ministry of Education of the Russian Federation under program No. FSZN-2020-0016.

Institutional Review Board Statement: Not applicable.

Informed Consent Statement: Informed consent was obtained from all subjects involved in the study.

Data Availability Statement: Not applicable.

Conflicts of Interest: The author declares no conflict of interest. The funders had no role in the design of the study; in the collection, analyses, or interpretation of data; in the writing of the manuscript; or in the decision to publish the results.

References

1. Oonk, S.; Slomp, C.P.; Huisman, J.D. Geochemistry as an aid in archaeological prospection and site interpretation: Current issues and research directions. *Archaeol. Prospect.* **2009**, *16*, 35–51. [CrossRef]
2. Entwistle, J.A.; Abrahams, P.W.; Dodgshon, R.A. Multi-Element Analysis of Soils from Scottish Historical Sites. Interpreting Land-Use History Through the Physical and Geochemical Analysis of Soil. *J. Archaeol. Sci.* **1998**, *25*, 53–68. [CrossRef]
3. Wilson, C.A.; Davidson, D.A.; Cresser, M.S. An evaluation of multielement analysis of historic soil contamination to differentiate space use and former function in and around abandoned farms. *Holocene* **2005**, *15*, 1094–1099. [CrossRef]
4. Aston, M.; Martin, M.; Jackson, A. The use of heavy metal soil analysis for archaeological surveying. *Chemosphere* **1998**, *37*, 456–477. [CrossRef]
5. Cabala, J.; Smieja-Kroól, B.; Jablonska, M.; Chrost, L. Mineral components in a peat deposit: Looking for signs of early mining and smelting activities in Silesia–Cracow region (Southern Poland). *Environ. Earth Sci.* **2013**, *69*, 2559–2568. [CrossRef]
6. Lutz, H.J. The concentration of certain chemical elements in the soils of Alaskan archeological sites. *Am. J. Sci.* **1951**, *249*, 925–928. [CrossRef]
7. Schlezinger, D.R.; Howes, B.L. Organic Phosphorus and Elemental Ratios as Indicators of Prehistoric Human Occupation. *J. Archaeol. Sci.* **2000**, *27*, 479–492. [CrossRef]
8. Terry, R.E.; Hardin, P.J.; Houston, S.D.; Nelson, S.D.; Jackson, M.W.; Carr, J.; Parnell, J.J. Quantitative phosphorus measurement: A field test procedure for archaeological site analysis at Piedras Negras, Guatemala. *Geoarchaeol. Int. J.* **2000**, *15*, 151–166. [CrossRef]
9. Tian, X.S.; Zhu, C.; Sun, Z.B.; Shui, T. An evaluation of heavy metal pollution within historic cultural strata at a specialized salt production site at Zhongba in the Three Gorges Reservoir region of the Yangtze River, China. *Environ. Earth Sci.* **2013**, *69*, 2129–2138. [CrossRef]
10. Wells, E.C.; Terry, R.E.; Parnell, J.J.; Hardin, P.J.; Jackson, M.W.; Houston, S.D. Chemical analyses of ancient anthrosols in residential areas at Pieda Negras, Guatemala. *J. Archaeol. Sci.* **2000**, *27*, 449–462. [CrossRef]

11. Sanchez, C.A. Phosphorus. In *Handbook of Plant Nutrition*; Barker, A.V., Pilbeam, D.J., Eds.; CRC Press: Boca Raton, FL, USA, 2007; pp. 51–90. [CrossRef]
12. Marwick, B. Element concentrations and magnetic susceptibility of anthrosols: Indicators of prehistoric human occupation in the inland Pilbara, Western Australia. *J. Archaeol. Sci.* **2005**, *32*, 1357–1368. [CrossRef]
13. Lima da Costa, M.; Kern, D. Geochemical signatures of tropical soils with archaeological black earth in the Amazon, Brazil. *J. Geochem. Explor.* **1999**, *66*, 369–385. [CrossRef]
14. Wells, C. Investigating Activity Patterns in Prehispanic Plazas: Acid-extraction ICP-AES Analysis of Anthrosols at Classic Period El Coyote, Northwestern Honduras. *Archaeometry* **2004**, *46*, 67–84. [CrossRef]
15. Proudfoot, B. The analysis and interpretation of soil phosphorus in archaeological contexts. In *Geoarchaeology*; Davidson, D.A., Shackley, M.L., Eds.; Cambridge University Press: Duckworth, London, UK, 1976; pp. 93–113.
16. Stevenson, F.J. *Cycles of Soil: Carbon, Nitrogen, Phosphorus, Micronutrients*; Wiley Interscience: New York, NY, USA, 1986; p. 380.
17. Middleton, W.D.; Price, T.D. Identification of Activity Areas by Multi-element Characterization of Sediments from Modern and Archaeological House Floors Using Inductively Coupled Plasma-atomic Emission Spectroscopy. *J. Archaeol. Sci.* **1996**, *23*, 673–687. [CrossRef]
18. Kulkova, M.A. Metodi priklanih paleolandshaftnih geokhimicheskih issledovany. In *Methods of Applied Paleolandscape Geochemical Investigations*; Herzen State University Publisher: St. Petersburg, Russia, 2012; p. 152. ISBN 978-5-8064-1678-1. (In Russian)
19. Cook, S.F.; Heizer, R.F. *Studies on the Chemical Analysis of Archaeological Sites University of California Publications in Anthropology 2*; University of California Publications: Berkeley, CA, USA; Los Angeles, CA, USA, 1965; p. 102.
20. Keeley, H.C.M. Recent work on soil phosphorus analysis in archaeological prospection. *Rev. D'archéométrie* **1981**, *95*, 89–95. [CrossRef]
21. Bethell, P.H.; Smith, J.U. Trace element analysis of an inhumation from Sutton Hoo, using inductively coupled plasma emission spectrometry: An evaluation of the techniques applied to analysis of organic residues. *J. Archaeol. Sci.* **1989**, *16*, 47–55. [CrossRef]
22. Barba, L.; Ortiz, A.; Link, K.; Lopez-Lujan, L.; Lazos, L. The chemical analysis of residues in floors and the reconstruction of ritual activities at the Templo Mayor, Mexico. In *Archaeological Chemistry: Organic, Inorganic and Biochemical Analysis*; Orna, M.V., Ed.; American Chemical Society: Washington, DC, USA, 1996; pp. 139–156.
23. Knudson, K.J.; Frink, L.; Hoffman, B.W.; Price, T.D. Chemical characterization of Arctic soils: Activity area analysis in contemporary Yupik fish camps using ICP–AES. *J Archaeol. Sci.* **2004**, *31*, 443–456. [CrossRef]
24. Fernandez, F.G.; Terry, R.E.; Inomata, T.; Eberl, M. An ethnoarchaeological study of chemical residues in the floors and soils of Qéqchì Maya houses at Las Pozas, Guatamala. *Geoarchaeology* **2002**, *17*, 487–519. [CrossRef]
25. Parnell, J.J.; Terry, R.E.; Golden, C. Using in-field phosphate testing to rapidly identify middens at Piedras Negras, Guatemala. *Geoarchaeology* **2001**, *16*, 855–873. [CrossRef]
26. Zimmermann, W.H. Die Siedlungen des 1. bis 6. Jahrhunderts nach Christus von Flögeln-Eekhöltjen, Niedersachsen: Die Bauformen und ihre Funktionen. *Probl. Der Küstenforschung Im Südlichen Nordseegebiet* **1992**, *19*, 360S.
27. Manzanilla, L. 1996. Soil analyses to identify ancient human activities. *Can. J. Soil Sci.* **1996**, *76*, 107–108. [CrossRef]
28. Entwistle, J.A.; Abrahams, P.W.; Dodgshon, R.A. The geoarchaeological significance and spatial variability of a range of physical and chemical soil properties from a former habitation site, Isle of Skye. *J. Archaeol. Sci.* **2000**, *27*, 287–303. [CrossRef]
29. Entwistle, J.A.; Dodgshon, R.A.; Abrahams, P.W. An investigation of former land-use activity through the physical and chemical analysis of soils from the Isle of Lewis, Outer Hebrides. *Archaeol. Prospect.* **2000**, *7*, 171–188. [CrossRef]
30. Jenkins, D.A. Trace element geochemistry in archaeological sites. *Environ. Geochem. Heatlth* **1989**, *11*, 57–62. [CrossRef] [PubMed]
31. Maskall, J.E.; Thornton, I. Chemical partitioning of heavy metals in soils, clays and rocks at historical lead smelting sites. *Water Air Soil Pollut.* **1998**, *108*, 391–409. [CrossRef]
32. Pyatt, F.B.; Amos, D.; Grattan, J.P.; Pyatt, A.J.; Terrell-Nield, C.E. Invertebrates of ancient heavy metal spoil and smelting tip sites in southern Jordan: Their distribution and use as bioindicators of metalliferous pollution derived from ancient sources. *J. Arid. Environ.* **2002**, *52*, 53–62. [CrossRef]
33. Monna, F.; Galop, D.; Carozza, L.; Tual, M.; Beyrie, A.; Marembert, F.; Chateau, C.; Dominik, J.; Grousset, F.E. Environmental impact of early Basque mining and smelting recorded in a high ash minerogenic peat deposit. *Sci. Total Environ.* **2004**, *327*, 197–214. [CrossRef]
34. Ottaway, J.H.; Matthews, M.R. Trace element analysis of soil samples from a stratified archaeological site. *Environ. Geochem. Health* **1988**, *10*, 105–112. [CrossRef]
35. Kulkova, M.A.; Gusentsova, T.M.; Sapelko, T.V.; Nesterov, E.M.; Sorokin, P.E.; Ludikova, A.V.; Ryabchuk, D.V.; Markova, M.A. Geoarcheological investigations on the development of the Neva River delta (Gulf of Finland) during the Holocene. *J. Mar. Syst.* **2014**, *129*, 19–34. [CrossRef]
36. Kulkova, M.; Mazurkevich, A.; Gerasimov, D. Stone Age archaeological sites and environmental changes during the Holocene in the NW region of Russia. In *Geology and Archaeology: Submerged Landscapes of the Continental Shelf*; Harff, J., Bailey, G., Lüth, F., Eds.; Geological Society of London Special Publications 411: London, UK, 2015; pp. 27–49.
37. Mazurkevich, A.N.; Dolbunova, E.V.; Kulkova, M.A.; Savelieva, L.A. *Geoarchaeological Issues of the Upper Dnieper–Western Dvina River Region (Western Russia): Fieldtrip guide*; Smolensk Universum: Moscow, Russia, 2012; p. 105. (In Russian)
38. Kulkova, M.A.; Gusentsova, T.M.; Nesterova, L.A.; Nesterov, E.M. The reconstruction of functional zones at Neolithic to Early Iron Age sites in the Neva river basin (Russia) by means of geochemical markers. In *Monographs of the Archaeological Society of Finland*

7. *Proceedings of the 11th Nordic Conference on the Application of Scientific Methods in Archaeology*; Mannermaa, A., Manninen, M.A., Pesonen, P., Seppänen, L., Eds.; Archaeological Society of Finland: Helsinki, Finland, 2019; pp. 72–92. ISBN 978-952-68453-5-7.
39. Shahknovich, M.M.; Kulkova, M.A. "Bratja" poluostrova Sredny Murmanskogo poberezhija Barenzeva morja. Opit identificazii saamskih sakral'nih objektov Russkoi Laplandii ["Bratja" of Sredny Isthmus of the Murmansky shore of Barents Sea. The experience of identification of Sami sacral objects of the Russian Lapland]. *Archaeol. Arctic.* **2020**, *7*, 300–326. (In Russian)
40. Ryabchuk, D.; Zhamoida, V.; Amantov, A.; Sergeev, A.; Gusentsova, T.; Sorokin, P.; Kulkova, M.; Gerasimov, D. Development of the coastal systems of the easternmost Gulf of Finland, and their links with Neolithic–Bronze and Iron Age settlements. *Geol. Soc.* **2016**, *411*, 51–76. [CrossRef]
41. Gesentsova, T.; Kulkova, M. The subsistence strategy and paleoenvironment on the Stone Age site Podolye 1 in the southern Ladoga Lake region (Eastern Baltic). *Quat. Int.* **2020**, *541*, 41–51. [CrossRef]
42. Sapelko, T.V.; Gusentsova, T.M.; Kulkova, M.A.; Ludikova, A.V.; Denisenkov, V.P.; Korneenkova, N. Ladozhskaya transgressiya I landshafti vtoroi polovini golozena v Yuzhnom Prilagozhie (po dannim izucheniya arheologicheskogo pamyatnika Podolye) [Ladoga transgression and landscapes of the second half of Holocene in the Southern Ladoga Lake basin (on the data of study of archaeological site Podolye 1)]. Izvestiya RAS. *Geogr. Ser.* **2019**, *5*, 78–95. (In Russian) [CrossRef]
43. Kashuba, M.T.; Smekalova, T.N.; Kulkova, M.A.; Gurov, E.Y. New Results of Interdisciplinary Study of Bronze Age Settlements in Northwestern Crimea. *Vestn. St. Petersburg Univ. Hist.* **2021**, *66*, 1270–1295. [CrossRef]
44. Khlopachev, G.A.; Kulkova, M.A.; Gribchenko, Y.N. The characteristics of settlement structure formation on the Upper Paleolithic site of Yudinovo according to archaeological and geochemical investigations of 2015–2018. *Strat. Plus* **2022**, *1*, 239–258. [CrossRef]
45. Nielsen, N.H.; Kristiansen, S.M. Identifying ancient manuring: Traditional phosphate vs. multi-element analysis of archaeological soil. *J. Archaeol. Sci.* **2014**, *42*, 390–398. [CrossRef]
46. Barba, L. Chemical residues in lime-plastered archaeological floors. *Geoarchaeology* **2007**, *22*, 439–452. [CrossRef]
47. Pecci, A. Almost ten years of plasters residue analysis in Italy: Activity areas and the function of structures. *Period. Di Mineral.* **2013**, *82*, 393–410.
48. Bintliff, J.; Degryse, P.; van Zwienen, J. The long-term programme of trace metal analysis at the ancient city of Hyettos. *J. Archaeol. Sci. Rep.* **2022**, *43*, 103432. [CrossRef]
49. Bintliff, J.; Degryse, P. A review of soil geochemistry in archaeology. *J. Archaeol. Sci. Rep.* **2022**, *43*, 103419. [CrossRef]

minerals MDPI

Article

Features of Arsenic Distribution in the Soils of Potash Mines

Tatiana Karavaeva , Elena Menshikova , Pavel Belkin * and Vyacheslav Zhdakaev

Natural Science Institute, Perm State University, Genkel st.4, Perm 614068, Russia
* Correspondence: pashabelkin@mail.ru

Abstract: The purpose of the present study is to analyse the distribution of arsenic in the soils of the Verkhnekamskoe potassium salt deposit (Perm Krai, Russia). The danger of arsenic pollution is determined by its high toxicity and carcinogenic hazard. Being a technophilic element, arsenic enters the environment primarily as a result of mining activities. Mining and processing sites for arsenic-containing ores are the most prone to technophilic arsenic accumulation. Solid wastes from potash production also contain elevated concentrations of arsenic. The content of arsenic in soils was determined by inductively coupled plasma mass spectrometry (ICP-MS). Statistical methods were used to analyse the features of arsenic distribution in soils of background areas and potash mining areas near production facilities. Three types of landscapes were studied within each territory, which were each distinguished by the leading processes of substance migration. Arsenic concentrations in both the background areas and the potash mining territories vary considerably, ranging from $n \times 10^{-1}$ to $n \times 10$. The study found no statistically significant differences in arsenic concentrations in soils of potash mining areas and background areas. Arsenic concentrations in soils from various types of landscapes also do not differ statistically. Arsenic concentrations in soils of saline areas were found to be higher than in the rest of the territories. Outside of saline areas, the identified patterns of arsenic distribution in the soils of the Verkhnekamskoe potassium salt deposit indicate that potash operations are not a determinant in the technophilic accumulation of arsenic.

Keywords: geoecology; pollution; arsenic; soil contamination; condition assessment; trace elements; potassium salt deposits; Verkhnekamskoe deposit

Citation: Karavaeva, T.; Menshikova, E.; Belkin, P.; Zhdakaev, V. Features of Arsenic Distribution in the Soils of Potash Mines. *Minerals* **2022**, *12*, 1029. https://doi.org/10.3390/min12081029

Academic Editor: Juan Antelo

Received: 15 July 2022
Accepted: 13 August 2022
Published: 16 August 2022

1. Introduction

Mining areas have an increased geochemical technogenic load due to mining and processing. In the hypergenesis zone, primary minerals are transformed, and toxic elements and their compounds are released into the environment [1]. Studies of natural environment components' contamination with toxic trace elements in mining-affected areas are of particular interest.

Arsenic is a chemical element of the first class of environmental hazard with high toxicity and carcinogenic properties that pose a serious risk to humans [2,3]. Arsenic can enter the human body through the consumption of arsenic-contaminated water or agricultural products grown on arsenic-contaminated soils. In India, Bangladesh, Nepal, China, Taiwan, Thailand, Argentina, Mexico, and other countries, arsenic contamination of groundwater used for drinking is a concern [3,4]. In this regard, effective methods for detecting [5] and removing arsenic from drinking water are being actively developed [6–8].

As a technophilic element, arsenic enters the environment primarily as a result of the mining and processing of minerals, where it is a major component in the ore or is present as an impurity as well as in the composition of pesticides used in agriculture. Arsenic compounds are found in small quantities in ores, hydrocarbon feedstocks, industrial clays, etc. During the mining and processing of minerals, arsenic is emitted into the atmosphere with inorganic pollutants, discharged into sewage, deposited in solid waste dumps, and washed out by atmospheric precipitation, polluting surface water and groundwater. Localised areas of ecologically hazardous pollution are formed as a result of arsenic distribution in the natural environment [9].

The assessment of arsenic contamination of soils in mining areas has received a lot of attention in China, which holds 70% of the world's arsenic reserves [10]. Arsenic concentrations of up to 76,400 mg/kg with an average concentration of 1205.97 mg/kg have been found in soils near waste sites of mining and processing of arsenic-containing ore in Yunnan province, southwest China [10].

Negative environmental impacts, such as increased concentrations of a variety of trace elements in the natural environment, can also occur after mining has ceased, e.g., as a result of tailings material dispersion into the natural environment [11], or a lack of mothballing of abandoned mining sites [12]. The following elements dominate in the geochemical series in the surface element concentrations within the areas affected by the Dalnegorsk and Krasnorechensk tailings (Dalnegorsk district, Primorsky Krai, Russia): manganese, zinc, and arsenic; zinc, manganese, lead, and arsenic, respectively [11]. Arsenic and other trace element concentrations in mining wastes reach levels that classify tailings sites as technogenic deposits. The arsenic content in tailings of tin ore deposits in the Kavalerovsky region (Primorsky Krai, Russia) reaches 0.01–0.05% [12]. Arsenic has the highest concentrations in technogenic soils of tailings among toxic elements and exceeds average concentrations in the earth's crust by 20–886 times in gold ore tailings, 152–5340 times in polymetallic tailings, and 1.2–172 times in rare-metal tailings [13].

Pollutants are carried out of the technogenic system via air and water streams, causing changes in the geochemical background of adjacent territories. The mining plant "Khrustalnensky GOK" (Primorsky Krai, Russia) ceased its operations at the beginning of the 21st century. In different soil horizons up to the depth of 45 cm within the area affected by tailing dumps of the mining plant, arsenic concentrations are 317.27–377.86 mg/kg [1], while the established approximate allowable concentrations are 2–10 mg/kg, depending on the grain size composition and acidity of soils. The arsenic content in tin ore processing dust ranges from 16.04 to 28.3 mg/kg [14]. In soils near an arsenic-containing ore (As_4S_4) mining facility that closed in 2011 (Hunan Province, China), arsenic is the main pollutant, with average concentrations of 394 mg/kg, exceeding background values by 23 times. Arsenic contamination of soil was detected at a relatively close distance—about 500 m from the sources of exposure [15].

According to [16], the majority of the As found in soils at abandoned mine sites (Rita Mine, Tres Amigos Mine, Las Viescas Mine (northern part of Castilla-León, Spain))is in the so-called "residual fraction", i.e., in grains of specific As minerals that come from wastes and are later integrated within the soil mineral fraction. Mechanical dispersion is thus quantitatively greater than chemical dispersion.

Researchers consider arsenic to be the most dangerous of all mining waste pollutants due to the prolonged activity (chemical transformation and migration) of arsenic technogenic formations in the natural environment [12,17] and the high bioavailability of arsenic [11,18].

The geochemical stress caused by the potassium industry (extraction and processing of fossil salts) is seen in an increase in the content of potassium and sodium chlorides in soils and natural waters near potash enterprises and waste disposal sites [19,20]. In addition to easily soluble compounds, fossil salts contain impurities of high-risk elements such as zinc, lead, copper, nickel, cadmium, and arsenic [21], which create areas of technogenic dispersion when extracted on the earth's surface.

Several studies [22–26] explore the mineral and trace element compositions of salt rocks, insoluble salt residues, and potassium production wastes, which determine the man-made transformation of the geochemical spectra of natural environment components.

Halite, sylvite, and carnallite dominate the mineral composition of the Verkhnekamskoe salt deposit (Perm Krai, Russia) [24] with varying percentages of their content in different beds. The main components of the chemical composition of productive formations of the deposit are NaCl (45.31–76.11 wt%), KCl (15.25–31.04 wt%), $MgCl_2$ (0.24–0.34 wt% in sylvinite formations; 5.97–12.48 wt% in carnallite formations), and $CaSO_4$ (1.15–2.68 wt%) [27], which determine the predominance of Na^+, K^+, Mg^+, and Ca^{2+} cations and Cl^- and SO_4^{2-} anions in salt compositions. At the same time, approximately 30 trace elements were found in the ores of the deposit [22], with the metals of greatest ecological interest being zinc, lead, copper, nickel, cadmium, and arsenic, which is related to metalloids. These elements can be found as minerals on their own or as impurities that isomorphically replace the main cation. The mineral composition is very diverse, containing both soluble and slightly soluble compounds. Most researchers believe that the majority of these elements are found in the insoluble ore residue [22], which is mainly represented by carbonates (10–20%), sulphates (5–30%), and aluminosilicates (42–51%), the mineral composition of which is dominated by dolomite and magnesite, anhydrite and gypsum, hydroslides and feldspars, respectively [27].

The content of insoluble residue in different strata of the potassium deposit as well as potassium production wastes is as follows (in wt%): 2.6–6 (in sylvinite stratum), up to 6 (in carnallite rock), up to 10–12 (in interstitial rock salt), 15–40 (in clay and anhydrite interlayers), up to 4 (in solid potash waste), 15–70 (in the solid phase of clay-salt pulp), and 5–28 (in slimes) [24,27]. Cu-As-Sb is present in the composition of minerals–micro impurities of insoluble residue of sylvinite, carnallite, and rock salt [28]. The content of arsenic in the insoluble salt rock residue reaches 6.1 μg/g [24], while solid wastes of potassium production contain 0.48–30.7 mg/kg of arsenic [26].

The results of determining the background content of a number of macrocomponents and trace elements, including arsenic, in soils of six natural zones of Perm Krai (Russia) are presented in [29]. In general, Perm Krai has an average regional content of arsenic, which is 7.52 ± 0.25 mg/kg. In the soils of the middle and southern taiga, in the contact zone of which the potash industry facilities are located, the average arsenic content is 5.77 ± 0.62 and 7.83 ± 0.62 mg/kg, respectively. According to A.P. Vinogradov, arsenic accumulation relative to its average content in the earth's crust was noted in all natural areas of Perm Krai. The following concentration factors were found: 4.42 for Perm Krai, 3.39 for the Middle Taiga natural area, and 4.61 for the Southern Taiga natural area. Relative to the calculated regional average content [29], the Middle Taiga area is characterised by element dispersion (the dispersion factor is 1.3). The Southern Taiga area is distinguished by insignificant accumulation (the concentration factor is 1.04). It should be noted that in [29], the distribution of arsenic and other determined elements is estimated by natural areas without regard to soil type, composition, or physical and chemical properties.

The goal of this research is to analyse the specifics of arsenic distribution in soils near potash mines in Perm Krai (Russia), where the Verkhnekamskoe salt deposit is being developed. The studies included: (1) an analysis of the landscape structure of the study area; (2) a comparative analysis of arsenic content in soils of background territories and potash mining territories near production facilities, with a detailed range of arsenic concentrations determined by saline soil studies. The obtained results provide useful information on the arsenic content in the mining-affected area. The findings can be used to assess the role of salt mining in the formation of territorial environmental situations.

2. Materials and Methods

The study considers the soils of the Verkhnekamskoe potassium salt deposit (Perm Krai, Russia). The potash industry in Perm Krai is associated with the development of one of the world's largest deposits—the Verkhnekamskoe salt deposit, whose development started in the 1930s. In addition to the potash facilities, other major regional industrial enterprises in metallurgy, chemistry, and oil production, as well as residential and agricultural development, have all greatly contributed to the transformation of the natural environment in the area.

The confinement of the territory to the taiga zone in humid climate conditions has determined the dominance of typical taiga soils with a clear morphogenetic differentiation of the profile. A clarified and silt-lightened eluvial horizon forms under the humus horizon.

This horizon is underlain by a median horizon with morphologically and analytically pronounced illuvial accumulation. The local territorial differentiation of soil formation factors leads to the development of intrazonal soil processes in river valleys under constant moistening. It determines the development of regenerative conditions in the environment. Economic activity alters the profile structure, acidity range, and chemical composition of natural soils.

Methods of system analysis and generalisation of theoretical and experimental research, statistical processing of empirical data, and modern instrumental and chemical methods are used in this work. The findings of geo-ecological soil studies from 2012 to 2021 are summarised and analysed.

Field studies included a route reconnaissance survey of the territory, the laying and description of soil trenches with soil taxonomic identification, and soil sampling from the upper humus horizon (0–10 to 0–20 cm) for subsequent laboratory tests. The soil was sampled from the background areas remote from economic activities (the planned development area of the Verkhnekamskoe salt deposit), developed areas of the deposit (adjacent to production facilities), and saline soil areas affected by potash industry facilities. Within the background area, sampling was conducted at three sites: (1) eluvial and transit landscapes occupied by typical forest communities on zonal soils; (2) eluvial and transit former agricultural landscapes; (3) transaccumulative and accumulative landscapes within wetland ecotopes and small river floodplains. A total of 81 soil samples were collected within the background areas. In addition, 64 soil samples were also collected within potash mining areas from (1) eluvial and transit landscapes occupied by forest vegetation on conditionally natural zonal soils; (2) eluvial and transit former agricultural landscapes; (3) transaccumulative landscapes within small river valleys. Furthermore, soils were sampled at sites with high concentrations of water-soluble salts and chloride-ion concentrations in soils ranging from 1.49 to 36.35 g/kg. Salinisation areas are distributed locally, located near salt waste disposal sites, and have no direct impact on the soil structure of the area. Seven soil samples were collected at the salinisation sites. The granulometric composition of soils was determined in the field using the rolling cord method according to N.A. Kachinsky [30]. Three to four grams of soil were moistened until they formed a thick paste (no water was squeezed out of the soil). The soil was well kneaded and mixed by hand before being rolled out in the palms into a cord (about 3 mm thick) and then rolled up into a ring (about 3 cm in diameter). When rolled, the cord takes on a different appearance depending on the granulometric composition of the soil. If no cord is formed, the soil has sandy composition (sand); rudiments of the cord are formed—the soil has loamy–sandy composition (loamy sands); the cord crumbles when rolled—the soil has light loamy composition (light loam); the cord is continuous, the ring is breaking up when rolled—the soil has medium loamy composition (medium loamy loam); the cord is continuous, the ring has cracks—the soil has heavy loamy composition (heavy loamy loam); the cord is continuous, the ring is continuous without cracks—the soil has clayey composition (clay).

Laboratory tests were conducted in the Nanomineralogy Sector of the Perm State University's Collaborative Use Centre and the Hydrochemical Analysis Laboratory of the Geology Department of Perm State University. Analytical studies were conducted using unified methods.

The As concentrations were measured using the Aurora M90 ICP-MS spectrometer (Bruker, Fremont, CA, USA). Autoclave digestion was used to dissolve the sample prior to ICP-MS measurements. To achieve an efficient digestion, sediment was used with various acids or mixtures, such as concentrated HNO_3 or other acids (HCl, $HClO_4$, and H_2SO_4) or H_3BO_3 solution diluted with deionized water. For the analysis, 0.1 g sample weights were used. Control samples (blank samples) and one standard sample were decomposed together with the analysed samples. To ensure the accuracy of the sample analysis, standard samples from the Institute of Geochemistry, Siberian Branch of the Russian Academy of Sciences (Irkutsk, Russia) were used. The validity of the analytical methods was confirmed by the analysis of the standard reference material Gabbro Essexit STD-2A (GSO 8670-2005).The following are the typical measurement errors for the method used, depending on chemical element concentrations: <0.001 μg/dm^3—RSD >25%; 0.001–0.1 μg/dm^3—RSD 25–10%; 0.1–1 μg/dm^3—RSD 10–5%; >1 μg/dm^3—RSD 5%.

The pH was determined using the national standard method (GOST 26483-85 [31], Russia) by extracting soil samples with a potassium chloride solution prepared at 75 g of potassium chloride per 1000 cm^3 of solution, which was followed by a pH-meter measurement (ANION 4100, Infraspak-Analit, Novosibirsk, Russia). The measurement error was less than 0.1 pH.

The obtained results were statistically processed using STATISTICA 12 software (StatSoft. Inc., Tulsa, OK, USA) and Microsoft Office Excel 2019 (Microsoft, Redmond, DC, USA). The Cheddock scale was used to assess the correlation between arsenic content and soil pH_{KCl} with Spearman's rank correlation coefficient. The statistical significance of the Spearman's correlation coefficient was determined using the Student's t-criterion. The Mann–Whitney U-criterion with a 95% confidence probability was used to assess the reliability of differences in arsenic content in soils from background areas and potash mining territories. The Kruskal–Wallis H-criterion with a 95% confidence probability was used to assess the reliability of differences in arsenic content in soils from selected types of landscapes. The Kruskal–Wallis criterion is used to compare three or more samples; thus, it was used to evaluate the differences between the three studied landscape types within the background areas and the three types of landscapes within potash mining areas. The FactoMineR package in R was used to perform principal component analysis (PCA) to visualise the correlations [32]. Figures were made using the package "ggplot2" [33].

The contamination factor (CF) was calculated as the ratio between the metal concentrations and its background values:

$$CF = C^i / C^i_B \tag{1}$$

CF is the contamination factor;

C^i (mg/kg) is the concentration of a target element in a sampled soil;

C^i_B (mg/kg) is the background value of the element.

The criteria adopted to determine the extent of the contamination were as follows: no contamination/low contamination ($CF < 1$), moderate ($1 \leq CF < 3$), high ($3 \leq CF < 6$), and very high ($6 \leq CF$) [34].

Potential Ecological Risk Index, proposed by Hakanson [35], is a standard and widely used method in modern research [36] for assessing the ecological risk posed by potentially toxic elements in soils. The ecological risk factor (E^i_r) for individual elements, e.g., arsenic, was calculated using its toxicity factor (T^i_r) according to the formula:

$$E^i_r = T^i_r \left(C^i / C^i_B \right) \tag{2}$$

E^i_r is the ecological risk factor for individual elements;

T^i_r is the toxicity response factor. The toxicity coefficient of arsenic is 10 [36];

C^i (mg/kg) is the concentration of a target element in a sampled soil;

C^i_B (mg/kg) is the background value of the element.

For risk assessments, we adopted the following classification: $E^i_r \leq 40$ represented low risk; $40 < E^i_r \leq 80$ moderate risk; $80 < E^i_r \leq 160$ considerable risk; $160 < E^i_r \leq 320$ high risk; $320 < E^i_r$ very high risk [37].

3. Results and Discussion

The soils of the taiga zone, which include the studied soils, are characterised by the active development of the oxidogenesis processes in conditions of free access of oxidants to weathering products, resulting in a decrease in the migration activity of chemical elements, especially oxidised forms of iron and manganese. As a result, weathering products and soils of humid taiga landscapes become saturated with these elements.

Under oxidising conditions, the predominant form of arsenic among the dissolved forms is As(V), which is present in the form of arsenic acid oxyanions. Arsenic adsorption in soils occurs mainly on the surfaces of colloidal soil particles. These particles can be represented by clay, oxides or hydroxides of aluminium, iron and manganese, calcium carbonates, and organic matter. Because iron oxide and hydroxide are the best adsorbents, iron arsenates are the most common arsenic compounds in acidic soils [38]. Studies of the correlations between clay and arsenic content in the top layer of soils also show that oxygen iron compounds play a determining role in clay fractions. The absorption of arsenic by clays is determined by the content of oxide and hydroxide forms of iron. Purified quartz sand without clay fractions, for example, showed minimal adsorption of arsenic oxyanions [38]. The low arsenic content is typical of soils in the State of Pará (Brazilian Amazon) with a predominant sand fraction [39].

Researchers noted that arsenic adsorption has a strong pH dependence due to the variable charge of the adsorbent surface (iron oxides and hydroxides). The maximum adsorption of arsenic oxyanions is observed in acidic conditions, at pH values close to 3. In the pH range from 3 to 7, arsenic adsorption is reduced up to 95–85%. A sharp reduction in arsenic adsorption is observed at pH 7–10 with an average of about 8.5. In this pH range, iron oxides and hydroxides have a zero charge. A higher pH value promotes the formation of a total negative charge on the adsorbent surface, preventing the adsorption of arsenic oxyanions from the solution. At pH values of 9–10, arsenic adsorption is reduced to 40–50% [38]. Soils generally have a pH below 8.6, at which most iron oxide and hydroxide surfaces should be positively charged, facilitating the adsorption of arsenic oxyanions [38]. The results of experiments evaluating the adsorption of arsenic by various clay minerals indicate that kaolinite, montmorillonite, illite, halloysite, and chlorite have the highest adsorption of As(V) at pH around 7, and that it decreases with increasing pH [38].

Table 1 shows the statistical characteristics of the pH_{KCl} of the studied soils. In 90% of cases, the soils in the background areas had pH_{KCl} values ranging from 3 to 7. Acidic soils with pH_{KCl} values < 3 were observed in 10% of cases. Soils of eluvial and transit former agricultural landscapes had relatively higher pH_{KCl} values. In 100% of cases, pH_{KCl} values ranged from 3 to 7, and in 96% of cases, pH_{KCl} values were above 5. Zonal soils of eluvial and transit landscapes occupied by typical forest communities in 93% of cases had pH_{KCl} values in the range from 3 to 7. In 7% of cases, more acidic soils were found. Within the areas of transaccumulative and accumulative landscapes of wetland ecotopes and small river floodplains, the proportion of acidic soils with pH_{KCl} < 3 increased to 22%. In 78% of cases, the pH_{KCl} values ranged from 3 to 7.

Soils in potash mining areas generally had higher pH_{KCl} values. Soils with pH_{KCl} < 3 were not found in these areas. In 95% of cases, soils had pH_{KCl} values ranging from 3 to 7. pH_{KCl} values > 7 were observed in 5% of cases. In 100% of cases, soils of eluvial and transit former agricultural landscapes had pH_{KCl} values ranging from 3 to 7, as in background areas.

Zonal soils of the territories of eluvial and transit landscapes occupied by typical forest communities in 97% of cases had pH_{KCl} values ranging from 3 to 7. In 3% of cases, more alkaline soils were found. Within the transaccumulative landscapes of small river valleys, the proportion of soils with pH_{KCl} > 7 increased to 17%. In 83% of cases, pH_{KCl} values ranged from 3 to 7.

Saline areas had even higher pH_{KCl} values—all pH_{KCl} values were >7, but they had a minimal spread compared to the rest of the study area.

No correlation was found between arsenic content and the soil pH_{KCl} (Figures 1 and 2, Table 2). The lack of a strong correlation can be explained by the prevailing range of pH_{KCl} values. Only 2% of the total number of soil samples fell within the pH_{KCl} range from 7 to 10, which is characterised by a decrease in arsenic adsorption by soil colloids [38], while all values were close to its lower limit (pH_{KCl} = 7.4). The observed correlation between arsenic content and pH_{KCl} of soils in the saline areas cannot be considered reliable, as it is not statistically significant (Figure 3, Table 2).

Table 1. Variation of pH_{KCl} in soils.

Sampling	Min. Value	Max. Value	Average Value	Median
Background area relative to potash mining area:				
Entire territory (n = 81)	2.42	6.30	4.56	4.61
by landscape type:				
Areas of eluvial and transit landscapes occupied by typical forest communities on zonal soils (n = 27)	2.60	5.54	4.10	4.03
Eluvial and transit former agricultural landscapes (n = 27)	3.30	6.30	5.61	5.80
Transaccumulative and accumulative landscapes within wetland ecotopes and floodplains of small rivers (n = 27)	2.42	5.70	3.95	4.03
Potash mining areas outside of soil salinisation areas:				
Entire territory (n = 64)	3.10	7.40	5.03	4.90
by landscape type:				
Areas of eluvial and transit landscapes occupied by forest vegetation on conditionally natural zonal soils (n = 36)	3.10	7.40	4.55	4.20
Eluvial and transit former agricultural landscapes (n = 16)	4.10	6.40	5.40	5.80
Transaccumulative landscapes within small river valleys (n = 12)	4.20	7.40	5.98	6.20
Areas of soil salinisation in the zone affected by the potash industry:				
Chloride-type soil salinity areas—content Cl^- 1.49–36.35 g/kg (n = 7)	7.40	8.00	7.64	7.60

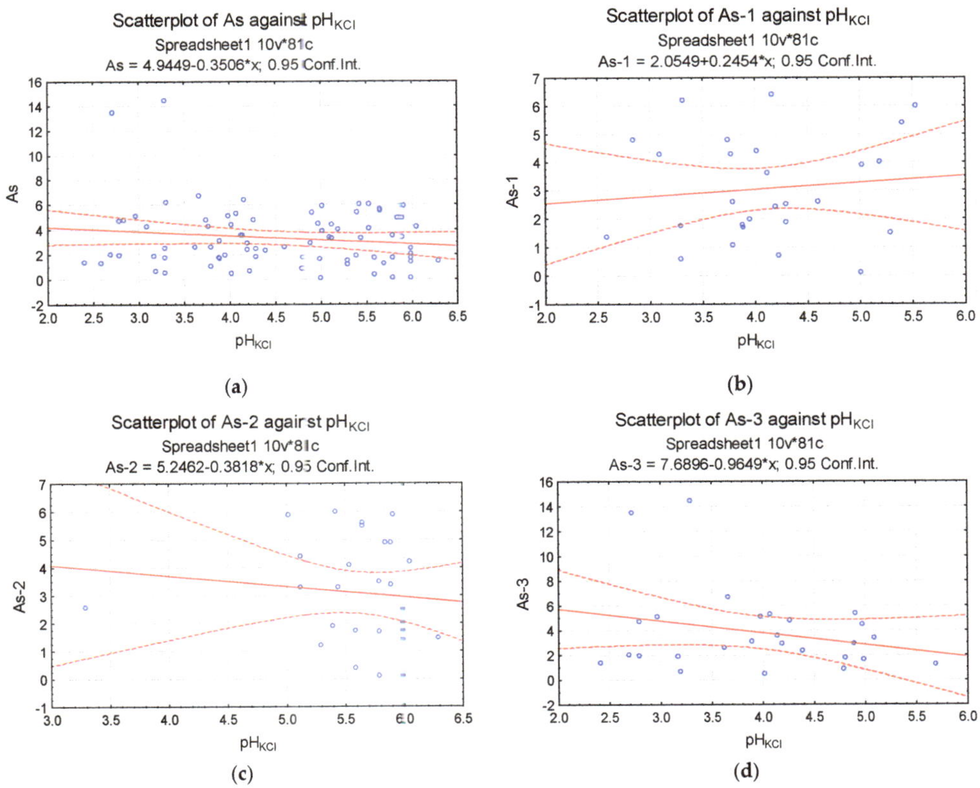

(a) (b) (c) (d)

Figure 1. Dependence of arsenic content in background soils on pH_{KCl}: (**a**) background area relative to potash mining area (As); (**b**) background areas of eluvial and transit landscapes occupied by typical forest communities on zonal soils (As-1); (**c**) background areas of eluvial and transit former agricultural landscapes (As-2); (**d**) background areas of transaccumulative and accumulative landscapes within wetland ecotopes and floodplains of small rivers (As-3).

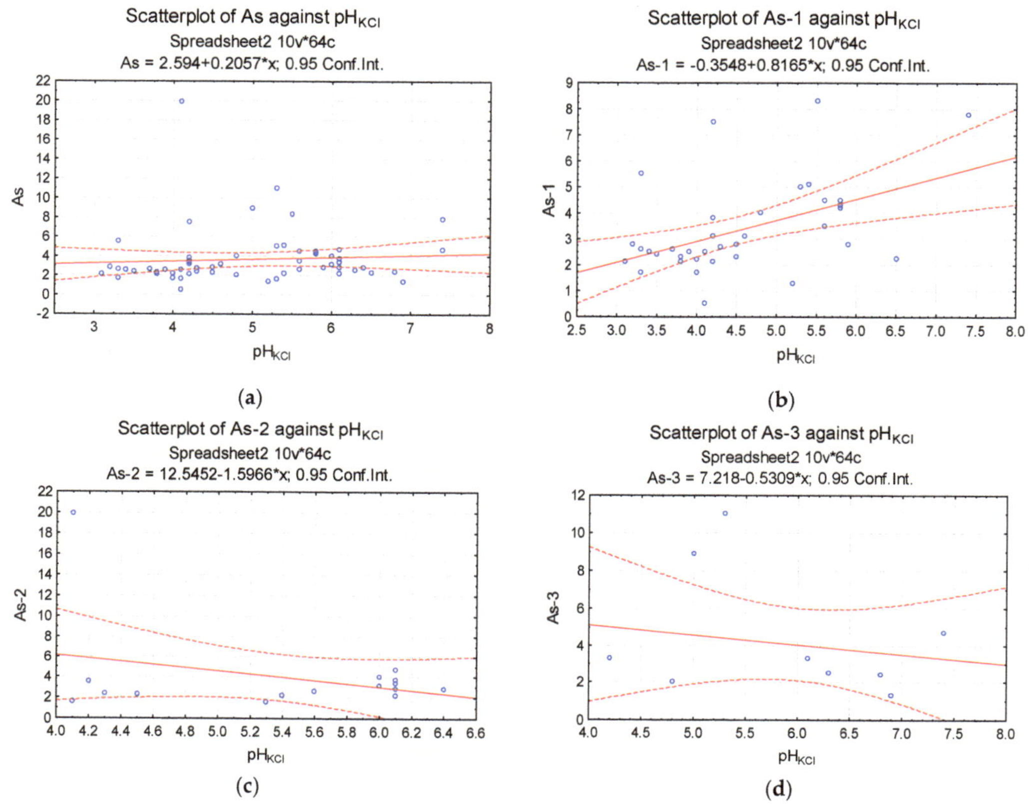

Figure 2. Dependence of arsenic content in soils of potash mining areas on pH_{KCl}: (**a**) potash mining areas outside of soil salinisation areas (As); (**b**) potash mining areas within eluvial and transit landscapes, covered with forest vegetation on conditionally natural zonal soils (As-1); (**c**) potash mining areas within eluvial and transit former agricultural landscapes(As-2); (**d**) potash mining areas within the transaccumulative landscapes of small river valleys (As-3).

Table 2. Assessment of the correlation between arsenic content and soil pH_{KCl} by Spearman's coefficient (ρ).

Sampling	**ρ ***	**Tightness of Correlation on the Cheddock Scale**	**Statistical Significance of the Spearman's Correlation Coefficient**		
			t (ρ)	$t_{critical}$ ($\rho_{critical}$)	**Significance Assessment**
Background area relative to potash mining area:					
Entire territory (n = 81)	−0.064	low	−0.566	1.991	the relationship is not statistically significant (p = 0.572786)
by landscape type:					
Areas of eluvial and transit landscapes occupied by typical forest communities on zonal soils (n = 27)	0.064	low	0.064	0.382	the relationship is not statistically significant (p > 0.05)

Table 2. *Cont.*

Sampling	ρ *	Tightness of Correlation on the Cheddock Scale	Statistical Significance of the Spearman's Correlation Coefficient		
			t (ρ)	$t_{critical}$ ($\rho_{critical}$)	Significance Assessment
Eluvial and transit former agricultural landscapes (n = 27)	−0.225	low	−0.225	0.382	the relationship is not statistically significant ($p > 0.05$)
Transaccumulative and accumulative landscapes within wetland ecotopes and floodplains of small rivers (n = 27)	−0.105	low	−0.105	0.382	the relationship is not statistically significant ($p > 0.05$)
Potash mines outside of soil salinisation areas:					
entire territory (n = 64)	0.276	low	2.262	1.999	the relationship is statistically significant ($p = 0.027297$)
by landscape type:					
Areas of eluvial and transit landscapes occupied by forest vegetation on conditionally natural zonal soils (n = 36)	0.485	moderate	0.485	0.33	the relationship is statistically significant ($p < 0.05$)
Eluvial and transit former agricultural landscapes (n = 16)	0.229	low	0.229	0.503	the relationship is not statistically significant ($p > 0.05$)
Transaccumulative landscapes within small river valleys (n = 12)	0.075	low	0.075	0.587	the relationship is not statistically significant ($p > 0.05$)
Areas of soil salinisation in the zone affected by potash industry:					
Chloride-type soil salinity areas—content Cl^- 1.49–36.35 g/kg (n = 7)	0.563	significant	0.563	0.786	the relationship is not statistically significant ($p > 0.05$)

* ρ—Spearman's correlation coefficient; $\rho_{critical}$—Spearman's criterion critical value; t—Student *t*-test; $t_{critical}$—critical value of Student's *t*-test.

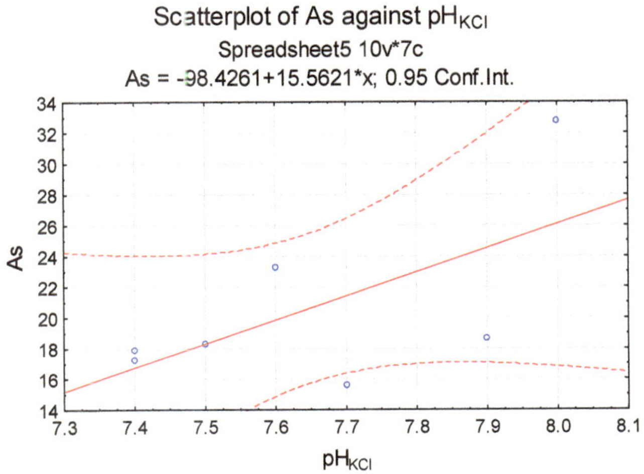

Figure 3. Dependence of arsenic content in soils of saline areas.

By analysing the features of arsenic distribution in different types of landscapes (Tables 3–5, Figures 4 and 5), the following patterns can be identified. Arsenic concentrations in both background and potash mining territories vary considerably (from n $\times$ 10^{-1} to n $\times$ 10), with 97% of values falling within the range from n $\times$ 10^{-1} to n. A smaller range of variation and the most uniform distribution were found for areas of eluvial and transit landscapes occupied by typical forest communities on zonal soils, both in background areas and potash mining areas, and for background areas of eluvial and transit former agricultural landscapes. There were no statistically significant differences in arsenic content in the soils of potash mining areas outside of saline areas and the background areas. The concentration factors, calculated as the ratio of average arsenic concentrations in the soils of potash mining territories to the average concentrations in the background areas, are 1.04–1.26. The contamination factor (*CF*) values for As are moderate. The ecological risk factor (E_r^i) values for As regarding the toxicity response factor (T_r^i) are 10.41–12.65, indicating low risk, both for the area as a whole and for each of the identified landscape types. There are also no statistically significant differences in arsenic content in the soils of the examined landscape types.

Table 3. Statistical characteristics of arsenic distribution in soils.

Sampling	Min. Value	Max. Value	Average Value mg/kg	Median	SD *	V %
Background area relative to potash mining area:						
Entire territory (n = 81)	0.10	14.48	3.35	2.96	2.45	73.17
by landscape type:						
Areas of eluvial and transit landscapes occupied by typical forest communities on zonal soils (n = 27)	0.10	6.40	3.06	2.60	1.82	59.55
Eluvial and transit former agricultural landscapes (n = 27)	0.10	6.00	3.10	3.30	1.86	59.98
Transaccumulative and accumulative landscapes within wetland ecotopes and floodplains of small rivers (n = 27)	0.50	14.48	3.88	2.96	3.35	86.42
Potash mines outside of soil salinisation areas:						
Entire territory (n = 64)	0.50	19.90	3.63	2.80	2.81	77.35
by landscape type:						
Areas of eluvial and transit landscapes occupied by forest vegetation on conditionally natural zonal soils (n = 36)	0.50	8.30	3.36	2.75	1.77	52.67
Eluvial and transit former agricultural landscapes (n = 16)	1.60	19.90	3.92	2.80	4.34	110.71
Transaccumulative landscapes within small river valleys (n = 12)	1.30	11.00	4.04	2.90	2.98	73.70
- on the granulometric composition of soils						
Sandy soils (n = 6)	1.30	3.30	2.30	2.35	0.64	27.91
Sandy loam soils (n = 9)	0.50	7.50	2.72	2.00	2.19	80.52
Light loamy soils (n = 32)	1.60	19.90	3.80	2.80	3.39	89.30
Soils of medium loamy composition (n = 10)	2.20	5.00	3.25	2.95	0.89	27.34
Heavy loamy soils (n = 7)	2.20	8.90	5.70	4.70	2.63	46.10
Areas of soil salinisation in the zone affected by the potash industry:						
Chloride-type soil salinity areas—content Cl^- 1.49–36.35 g/kg (n = 7)	15.56	32.75	20.51	18.27	5.89	28.72

* SD—Standard deviation; V—Coefficient of variation.

Table 4. Mann–Whitney U-test for significance of differences in arsenic content in soils, $p < 0.05$.

№	Sampling	n *	T *	U *	$U_{critical}$ *	Assessing the Credibility of Differences
1	Background areas of eluvial and transit landscapes occupied by typical forest communities on zonal soils	27	810	432	367	Differences between samples are not significant
	Potash mining areas within eluvial and transit landscapes, covered with forest vegetation on conditionally natural zonal soils	36	1206			
2	Background areas of eluvial and transit former agricultural landscapes	27	588	210	150	Differences between samples are not significant
	Potash mining areas within eluvial and transit former agricultural landscapes	16	358			
3	Background areas of transaccumulative and accumulative landscapes within wetland ecotopes and floodplains of small rivers	27	529	151	107	Differences between samples are not significant
	Potash mining areas within the transaccumulative landscapes of small river valleys	12	251			

* n—number of values in the sample; T—sum of ranks in the sample; U—criterion value; $U_{critical}$—critical importance.

Table 5. Assessment of the significance of differences in arsenic content in soils by Kruskal–Wallis H-criterion, $p < 0.05$.

№	Sampling	n *	T *	H *	*p*-Value	Assessing the Credibility of Differences
1	Background areas of eluvial and transit landscapes occupied by typical forest communities on zonal soils	27	1069	0.3485	0.84009	The result is not significant at $p < 0.05$
	Background areas of eluvial and transit former agricultural landscapes	27	1087			
	Background areas of transaccumulative and accumulative landscapes within wetland ecotopes and floodplains of small rivers	27	1165			
2	Potash mining areas within eluvial and transit landscapes, covered with forest vegetation on conditionally natural zonal soils	36	1171	0.076	0.96269	The result is not significant at $p < 0.05$
	Potash mining areas within eluvial and transit former agricultural landscapes	16	506			
	Potash mining areas within the transaccumulative landscapes of small river valleys	12	403			
3	Background relative to potash mines area	81	5759	19.8666	0.00005	The result is significant at $p < 0.05$
	Potash mines outside of soil salinisation areas	64	4831			
	Areas of soil salinisation in the zone affected by the potash industry	7	1038			
4	Potash mining areas with sandy soils	6	115.5	14.2319	0.00659	The result is significant at $p < 0.05$
	Potash mining areas with sandy loam soils	9	171			
	Potash mining areas with light loamy soils	32	1077.5			
	Potash mining areas with medium loamy soils	10	370.5			
	Potash mining areas with heavy loam soils	7	345.5			

* n—number of values in the sample; T—sum of ranks in the sample; H—criterion value.

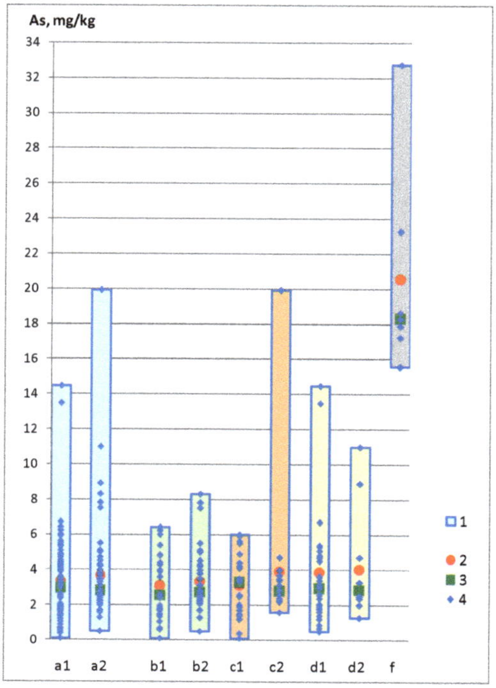

Abscissa axis:

a1—background area relative to potash mining area; **a2**—potash mines outside of saline soil areas; **b1**—background areas of eluvial and transit landscapes occupied by typical forest communities on zonal soils; **b2**—potash mining areas within eluvial and transit landscapes, covered with forest vegetation on conditionally natural zonal soils; **c1**—background areas of eluvial and transit former agricultural landscapes; **c2**—potash mining areas within eluvial and transit former agricultural landscapes; **d1**—background areas of transaccumulative and accumulative landscapes within wetland ecotopes and floodplains of small rivers; **d2**—potash mining areas within the transaccumulative landscapes of small river valleys; **f**—chloride saline areas

Legend:1—range; 2—mean; 3—median; 4—value (variant) of arsenic content

Figure 4. Change in arsenic content of soils.

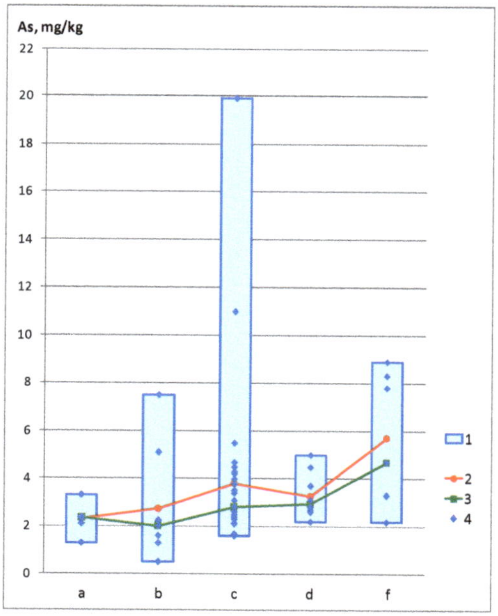

Abscissa axis:

a—sandy soils;

b—sandy loam soils;

c—light loamy soils;

d—soils of medium loamy composition;

f—heavy loamy soils

Legend:1—range; 2—mean; 3—median; 4—value (variant) of arsenic content

Figure 5. Change in arsenic content in soils of potash mining areas as a function of soil particle size distribution.

Dissolved sulphate, nitrate, and chloride in saline soil concentrations were shown to have little effect on arsenic adsorption [38]. Nonetheless, studies have found higher arsenic concentrations in saline soils than in the rest of the area. Arsenic concentrations in saline soils are comparable to those in solid wastes from potassium production, as shown in [26]. The concentration factors of average values are 6.12 in relation to background territories and 5.65 in relation to potash mining territories outside of saline areas. The contamination factor (CF) values for As are very high. The ecological risk factor (E_r^i) for As regarding the toxicity response factor (T_r^i) is 61.22, indicating moderate risk.

Outside of saline areas, the average arsenic content in soils of both background and potash mining territories is lower than the average regional contents shown in [29]. Relative to the calculated regional average content [29], soils are characterised by element dispersion: the factor of dispersion within background areas ranges from 1.94 to 2.46; within potash mining areas, it ranges from 1.86 to 2.24. Soils in saline areas relative to the calculated regional average content [29] are characterised by element accumulation: the concentration factor is 2.73.

Arsenic concentrations differ statistically in soils with different granulometric compositions (Tables 2 and 5, Figure 5). The average arsenic content in soils of heavy loam composition is 1.5–2.5 times higher than its average content in other soils. Minimum average values are typical of soils with prevailing sandy fractions in their granulometric composition. The identified characteristics support the role of clay particles as an adsorbent.

In environmental studies, permissible arsenic concentrations in soils are determined based on particle size distribution and pH_{KCl}. In Russia, the permissible concentration is 2 mg/kg for sandy and sandy loam soils, 5 mg/kg for sour ($pH_{KCl} < 5.5$) loamy and clayey soils, and 10 mg/kg for near neutral and neutral ($pH_{KCl} > 5.5$) loamy and clayey soils.

Soils of potash mining territories outside of saline areas in 20% of cases exceed permissible concentrations by 1.05–3.98 times. The ranges of exceedances in soils of different granulometric composition are similar: 1.05–3.75 in soils of sandy and sandy loam composition, and 1.1–3.98 in loamy and clayey soils. Despite their lower adsorption capacity, sandy soils consistently exceed the permissible values (69% of cases). This is due to the lower permissible arsenic concentrations in sandy and sandy loam soils.

The principal component analysis (PCA) was used to detect patterns and analyse linear dependencies in samples of arsenic and pH_{KCl} concentrations in background areas as well as non-saline and saline areas near potash enterprises(Figure 6). The first PC1 axis explains 70.05% of the total variance between samples and the second PC2 axis explains 29.95%.

Figure 6 shows the general distribution patterns of arsenic and pH_{KCl} for Samples A (background area) and B (potash mining areas outside of saline areas), with changes in acid–alkaline conditions playing the most essential role. Sample C (saline areas near potash mines) has considerably more variation in arsenic content.

The world's largest potash mining operations (Russia, Belarus, and Germany) are located in temperate latitudes in a humid climate zone [40,41]. Therefore, it can be assumed that the patterns of arsenic distribution in soils of potash mining areas in other countries with similar environmental conditions will be similar to those identified in this study.

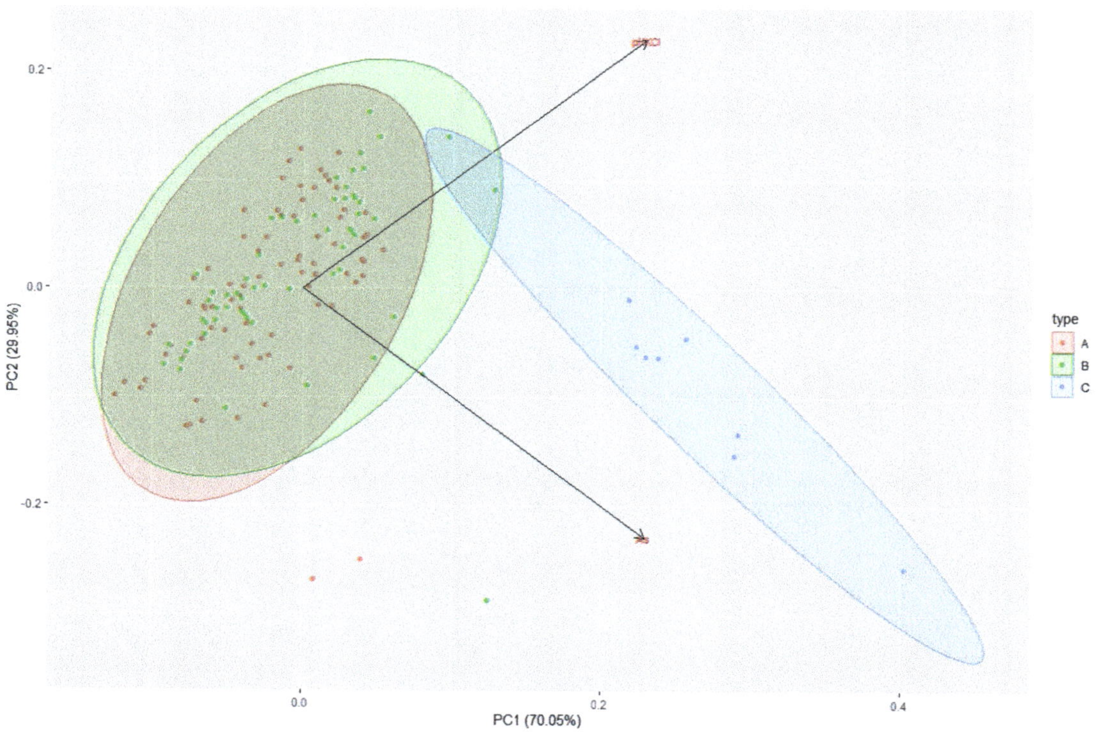

Figure 6. Analysis of the main components for samples A (background area), B (potash mining territories outside of salinity areas), and C (salinity areas).

4. Conclusions

The findings show a significant range of variation in arsenic concentrations in soils from both potash mining areas and background areas. In 97% of cases, arsenic concentrations range from $n \times 10^{-1}$ to n. No statistically significant differences in arsenic concentrations were found in the soils of potash mining territories, background territories, and soils of different types of landscapes. The soils of saline areas have higher arsenic concentrations than the rest of the territories. The arsenic concentrations in saline soils are comparable to those found in potash production solid waste. Saline areas are distributed locally, usually near salt waste disposal sites (salt dumps, sludge storages).

Despite high arsenic concentrations in the insoluble salt rock residue and solid waste from potash production, the identified patterns suggest that the activities of the potash companies operating in the Verkhnekamskoe potassium salt deposit are not determinants in the technophilic accumulation of arsenic in potash mining territories outside of saline areas. Arsenic concentrations in background soils can be used to adjust its regional background concentrations. Information on this toxic element is required by environmental studies for construction projects.

Author Contributions: Conceptualization, T.K. and E.M.; data curation, V.Z.; formal analysis, V.Z.; funding acquisition, T.K. and E.M.; investigation, T.K. and P.B.; methodology, T.K., E.M. and P.B.; project administration, T.K. and E.M.; resources, E.M.; software, E.M.; supervision, E.M.; validation, E.M.; visualization, T.K. and P.B.; writing—original draft, T.K. and V.Z.; writing—review and editing, T.K., E.M. and P.B. All authors have read and agreed to the published version of the manuscript.

Funding: The research was supported by the Perm Research and Education Centre for Rational Use of Subsoil, 2022.

Data Availability Statement: Not applicable.

Conflicts of Interest: The authors declare no conflict of interest.

References

1. Kolobanov, K.A.; Filatova, M.Y.; Bubnova, M.B.; Romashkina, E.A. Improving valuation of ecosphere pollution from mining waste using mathematical tools. *Min. Inf. Anal. Bull.* **2021**, *4*, 85–99. (In Russian) [CrossRef]
2. Lalmalsawmi, J.; Zirlianngura; Tiwari, D.; Lee, S. Low cost, highly sensitive and selective electrochemical detection of arsenic (III) using silane grafted based nanocomposite. *Environ. Eng. Res.* **2020**, *25*, 579–587. [CrossRef]
3. Gupta, A.D.; Giri, B.S.; Rene, E.R.; Chaturvedi, P.; Goswami, M.; Singh, H. Batch and continuous reactor studies for the adsorption of As(III) from wastewater using a hybrid biochar loaded with transition metal oxides: Kinetics and mass transfer analysis. *Environ. Eng. Res.* **2021**, *26*, 200438. [CrossRef]
4. Khosravi-Darani, K.; Rehman, Y.; Katsoyiannis, I.A.; Kokkinos, E.; Zouboulis, A.I. Arsenic Exposure via Contaminated Water and Food Sources. *Water* **2022**, *14*, 1884. [CrossRef]
5. Tiwari, D.; Jamsheera, A.; Zirlianngura; Lee, S. Use of hybrid materials in the trace determination of As(V) from aqueous solutions: An electrochemical study. *Environ. Eng. Res.* **2017**, *22*, 186–192. [CrossRef]
6. Jung, Y.; Kamimoto, Y. Arsenic removal properties by electrolyzed and calcined manganese dioxide. *Environ. Eng. Res.* **2020**, *25*, 735–741. [CrossRef]
7. Peng, Y.; Li, Y.; Tang, S.; Zhang, L.; Zhang, J.; Zhao, Y.; Zhang, X.; Zhu, Y. Dynamic Adsorption of As(V) onto the Porous α-Fe_2O_3/Fe_3O_4/C Composite Prepared with Bamboo Bio-Template. *Water* **2022**, *14*, 1848. [CrossRef]
8. Sun, Y.; Yu, F.; Han, C.; Houda, C.; Hao, M.; Wang, Q. Research Progress on Adsorption of Arsenic from Water by Modified Biochar and Its Mechanism: A Review. *Water* **2022**, *14*, 1691. [CrossRef]
9. Voronchikhina, E.A.; Zhdakaev, V.I. Arsenic in natural and technogenic geosystems of the Perm Territory. In Proceedings of the Sergeev Readings. Materials of the Annual Session of the Scientific Council of the Russian Academy of Sciences on the Problems of Geoecology, Engineering Geology and Hydrogeology, Perm, Russia, 2–4 April 2019; pp. 278–283. (In Russian)
10. Liu, G.; Shi, Y.; Guo, G.; Zhao, L.; Niu, J.; Zhang, C. Soil pollution characteristics and systemic environmental risk assessment of a large-scale arsenic slag contaminated site. *J. Clean. Prod.* **2020**, *251*, 119721. [CrossRef]
11. Mayorova, L.P.; Cherentsova, A.A.; Krupskaya, L.T.; Golubev, D.A.; Kolobanov, K.A. Assessment of manmade air pollution due to dusting at mine tailings storage facilities. *Min. Inf. Anal. Bull.* **2021**, *1*, 5–20. (In Russian) [CrossRef]
12. Krupskaya, L.T.; Zvereva, V.P.; Sklyarova, G.F.; Orlov, A.M. Aboveground mining waste storage as an ecosphere pollution source and waste exploitability in Russia's Far East. *Min. Inf. Anal. Bull.* **2021**, *2*, 5–21. (In Russian) [CrossRef]
13. Abramov, B.N. Toxicity assessment of mine tailings ponds in Transbaikalia. *Min. Inf. Anal. Bull.* **2021**, *11*, 136–145. (In Russian) [CrossRef]
14. Rastanina, N.K.; Golubev, D.A.; Shavrin, E.I. Ecosphere and health conditions within the influence zone of a closed mine in Primorye. *Min. Inf. Anal. Bull.* **2021**, *3*, 114–127. (In Russian) [CrossRef]
15. Ran, H.; Guo, Z.; Yi, L.; Xiao, X.; Zhang, L.; Hu, Z.; Li, C.; Zhang, Y. Pollution characteristics and source identification of soil metal(loid)s at an abandoned arsenic-containing mine, China. *J. Hazard. Mater.* **2021**, *413*, 125382. [CrossRef] [PubMed]
16. Álvarez-Quintana, J.; Álvarez, R.; Ordóñez, A. Arsenic in Soils Affected by Mining: Microscopic Studies vs. Sequential Chemical Extraction. *Int. J. Environ. Res. Public Health* **2020**, *17*, 8426. [CrossRef]
17. Kachor, O.L. Development of a model for the migration of arsenic along the soil profile from accumulated waste from the mining industry. *Earth Sci. Subsoil Use* **2019**, *42*, 144–150. (In Russian) [CrossRef]
18. Bailey, A.S.; Jamieson, H.E. Characterization of arsenic-hosting solid phases in tailings dust from giant mine: Implications for air quality monitoring at legacy mine sites. *Geol. Soc. Am. Abstr. Programs* **2017**, *49*, 6. [CrossRef]
19. Khayrulina, E.A.; Khomich, V.S.; Liskova, M.Y. Environmental issues of potash deposit development. *Izvestiya TulGU. Earth Sci.* **2018**, *2*, 112–126. (In Russian)
20. Khayrulina, E.; Mitrakova, N.; Poroshina, N.; Menshikova, E.; Perminova, A. Formation of solonchak in the area of the discharged ancient brine wells (Perm krai, Russia). *Front. Environ. Sci.* **2022**, *10*, 858742. [CrossRef]
21. Menshikova, E.A.; Voronchikhina, E.A.; Zhdakaev, V.I. Water migration technophilic elements in the extraction and processing of minerals salts: Problems and prospects of ecologizations. In Proceedings of the Modern Problems of Reservoirs and Their Watersheds: VII All-Russian Scientific and Practical Conference with International Participation, Perm, Russia, 2 June 2019; pp. 150–155. (In Russian).
22. Baboshko, A.Y.; Bachurin, B.A. Heavy metals in potash industry waste. *Min. Inf. Anal. Bull.* **2009**, *5*, 369–376. (In Russian)
23. Tchaikovsky, I.I.; Utkina, T.A.; Isaeva, G.A. Evolution of the mineral composition of the insoluble salt residue of the Verkhnekamskoye deposit. *Gorn. Echo* **2021**, *4*, 17–20. (In Russian) [CrossRef]
24. Smetannikov, A.F.; Onosov, D.V.; Sinegribov, V.A.; Novikov, P.Y. Noble metal extraction from K-Mg ore processing waste. *Min. Inf. Anal. Bull.* **2013**, *S4*, 216–231. (In Russian)
25. Bayandina, E.O.; Kudryashov, A.I. *Insoluble Salt Residue of the Verkhnekamsk Deposit*; Tipograf: Perm, Russia, 2015; pp. 21–36.
26. Khokhryakova, E.S. Ecogeochemical assessment of solid potash waste based on the content of trace elements. *Gorn. Echo* **2020**, *3*, 30–35. (In Russian) [CrossRef]

27. Belkin, P.A.; Kataev, V.N. Regularities of the chemical composition of technogenic transformation which groundwater undergoes in the areas of the potash deposits development. *News Ural. State Min. Univ.* **2018**, *2*, 55–64. (In Russian) [CrossRef]
28. Smetannikov, A.F.; Filippov, V.N. Some features of the mineral composition of salt rocks and products of their processing (on the example of the Verkhnekamsk salt deposit). In *Problems of Mineralogy, Petrography and Metallogeny. Scientific Readings in Memory of P.N. Chirvinsky*; PSU: Perm, Russia, 2010; Volume 13, pp. 99–113. (In Russian)
29. Dziuba, E.A. Determination of local background content of some macro- and microelements in the soils of the Perm region. *Geogr. Bull.* **2021**, *1*, 95–108. (In Russian) [CrossRef]
30. Opekunova, M.G.; Opekunov, A.Y.; Kukushkin, S.Y.; Ganul, A.G. Background contents of heavy metals in soils and bottom sediments in the north of Western Siberia. *Eurasian Soil Sci.* **2019**, *52*, 380–395. (In Russian) [CrossRef]
31. *Gost 26483-85*; Soils. Preparation of Salt Extract and Determination of its PH by CINAO Method. The National Standard. USSR State Committee for Standards: Moscow, Russia, 1985; 6p. (In Russian)
32. Lê, S.; Josse, J.; Husson, F. FactoMineR: An R package for multivariate analysis. *J. Stat. Softw.* **2008**, *25*, 1–18. [CrossRef]
33. Wickham, H. *ggplot2: Elegant Graphics for Data Analysis*; Springer: New York, NY, USA, 2009; pp. 2197–5736.
34. Ruiz-Roso, J.; García-Lorenzo, M.L.; Castiñeiras, P.; Muñoz-Martín, A.; Crespo-Feo, E. 3D Spatial Distribution of Arsenic in an Abandoned Mining Area: A Combined Geophysical and Geochemical Approach. *Minerals* **2020**, *10*, 1130. [CrossRef]
35. Hakanson, L. An ecological risk index for aquatic pollution control, a sedimentological approach. *Water Res.* **1980**, *14*, 975–1001. [CrossRef]
36. Pan, L.; Guan, X.; Liu, B.; Chen, Y.; Pei, Y.; Pan, J.; Zhang, Y.; Hao, Z. Pollution Characteristics, Distribution and Ecological Risk of Potentially Toxic Elements in Soils from an Abandoned Coal Mine Area in Southwestern China. *Minerals* **2021**, *11*, 330. [CrossRef]
37. Yuan, D.; Liu, Y.; Guo, X.; Liu, J. Characteristic contaminants in snowpack and snowmelt surface runoff from different functional areas in Beijing, China. *Environ. Sci. Pollut. Res.* **2018**, *25*, 36256–36266. [CrossRef] [PubMed]
38. Putilina, V.S.; Galitskaya, I.V.; Yuganova, T.I. *The Behavior of Arsenic in Soils, Rocks and Groundwater. Transformation, Adsorption/Desorption, Migration*; State Public Scientific and Technical Library of the Siberian Branch of the Russian Academy of Sciences: Novosibirsk, Russia, 2011; 249p. (In Russian)
39. Gonçalves, D.A.M.; Pereira, W.V.d.S.; Johannesson, K.H.; Pérez, D.V.; Guilherme, L.R.G.; Fernandes, A.R. Geochemical Background for Potentially Toxic Elements in Forested Soils of the State of Pará, Brazilian Amazon. *Minerals* **2022**, *12*, 674. [CrossRef]
40. Prakash, S.; Verma, J.P. Global Perspective of Potash for Fertilizer Production. In *Potassium Solubilizing Microorganisms for Sustainable Agriculture*; Meena, V., Maurya, B., Verma, J., Meena, R., Eds.; Springer: New Delhi, India, 2016. [CrossRef]
41. Garrett, D.E. *Potash: Deposits, Processing, Properties and Uses*; Springer Science & Business Media: Berlin/Heidelberg, Germany, 2012.

minerals

MDPI

Article

Paleoenvironmental and Bio-Sequence Stratigraphic Analysis of the Cretaceous Pelagic Carbonates of Eastern Tethys, Sulaiman Range, Pakistan

Shuja Ullah [1,*], Irfan U. Jan [1], Muhammad Hanif [1], Khalid Latif [1], Mohibullah Mohibullah [2], Mahnoor Sabba [1], Aqsa Anees [3], Umar Ashraf [3] and Hung Vo Thanh [4,*]

1 National Centre of Excellence in Geology, University of Peshawar, Peshawar 25120, Pakistan; irfan_nceg@uop.edu.pk (I.U.J.); mhanif_nceg@uop.edu.pk (M.H.); khalidlatif@uop.edu.pk (K.L.); mahnoorsabba57@gmail.com (M.S.)
2 Department of Geology, University of Baluchistan, Quetta 87300, Pakistan; mohibullah.kakar@um.uob.edu.pk
3 Institute for Ecological Research and Pollution Control of Plateau Lakes, School of Ecology and Environmental Science, Yunnan University, Kunming 650504, China; aqsaanees01@outlook.com (A.A.); umarash2010@hotmail.com (U.A.)
4 School of Earth and Environmental Sciences, Seoul National University, Seoul 08826, Korea
* Correspondence: shujageo@gmail.com or shuja_nceg@uop.edu.pk (S.U.); vothanhhung198090@gmail.com (H.V.T.)

Abstract: The Cretaceous pelagic carbonate succession, i.e., Goru Formation was studied in the Chutair Section, Sulaiman Range, representing part of the eastern Tethys for the paleoenvironment and bio-sequence stratigraphy. Eight planktonic foraminiferal biozones are identified which include: 1. *Muricohedbergella planispira* Interval Zone; 2. *Ticinella primula* Interval Zone; 3. *Biticinella breggiensis* Interval Zone; 4. *Rotalipora appenninica* Interval Zone; 5. *Rotalipora cushmani* Total Range Zone; 6. *Whiteinella archeocretacea* Partial Range Zone; 7. *Helvetoglobotruncana helvetica* Total Range Zone; and 8. *Marginotruncana sigali* Partial Range Zone representing Albian-Turonian age. The petrographic studies revealed five microfacies: 1. Radiolarians-rich wacke-packestone microfacies; 2. Radiolarians-rich wackestone microfacies; 3. Planktonic foraminiferal wacke-packestone microfacies; 4. Planktonic foraminiferal wackestone microfacies; and 5. Planktonic foraminiferal packestone microfacies; indicating deposition of the Goru Formation in outer-ramp to deep basinal settings. Based on the facies variations and planktonic foraminiferal biozones, the 2nd and 3rd order cycles are identified, which further include six transgressive and five regressive system tracts. The sea level curve of the Goru Formation showed fluctuation between outer-ramp and deep-basin, showing the overall transgression in the 2nd order cycle in the study area, which coincides with Global Sea Level Curve; however, the 3rd order cycle represents the local tectonic control during deposition of the strata.

Keywords: microfacies; biostratigraphy; Chutair Section; sequence stratigraphy; Sulaiman Range; Stratigraphic Analysis; eastern Tethys

Citation: Ullah, S.; Jan, I.U.; Hanif, M.; Latif, K.; Mohibullah, M.; Sabba, M.; Anees, A.; Ashraf, U.; Vo Thanh, H. Paleoenvironmental and Bio-Sequence Stratigraphic Analysis of the Cretaceous Pelagic Carbonates of Eastern Tethys, Sulaiman Range, Pakistan. *Minerals* **2022**, *12*, 946. https://doi.org/10.3390/min12080946

Academic Editors: Marianna Kulkova and Dmitry Subetto

Received: 8 June 2022
Accepted: 26 July 2022
Published: 27 July 2022

1. Introduction

The Sulaiman Range in Pakistan consists of thick Triassic to Pleistocene sedimentary succession [1,2]. Whilst these units represent predominantly marine succession, the post Eocene shows the fluvial depositional realms [2]. Among all marine succession, the pelagic carbonates of the Goru Formation, which is subject of the current study, was deposited in the Cretaceous time (i.e., mid-Cretaceous) [3]. The Early and Early Late Cretaceous reflect a period of significant paleoenvironmental changes, which manifested itself in the installation of widespread dys-aerobic to anaerobic conditions in outer-shelf and basinal settings on multiple stages [4,5]. The mid-Cretaceous period (i.e., 124–90Ma) saw a shift in the ocean-climate system's dynamic. Increased tectonic activity and changing paleogeography were responsible for the changes [6]. The planktons preserved in deep-sea marl, organic-rich shale, and pelagic carbonate from the mid-Cretaceous (Barremian–Turonian) period provide an important record of how the marine biosphere responded to short- and long-term changes in the ocean-climate system [7]. Dynamic paleo-oceanographic settings affected the evolution of planktonic foraminifera throughout the mid-Cretaceous [8]. A vital radiation in the evolution of planktonic foraminifera occurred during the Late Albian–Cenomanian (105–93 Ma), characterised by the appearance of new morphologies, such as single-keeled trochospiral and biserial taxa with supplementary apertures, and the development of complex wall textures, such as macroperforate, muricate, and costellate, which lasted until the end of the Cretaceous [9]. The mid-Cretaceous, in the marine plankton, was likewise a time of high radiation and turnover [10]. The southern high latitudes were likely the main sources of deep-water production during the mid-Cretaceous, with subtropical convection restricted to isolated basins [11].

The autotrophic calcareous nano-plankton and heterotrophic planktonic foraminifera and radiolarians, both of which have mineralized skeletons of calcium carbonate and silica, are both abundant and well-preserved in mid-Cretaceous marine strata [7]. At or near the Oceanic Anoxic events (OAEs), radiolarians show high rates of evolutionary turnover that is extinction and radiation [12]. During the Cenomanian–Turonian boundary, the first bioevent occurred and the isochronous extinction of rotaliporids and the evolution/diversification of *Praeglobotruncana*, *Dicarinella*, and *Marginotruncana* indicate the start of the Turonian [7,13,14].

Based on planktonic foraminifera biostratigraphy, the pelagic carbonates of the Goru Formations in the Southern Kirthar Ranges represent Cretaceous age and the fauna show an open-marine environment [15]. The Goru Formation in the Lower Indus Basin (i.e., Kirthar Ranges) comprised of yellowish to greenish grey, fine to coarse grain sandstone interbedded with siltstone and shale in lower part whilst dark gray-to-black shales with siltstone and clay interbeds in upper part [16,17]. Here, the Goru Formation represents the main hydrocarbon bearing reservoir zone [18–21]. Previously researchers have conducted only patchy work on the Goru Formation in the Lower Indus Basin with respect to deposition and biostratigraphy; authors in a study [22] assigned the Albian to Cenomanian age to the Goru Formation on the Nar River Section, near Goru village in the Kirthar Range. In another study, authors [23] assigned to it Albian to Cenomanian age based on calcareous nannofossils and planktonic foraminifera of NW Pakistan. The Goru Formation in the Kirthar Range was allotted as an Lower Aptian to Coniacian age [24]. The Goru Formation has not been studied in detail in the study area for biostratigraphy, paleoenvironments and sequence stratigraphy. Therefore, the aims of this study are to determine the biostratigraphy, paleoenvironments and sequence stratigraphy of the Goru Formation.

2. Tectonic and Stratigraphic Settings

The Sulaiman Range is a structurally and tectonically active thin-skinned fold and thrust belt that developed from an oblique collision between Eurasia and India [25]. The Chaman sinistral strike slip fault and the Zhob Ophiolites (i.e., Muslim Bagh Ophiolites) form the northern and western boundaries of this thrust belt (Figure 1). It resulted by transpression of left lateral Chaman Fault as well as the southward thrusting of the western end of the India [26–28]. The Sulaiman Range has a Triassic to Oligocene stratigraphy. The Chutair Section is a part of the Sulaiman Range near Ziarat. The stratigraphic succession exposed in the study area ranges from the Triassic Wulgai Formation to the Oligocene conglomerate [3] (Figure 2), having being deposited on a broad shelf opening in the westward terminations into the southerly expansion of the Tethys during the Early Cretaceous [3]. The Goru Formation in the western part of the Sulaiman Range, where the study area is located, is primarily composed of thinly bedded limestone. The limestone has a grey hue in the lower part of the formation, a cream color in the middle, and a red and brown color in the upper part (Figure 2).

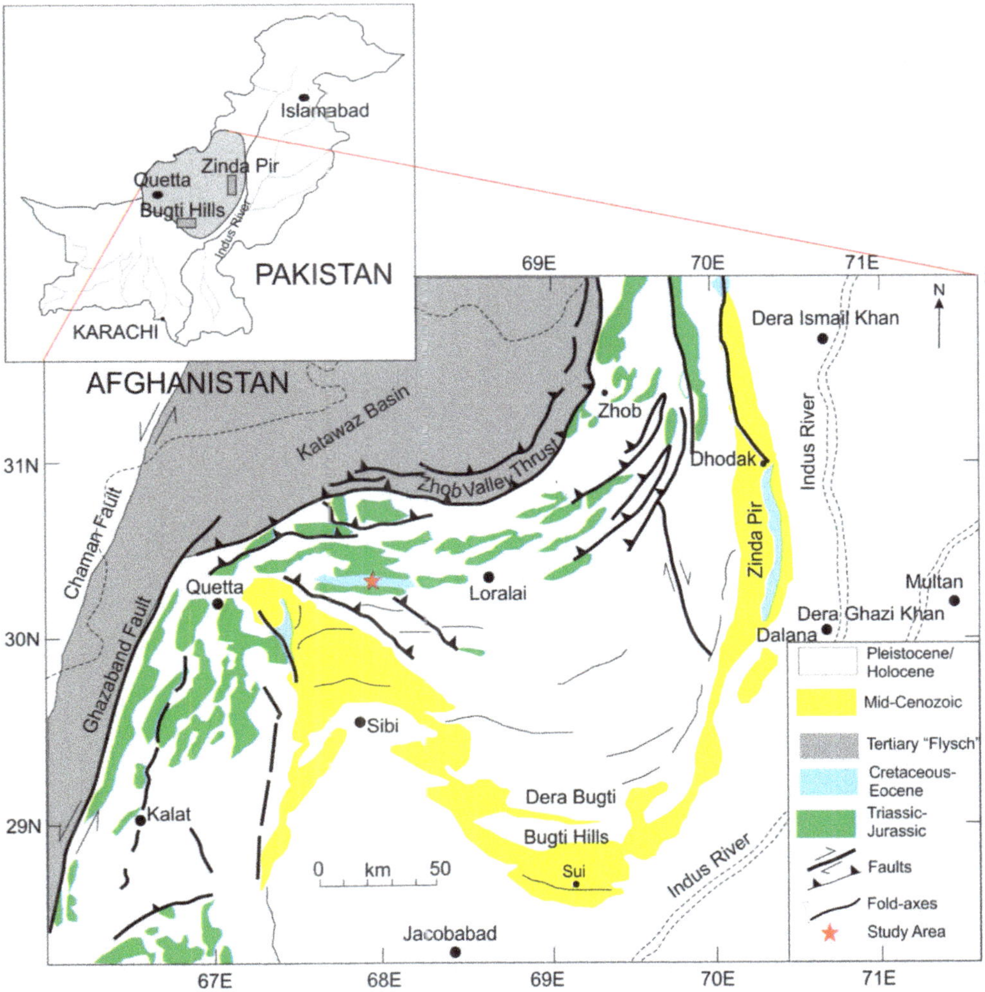

Figure 1. Map showing the study area and distribution of Triassic to Holocene rocks in the Sulaiman Range. (After [29]).

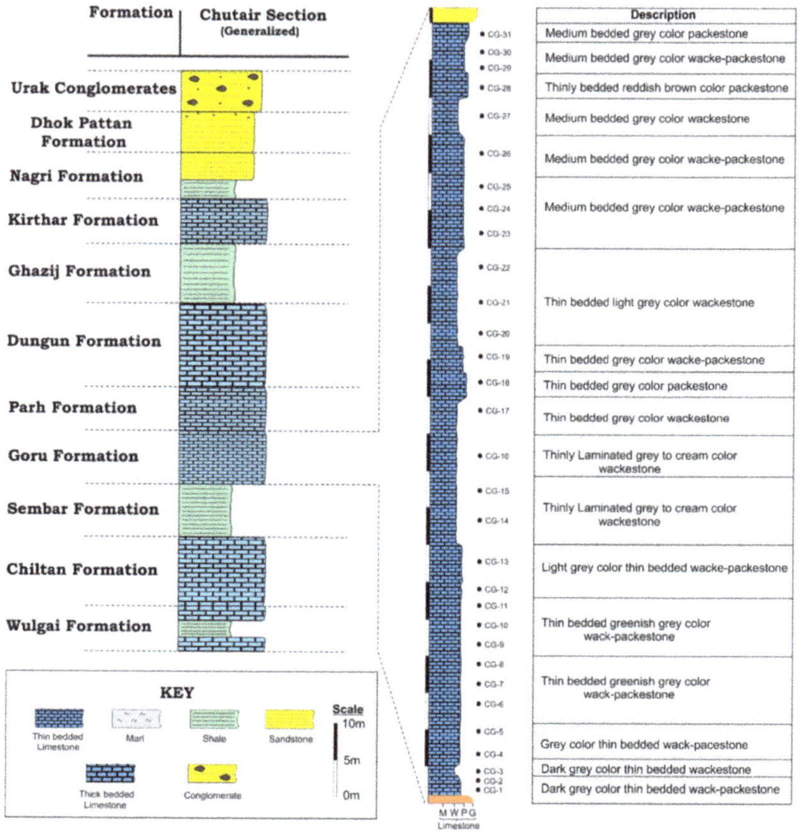

Figure 2. Generalized stratigraphy and lithologic log of Goru Formation at the study area, i.e., Chutair Section (After [3]).

3. Methodology

The Cretaceous pelagic carbonates of the Goru Formation exposed at Chutair Section (at Lat: 30°20′38.02″ N; and Long: 67°54′11.26″ E), Sulaiman Range is sampled for the determination of depositional environment, biostratigraphy, and sequence stratigraphy. The section is logged and 31 samples are collected on the basis of facies variation for the petrographic study [30]. The important features at the outcrop scale are noted. The field samples were cut into slabs and thin sections were generated at thin section laboratory of the National Centre of Excellence in Geology (NCE in Geology), University of Peshawar. The Nikon eclipse LV100ND polarizing microscope having dispersion staining observation at up to 400× magnification was used to examine and photograph the thin section at the Sedimentology and Paleontology Laboratory, NCE in Geology, University of Peshawar. The petrographic studies are carried out for the biostratigraphy and paleoenvironment. The morphological characteristics of planktonic foraminifera are used in association with published literature for identification [9,31,32]. For the sequence stratigraphic studies, the microfacies and biostratigraphy are used to define the system tracts and order of cycles, by using the sequence models from the literature [33,34].

4. Results and Discussions

4.1. Planktonic Foraminiferal Biostratigraphy

The planktonic foraminiferal studies are carried out for the determination of biostratigraphy of Goru Formation in the study area. Forty-four species of planktonic foraminifera that belonged to different genera are identified from thin section examination (Figure 3). Based on taxa's stratigraphic distribution within the Goru Formation in the study area (i.e., Chutair Section; Figure 1), 8 planktonic foraminiferal biozones are documented (Figure 4). These biozone are compared with literature [32,35–37]. The documented planktonic foraminifera are correlated with global biozonal schemes. All foraminiferal species' thin section photomicrographs are shown in Figure 3. The biozones discussed from the Albian–Turonian age of the pelagic carbonates of the Goru Formation are given as: *Muricohedbergella planispira* Interval Zone; *Ticinella primula* Interval Zone; *Biticinella breggiensis* Interval Zone; *Rotalipora appenninica* Interval Zone; *Rotalipora cushmani* Total Range Zone; *Whiteinella archeocretacea* Partial Range Zone; *Helvetoglobotruncana helvetica* Total Range Zone; and *Marginotruncana sigali* Partial Range Zone. These zones are discussed in detail as follows:

Muricohedbergella Planispira **Interval Zone (Moullade, [38])**

Age: Early Albian

Definition: Partial range zone of *Muricohedbergella planispira* from the transition to the First Appearance Datum (FAD) of *Ticinella primula.*

Assemblage: *Muricohedbergella delrioensis, M. planispira* (Figures 3 and 4).

Remarks: This zone is marked based on the FAD of the *Ticinella primula,* and is characterized by the low-diversity assemblages. The zonal marker's FAD is recorded at the bottom of the section, and it may extend deeper into the unexposed stratigraphic depth, casting doubt on the Late Aptian age, although the base of the Albian is marked by the presence of *Microhedbergella renilaevis* [39] and in this part of the basin this particular species have not been identified (Figure 4).

Ticinella Primula **Interval Zone (Moullade, [38])**

Age: Middle Albian

Definition: The zone ranges from the FAD of the *Ticinella primula* to the FAD of *Biticinella breggiensis*

Assemblage: *Muricohedbergella delrioensis, M. planispira, M. rischi, M. simplex, Ticinella praeticinensis, T. primula, T. roberti* (Figures 3 and 4).

Remarks: The planktonic foraminifera gradually increase in this zone. Some of the planktonic species have their FADs in this zone including *M. rischi, M. simplex, and Ticinella roberti.* These taxa FADs have also been reported from this zone of the Parh Formation of Mughal Kot Section [40]. However, elsewhere in the world, the *Ticinella roberti* has been reported from the older Zone, i.e., *Ticinella bejaouaensis* zone [32,36]. This zone is correlated with the *Ticinella primula* zone of Premoli Silva and Verga [9].

Biticinella Breggiensis **Interval Zone (Moullade, [38])**

Age: Middle Albian

Definition: This zone started from the FAD of *Biticinella breggiensis* to the FAD of *Rotalipora appenninica* and *Planomalina buxtorfi.*

Assemblage: *Biticinella breggiensis, B. subbreggiensis, Macroglobigerinelloides bentonensis, Muricohedbergella delrioensis, M. planispira, M. rischi, M. simplex, Ticinella praeticinensis, T. primula, T. raynaudi* (Figures 3 and 4).

Remarks: The upper boundary of this zone is marked on the FAD of *Rotalipora appenninica* and *Planomalina buxtorfi* instead of *Rotalipora ticinensis,* which is the defined boundary elsewhere in the world [32,38]. The diversity in the planktonic foraminifera increase in this zone such as *M. bentonensis, T. praeticinensis, and T. raynaudi,* start to appear in this zone.

Rotalipora Appenninica **Interval Zone (Bronnimann, [41])**

Age: Late Albian

Definition: This interval spans from the FAD of *Rotalipora appenninica* to the FAD of *Rotalipora globotruncanoides*.

Assemblage: *Heterohelix moremani, H. reussi, Macroglobigerinelloides bentonensis, M. caseyi, Muricohedbergella delrioensis, M. planispira, M. rischi, M. simplex, Planomalina buxtorfi, P. praebuxtorfi, Praeglobotruncana delrioensis, Rotalipora appenninica, R. balernaensis, Ticinella madecassiana, T. primula* (Figures 3 and 4).

Remarks: The planktonic foraminiferal diversity and abundance increase. The keeled forms mostly appeared in this zone, i.e., *Rotalipora appenninica, Planomalina praebuxtorfi, P. buxtorfi. The Heterohelix moremani, H. reussi, Macroglobigerinelloides caseyi*, and *Praeglobotruncana delrioensis*, also appeared at the base of this zone (Figure 4). Some of the species disappeared in this zone, i.e., *Biticinella breggiensis, B. subbreggiensis, Ticinella primula, T. roberti*. The lower zones that were identified [9,32,36] are missing in this succession which with other factors could also be attributed to the coarse sampling in the field.

Rotalipora Cushmani **Total Range Zone (Borsetti, [42])**

Age: Middle to Late Cenomanian

Definition: The Total range zone of *Rotalipora cushmani*.

Assemblage: *Heterohelix reussi, H. moremani, Muricohedbergella delrioensis, M. planispira, M. rischi, M. simplex, Macroglobigerinelloides bentonensis, Praeglobotruncana delrioensis, P. stephani, Rotalipora balernaensis, R. cushmani, R. globotruncanoides, R. greenhornensis, R. montsalvensis, Whiteinella baltica, W. praehelvetica* (Figures 3 and 4).

Remarks: Most of the keeled taxa appeared in this zone. The test sizes showed increase towards the middle part and gradually started disappearing at the top. The *Whiteinella* and *Dicarinella* have their FADs in this zone (Figure 4). The *rotaliporids* are dominant in this zone and are reported in other parts of the world [9,32,42]. The *rotaliporids* are diversified and having the gradual disappearance of older species that is balanced by the younger species and then by Whiteinella and Dicarinella species. The *rotaliporids* have Last appearance datums (LADs) at the top of this zone that is also reported from elsewhere with the appearance of black shale, i.e., Ocean anoxic event-2 (OAE2) [14,36]. On the basis of bio-stratigraphic constraints, the black shale is correlated with the global bonarelli event (i.e., OAE2). Such black shales have also been reported from Parh Formation by Khan ([43]) from the Northern Suleman Range, Mughal Kot Section. However, in this study the black shale is not reported from this part of the basin in the southern Suleman Range. The planktonic foraminifera disappeared at the onset of these black shale deposition. Below this zone, some of the biozones are missing.

Whiteinella Archeocretacea **Partial Range Zone (Bolli, [44])**

Age: Late Cenomanian to Early Turonian

Definition: This zone is the partial Range Zone of LAD of *Rotalipora cushmani* to the FAD of *Helvetoglobotruncana helvetica*.

Remarks: At the onset of black shale deposition, almost all the planktonic foraminifera disappeared. Consequently, this zone is only named based on the LAD of *Rotalipora cushmani* to the FAD of *Helvetoglobotruncana helvetica* as this zone is barren of foraminifera (Figure 4). The high organic rich black shale deposition corresponds to this zone [14,36]. Such black shales of OAE2 are not present in this part of the basin in the Goru Formation.

Helvetoglobotruncana helvetica **Total Range Zone (Dalbiez, [45])**

Age: Early to middle Turonian

Definition: The total Range zone of *Helvetoglobotruncana helvetica*.

Assemblage: *Dicarinella algeriana, D. canaliculate, D. imbricata, Helvetoglobotruncana helvetica, Heterohelix moremani, H. reussi, Macroglobigerinelloides bentonensis, Muricohedbergella delrioensis, M. planispira, M. rischi, M. simplex, Marginotruncana coronata, M. marianosi, M. pseudolinneina, M. schneegansi, M. sigali, M. renzi, Praeglobotruncana gibba, P. stephani, W. aprica, W. baltica, W. brittonensis, W. paradubia, and Whiteinella praehelvetica (Figures 3 and 4).*

Remarks: This zone started above the *Whiteinella archeocretacea* zone by the appearance of zonal marker that is FADs of *Helvetoglobotruncana helvetica* and extinction of *rotaliporids* (Figure 4). Some of the species survived this extinction like *Macroglobigerinelloides bentonensis*, and *Whiteinellids*. Most of the robust and large planktonic foraminifera have their FADs and diversification within this zone, i.e., *Marginotruncanids* like *Marginotruncana pseudolinneina, M. renzi, M. schneegansi, and M. sigali.*

Marginotruncana Sigali **Partial Range Zone (Barr, [46])**

Age: Late Turonian

Definition: This zone started from the extinction of *Helvetoglobbotruncana helvetica.*

Assemblage: *Dicarinella canaliculate, D. concavata, D. imbricata, Heterohelix moremani, H. reussi, Muricohedbergella delrioensis, M. flandrini, M. planispira, M. rischi, M. simplex Macroglobigerinelloides bentonensis, Marginotruncana coronata, M. pseudolinneina, M. marianosi, M. renzi, M. schneegansi, M. sigali, and M. undulata* (Figures 3 and 4).

Remarks: The base of this zone is marked by the LADs of *Praeglobotruncana* spp. and the extinction of *Helvetoglobotruncana helvetica* and FADs of *Muricohedbergella flandrini* and *Dicarinella concavata* (Figure 4). The upper boundary of this zone is not identified here. This zone is the last biozone that is encountered in the Goru Formation in the study area and is also observed elsewhere [32].

Biostratigraphic Discussions

The Goru Formation's planktonic foraminiferal biostratigraphy suggests Mid-Cretaceous (Early Albian to late Turonian) age in the study area (Figure 4). The planktonic foraminifera evolution in the mid-Cretaceous was defined by periods of high and low turnover. The latest Albian, the mid-Cenomanian, and the Cenomanian/Turonian boundary had the highest turnover rates [7]. All across the Cenomanian, the diversity remained high, with a wide range of morphologies [47]. The same type of conditions are also prevailing from vertebrate fauna of the peninsular India, and Cauvery Basin, India [48,49]. Almost all the identified zones in the Goru Formation are global in nature as was identified elsewhere [32,36]. The base of the deposition of pelagic carbonates of the Goru Formation started with *Muricohedbergella planispira* Zone that indicates the Early Albian age. This zone is marked based on the FAD of *Ticinella primula.* This is followed by the *Ticinella primula* Zone of Early Albian age in which the planktonic foraminifera become larger and abundant represented by the FAD of *Macroglobigerinelloides bentonensis and Muricohedbergella rischi.* In the same stage in Middle Albian time, the *Biticinella* appeared and have FAD in the *Biticnella breggiensis* Zone. Here, the notable increase in the taxa's diversity, and abundance and gradual transition in morphology from unkeeled to keeled taxa is observed. This zone is followed by the *Rotalipora appenninica* Zone of the latest Albian time in the study area. Some of the global biozone below *R. appenninica* Zone of the Late Albian are missing. Diversification of the *rotaliporids* occurs at the top of this zone where some species appeared. Some of the species become extinct at this zone that is *Ticinella's* and *biticinella's.* Above this zone, is the Total Range zone of *Rotalipora cushmani* Zone of Middle to Late Cenomanian age (Figure 4). Below this zone, some of the global biozone of Early Cenomanian age are missing in the Goru Formation that may be because of the coarse sampling intervals. At the top of this zone, is the zone OAE2 (Bonarelli event), but OAE2 black shale has not been identified in this part of the Suleman Range, despite the fact that the OAE2 black shale was identified in the Parh Formation of the Mughal Kot Section, northern Sulaiman Range [40]. This is correlated with the arbitrary biozone that is *Whiteinella archeocretacea* Biozone that is of the Latest Cenomanian to Early Turonian. Here, in the Goru Formation, the C/T (Cenomanian/Turonian) boundary is present in the pelagic carbonates. This arbitrary zone is marked on the LAD of *Rotalipora cushmani* to the FAD of *Helvetoglobotruncana helvetica.* With FAD of *Helvetoglobotruncana Helvetica, the Helvetoglobotruncana helvetica* Total Range Zone of the Early to Middle Turonian started where the large, robust *Marginotrancanids* appeared. After this, the *Marginotreuncana sigali* Zone of Late Turonian age started with the extinction of the *Helvetoglobotruncana helvetica.* Here, the *Praeglobotruncana* disappear and *Muricohedbergella flandrini* appeared. The Goru Formation deposition ends with this biozone in the Chutair Section, Sulaiman Range. The biozones identified in this study have been compared to the main Tethyan biozonation system on a global scale [9]. The Goru Formation's biostratigraphy also suggests that Cretaceous sedimentation in the Eastern Tethys was nearly complete.

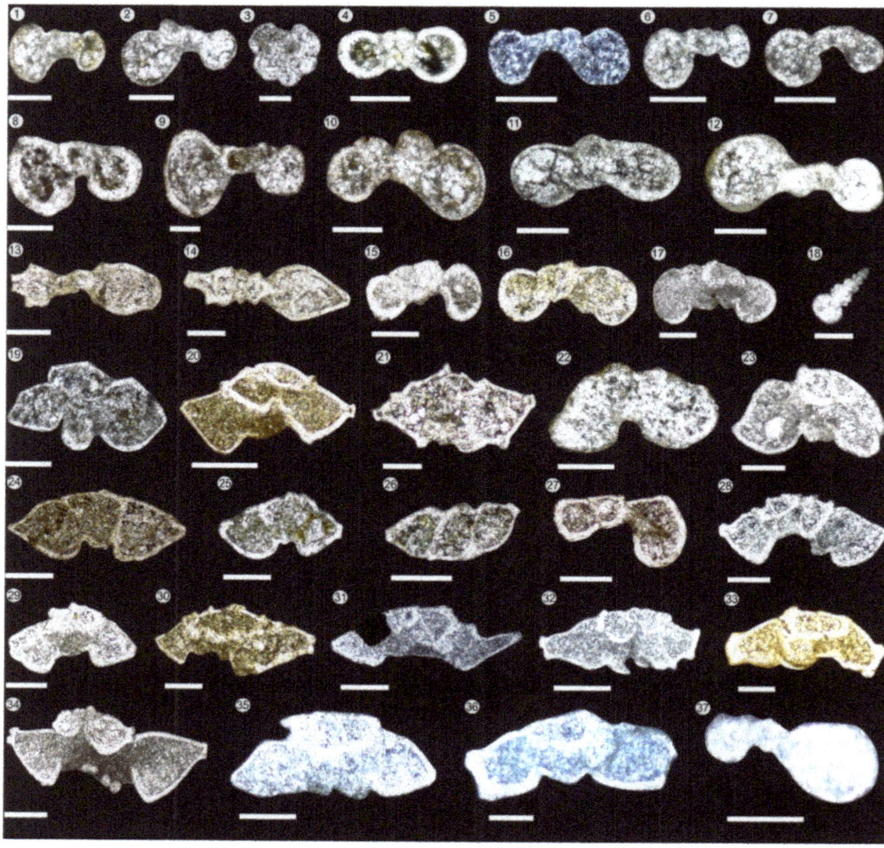

Figure 3. (**1**) Muricohedbergella planispira; (**2**) Muricohedbergella delrioensis; (**3**) Ticinella primula; (**4**) Ticinella roberti; (**5**) Muricohedbergella simplex; (**6**) Muricohedbergella rischi; (**7**) Ticinella praeticinensis; (**8**) Biticnella subbreggiensis; (**9**) Biticnella breggiensis; (**10**) Ticinella madecassiana; (**11**) Ticinella raynaudi; (**12**) Macroglobigerinelloides bentonensis; (**13**) Planomalina praebuxtorfi; (**14**) Planomalina buxtorfi; (**15**) Whiteinella praehelvetica; (**16**) Whiteinella aprica; (**17**) Whiteinella brittonensis; (**18**) Heterohelix reussi; (**19**) Rotalipora cushmani; (**20**) Rotalipora globotruncanoids; (**21**) Rotalipora greenhornensis; (**22**) Praeglobotruncana delrioensis; (**23**) Praeglobotruncana gibba; (**24**) Rotalipora appenninica; (**25**) Rotalipora balernaensis; (**26**) Rotalipora montsalvensis; (**27**) Helvetoglobotruncana helvetica; (**28**) Marginotruncana sigali; (**29**) Marginotruncana Renzi; (**30**) Marginotruncana schneegansi; (**31**) Marginotruncana marianosi; (**32**) Marginotruncana coronata; (**33**) Marginotruncana pseudolinniena; (**34**) Dicarinella concavata; (**35**) Dicarinella imbricata; (**36**) Dicarinella canaliculata; (**37**) Muricohedbergella flandrini. (Scale = 100 μm).

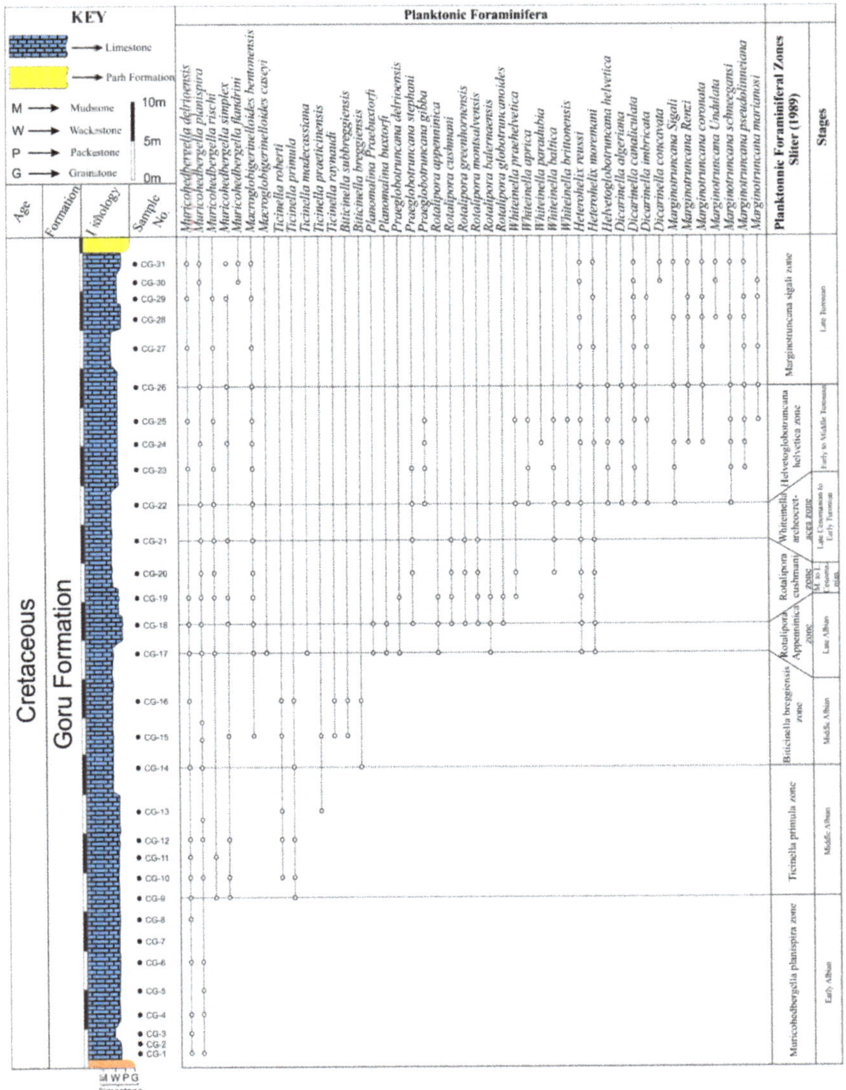

Figure 4. Biostratigraphic range chart of the Goru Formation exposed in the Chutair Section, Sulaiman Range.

4.2. Microfacies

The comprehensive petrographic analyses were carried out to evaluate the microfacies based depositional environment. Five microfacies were identified in the pelagic carbonates of the Goru Formation at Chutair Section based on texture, grain type, and other petrographic features discussed as follows:

4.2.1. Radiolarians-Rich Wacke-Packestone Microfacies (MF-1)

Description

The Radiolarians-rich wacke-packestone microfacies (MF-1) in the outcrop shows thin bedded dark grey color limestone. Petrographic observation reveals the presence of matrix and allochems. The planktonic foraminifera, radiolarians, calcispheres, and undifferentiated bioclasts make up the majority of the allochems. The matrix is neomorphosed and ferruginous. (Figure 5).

Interpretation

The MF-1 microfacies is mostly comprised of the deeper fauna such as radiolarians and calcisphere. The radiolarians' dominance indicates that energy levels are too low and that they are found deeper in nature [50]. An environment with low-energy levels is reflected by the extensive carbonate mud. [50]. The combination of mud, radiolarians, and planktonic foraminifera indicate that the deposition took place in a low-energy environment [50,51]. The dominant radiolarians show deposition in the outer-ramp to deep basinal settings [52]. The MF-1 is deposited in deep basinal setting of the ramp depositional environment (Figure 6).

4.2.2. Radiolarians-Rich Wackestone Microfacies (MF-2)

Description

The Radiolarians-rich wackestone microfacies (MF-2) displays medium-bedded, dark grey to light grey limestone. The petrographic studies indicate the presence of grains and matrix. The radiolarian, calcisphere, and planktonic foraminiferal grains are present. The micrite is a fine bioclastic matrix that is ferruginous. (Figure 5).

Interpretation

The MF-2 has similar type of fauna to that of MF-1 but different texture. The radiolarians-rich deposit is found in deeper outer to basinal environments [53]. The presence of radiolarians and planktonic foraminifera together indicates a deeper energy condition. [54]. Hence, the MF-2 is deposited in the deeper basinal setting (Figure 6).

4.2.3. Planktonic Foraminiferal Wacke-Packestone Microfacies (MF-3)

Description

The Planktonic foraminiferal wacke-packestone microfacies (MF-3) shows thin to medium bedded grey to light grey color limestone. Planktonic foraminiferal allochems, mollusk bioclasts, and micritic matrix are the main components of this microfacies. The planktonic genera's includes the *Dicarinella, Macroglobigerinelloides,* Marginotruncana, *Muricohedbergella, Rotalipora, Ticinella* and *Whiteinella*. The other bioclasts include the mollusks, pelagic bivalves and some are undifferentiated. The radiolarians and calcisphere are in minor amount. This microfacies contains pyrite in the form of lenses and along the stylolites. The foraminiferal grains are concentrated at some places showing the tempestites nature (Figure 5).

Interpretation

The combined occurrences of the planktonic foraminifera, bioclasts of mollusks and radiolarians represents that the energy condition was low [50]. The concomitant of the radiolarians and planktonic foraminifera shows the deposition in outer ramp [54]. The pyrite rich matrix is present in most of the microfacies but here it is in more concentration, which is related to the OAEs of the cretaceous time [55,56]. Its presence represents the deeper water conditions. The tempestites in the microfacies represent the deposition near the stormy weather wave base (SWWB) [50]. The microfacies is deposited in proximal part of outer ramp setting (Figure 6).

4.2.4. Planktonic Foraminiferal Wackestone Microfacies (MF-4)

Description

The Planktonic foraminiferal wackestone microfacies (MF-4) in outcrop shows medium bedded grey limestone. The MF-4's major components are micritic matrix and allochems. Bioclasts are present and undifferentiated, while also belonging to bivalves, ostracodes, and echinoderms. The major planktonic foraminifera in MF-4 include *Dicarinella, Heterohelix, Macrglobigerinelloides Marginotruncana, Muricohedbergella,* and *Whiteinella* (Figure 5).

Interpretation

There is a lot of lime mud in the MF-4. Lime mud's presence suggests low-energy conditions beneath the fair-weather wave base. [50]. The presence of bioclasts, as well as planktonic foraminifera and lime mud, support low-energy conditions [57]. Planktonic foraminifers show sediment deposition in an outer-ramp setting [30]. The micorfacies suggest deposition in the proximal outer ramp environment (Figure 6).

4.2.5. Planktonic Foraminiferal Packestone Microfacies (MF-5)

Description

The Planktonic foraminiferal packestone microfacies (MF-5) is identified by the presence of a micritic matrix, foraminiferal grains, and pyrite. The micritic matrix is fine bioclastic matrix. *Dicarinella, Heterohelix, Marginotruncana, and Muricohedbergella, and Whiteinella* are the dominant foraminifera present in MF-5. The foraminiferal chambers are filled with pyrite. The micritic matrix contains a significant amount of brown pyrite. The pyritization also occurred along the stylolites. Additionally, there are radiolarians, a small number of calcispheres, and sponge spicules. (Figure 5).

Interpretation

The pyrite is in the form of matrix and suture boundaries is the dominant feature of this microfacies. The lack of neritic fauna and dominance of the pyritic matrix shows deposition in outer ramp setting [43]. The coexistence of radiolarians and planktonic foraminifera supports deposition in outer ramp environments [54]. In the light of the above-mentioned biota and other features, distal outer ramp environment is suggested for this microfacies (Figure 6).

4.2.6. Depositional Environment

The geoscientists need to understand the depositional environment since it determines the architecture, heterogeneity, and ultimately the quality of any reservoirs [58–61]. The Goru Formation is the study area and is comprised of pelagic thin bedded carbonates. Based on petrographic studies, the texture is wackestone, wacke-packestone, and packestone. Allochems and micritic matrix are the most important constituents. The grains are of radiolarians, calcispheres, planktonic foraminifera, whilst others are bioclasts. The majority of the bioclasts are undifferentiated, although some are planktonic foraminifera and mollusks (bivalves), ostracodes and echinoderms. The micritic matrix is fine and mud dominated. The limestone's mud-dominated texture indicates that the mud was deposited from a suspended load in a low-energy environment [62]. A deep marine outer ramp energy setting below the storm wave base is suggested by the presence of planktonic foraminifera in the limestone unit [30]. The other grains such as radiolarians and calcisphere indicate the deeper water conditions [54]. Radiolarians are a key paleoenvironmental indicator that accumulate on the seabed because of suspended load, but they may also be found in the outer ramp setting [63]. The high content of radiolarians, planktonic foraminifera together with pelagic lime mud show outer ramp setting. Based on the microfacies details, the sediments of the Goru Formation are pelagic in nature and were deposited in the distal middle ramp via distal outer ramp to deep basinal setting in low-energy conditions (Figure 6). The ramp depositional environment was suggested for these carbonates because there is nothing about the deposition of resedimented deposits, i.e., turbidities, and also there is nothing about the grain stone belts and reefal offshore facies. All this evidence suggests a low-energy gradient slope that is the feature of ramp rather than a shelf [64].

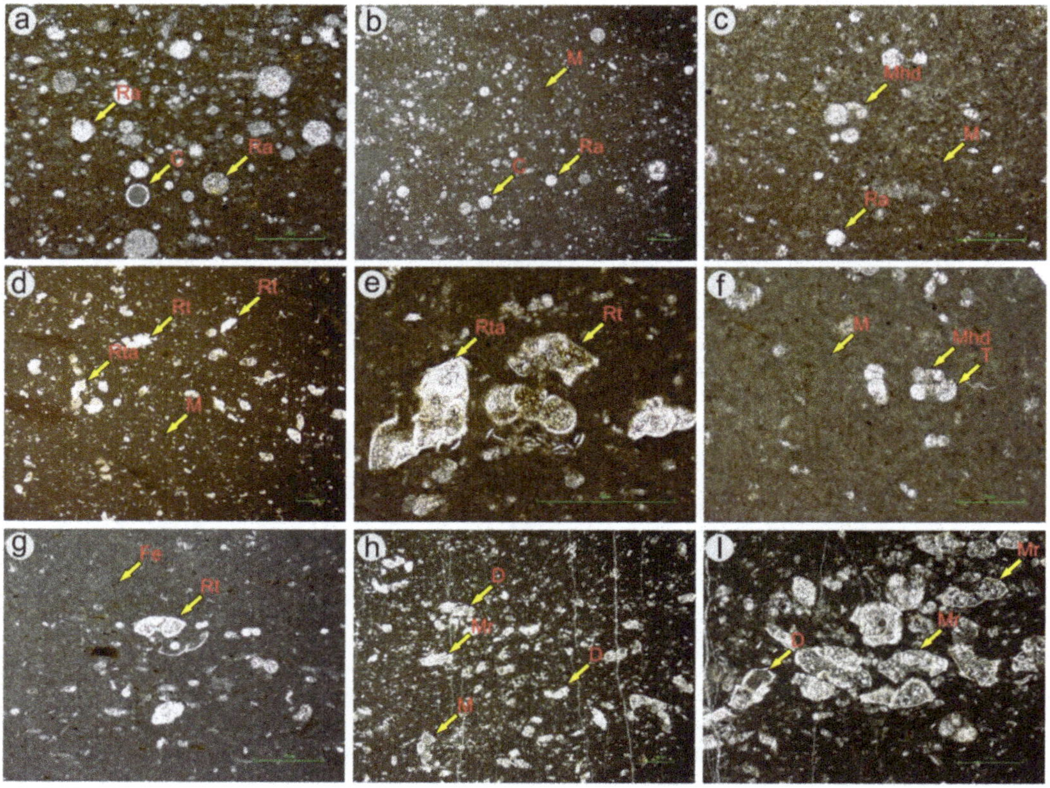

Figure 5. (**a**) Photomicrograph of the Radiolarians rich wacke-packestone microfacies (MF-1) showing radiolarians (Ra) and Calcisphere (C); (**b**) Photomicrograph of the MF-1 (zoomed view) showing matrix (M), radiolarians (Ra) and Calcisphere (C); (**c**) Photomicrograph of the Radiolarians rich wackestone microfacies (MF-2) showing radiolarians (Ra), matrix (M) and *muricohedbergella* (Mhd); (**d**) Photomicrograph of the Planktonic foraminiferal wacke-packestone microfacies (MF-3) showing *rotalipora* sp. (Rt), *rotalipora appenninica* (Rta), and matrix (M); (**e**) Photomicrograph of the MF-3 (zoomed view) showing *Rotalipora appenninica* (Rta), and *rotalipora* sp. (Rt); (**f**) Photomicrograph of the Planktonic foraminiferal wackestone microfacies (MF-4) having *muricohedbergella* (Mhd), *Ticinella* (T) and matrix (M); (**g**) Photomicrograph of the MF-4 (zoomed view) having iron filaments (Fe) and *rotalipora* (Rt); (**h**) Photomicrograph of the Planktonic Foraminiferal Packestone Microfacies (MF-5) showing *marginotruncana* (M) and *dicarinella* (D); (**i**) Photomicrograph of the MF-5 (zoomed view) having *dicarinella* (D); and *marginotruncana* (M).

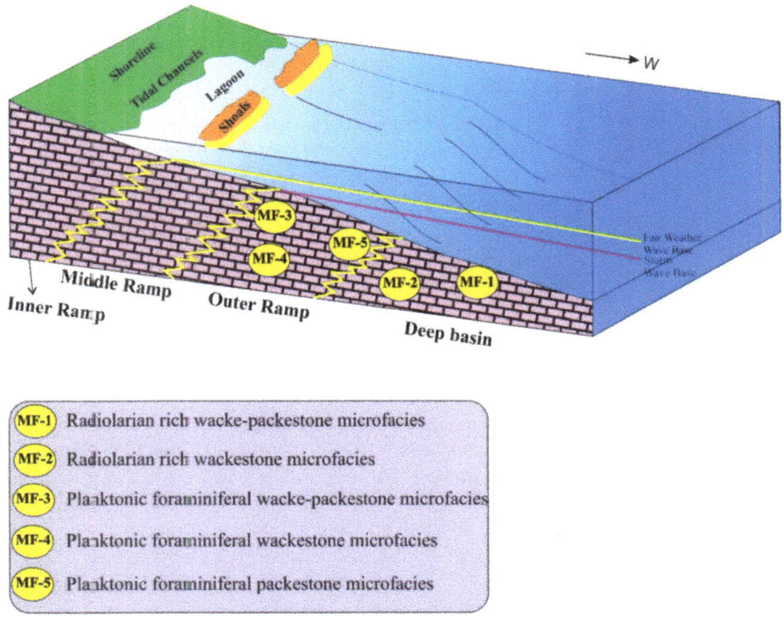

Figure 6. Proposed depositional model of the Goru Formation at Chutair Section, Sulaiman Range.

4.3. Sequence Stratigraphy

The microfacies studies along with bio-stratigraphic investigations of the pelagic carbonates of the Goru Formation are done, which indicates the outer ramp to deep basinal environment of deposition for the pelagic carbonates of Goru Formation in Albian to Turonian time (Figure 4). It is essential to classify depositional sequences according to their cyclicity orders because tectonics controls the eustatic rise and fall that might occur throughout a series of time periods [65]. Each sequence is the result of a specific tectonic or eustatic cycle order. The stratigraphic cycle is usually demonstrated in the following four orders of million years' time span [66]. That is, the first order (greater than 50 Ma), second order (3–50 Ma), third order (0.5–3 Ma), and fourth order (0.1–0.5 Ma) order of cycles [65,67]. The current bio-stratigraphic results show an Albian to Turonian age (113–89.8 Ma) for the deposition of Goru Formation in the study area (Figure 3). This represents an overall time span of 23.2 Ma for the deposition of Goru Formation. This time indicates that the Goru Formation is deposited in Second (2nd) order cycle. For the third (3rd) order cyclicity, the Transgressive-Regressive (T-R) sequence model [33] is used. The T-R sequence model employs the unconformable portion of the boundary as sub-aerial unconformity over the basin margin, and the correlative conformity as the maximum regressive surface (MRS) farther seaward. This model provides a special technique of classifying strata into sequences. T-R sequences are classified as transgressive systems tract (TST) or regressive systems tract (RST) based on Maximum Flooding Surfaces (MFS) (Figure 7). Six depositional sequences are identified that represent system tracts [68,69]. Two types of system tracts that is TST and RST are identified by using the T-R sequence model in a measured section of the Goru Formation in the Chutair Section.

4.3.1. Depositional Sequences

Depending on the sequence model used, depositional sequence corresponds to the depositional product of a whole cycle of base-level changes or shoreline movements [68]. The consequence of the interactions between Eustasy, climate, and tectonic, the depositional sequence constitutes a full cycle of deposition [65,70]. In the Goru Formation, the identified sequences contain total of eleven system tracts, which are the following.

4.3.2. System Tracts

Transgressive system tract (TST)

Six transgressive system tracts are identified comprising of the deeper pelagic carbonates of the Goru Formation. All the transgressive system tracts (TST1-TST6) represent a retro gradational stacking pattern that is the shallower microfacies are overlain by deeper microfacies (Figure 7). The TST1 and TST2 contains the deep basinal microfacies that are overlain by the proximal outer ramp microfacies. The TST3-TST5 are represented by the distal outer ramp microfacies that are overlain by the proximal outer ramp microfacies. The TST6 shows distal outer microfacies that is present at the top of the Formation and also overlain by the proximal outer ramp microfacies. Stratigraphically, the TST1, TST2, and TST3 are present in the Albian time. The TST4 in late Cenomanian to Early Turonian and TST5, TST6 in the Late Turonian time (Figure 7).

Regressive System tract (RST)

Five regressive system tracts are identified comprising of outer ramp carbonates of the Goru Formation. The regressive system tracts are characterized by the progradational facies pattern that is the deeper facies are overlain by the shallower facies (Figure 7). The RST1 is represented by the proximal outer ramp microfacies that is underlain and overlain by of deep basinal microfacies. The RST2 contains the same microfacies as that of RST1 that is the proximal outer ramp microfacies. The RST2 is underlain by deeper basinal microfacies and overlain by the distal outer ramp microfacies. The RST3, RST4, and RST5 is represented by proximal outer ramp microfacies and overlain by distal outer ramp microfacies. Similarly, the bio stratigraphic position of the RST1, and RST2 are present in Albian time. The RST3 in late Cenomanian to Early Turonian time and RST4, RST5 in Middle to Late Turonian time (Figure 7).

4.3.3. Comparison of Local with Global Sea Level Curve

The cretaceous is defined by the long- and short-term sea level oscillations [71]. The overall trend of sea level in the Cretaceous is characterized by a fall in the Cenomanian, a maximum rise in the early Turonian, a prolonged fall from the middle Turonian to the Santonian, a rise in the late Santonian to the early Campanian, and then a subsequent fall in the Middle Campanian that lasts until the Danian [72]. Long-term sea-level variations can be connected to paleoclimate, for example, the boundary between the late Cenomanian and early Turonian is documented as a worldwide warming event of OAE2 defined by positive $\delta^{13}C$ excursion, high organic burial and global sea level rise [4,56,73]. The Cretaceous' maximum sea-level increase occurred during the Earliest Turonian (93.5), and is believed to have been 240–250 m. above present-day mean sea level (or 180–190 m without the existing ice cap, which is anticipated to store water equal to another 60 m of sea-level rise) [71]. Therefore, the sea level curve of the Goru Formation is constructed on the basis of microfacies types under the bio stratigraphic framework. This local sea level curve of the formation is compared with the global curve of Haq et al., [34] to infer about the global and local tectonic on the Goru Formation deposition. The Long-term sea-level variations in Goru Formation have continued to deposit a second-order composite transgressive systems tract, which is comparable with Haq et al.'s, [34] long-term sea-level curve (Figure 7). The short-term 3rd order sea level fluctuations show six episodes of rise and fall, whilst the Haq et al., [34] curve shows almost double to these conditions (Figure 7). The difference in the short-term pattern of sea-level variation in the study area from the global short-term sea-level curve at the time of deposition is attributed to the local tectonics in the study area.

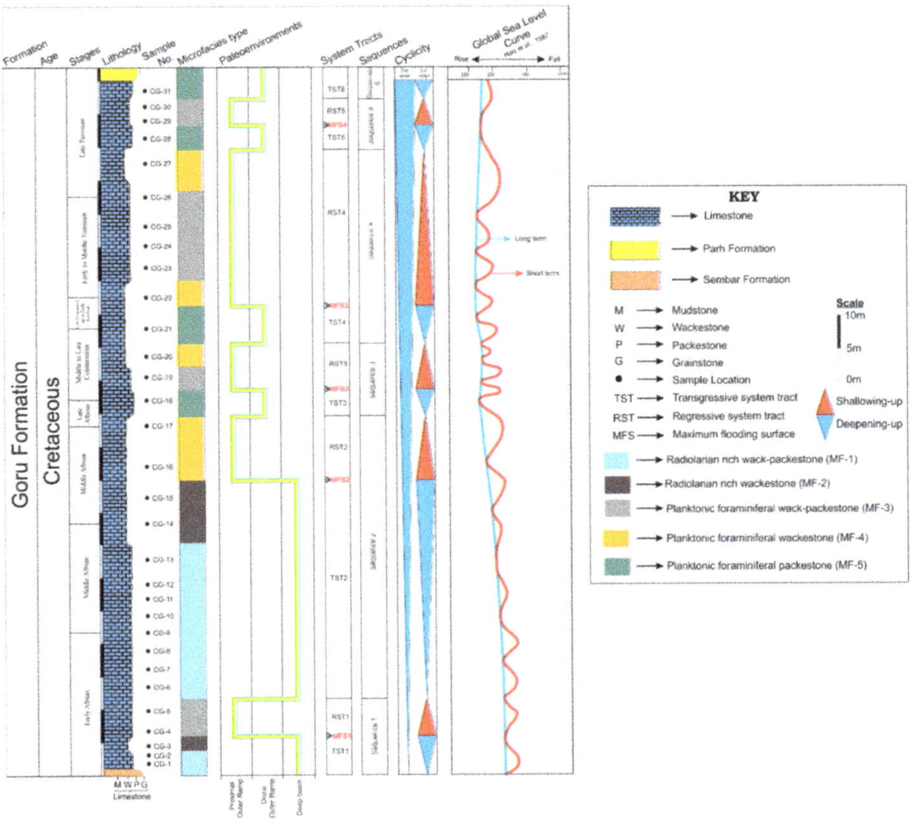

Figure 7. Showing the vertical distribution of microfacies, paleoenvironment, vertical distribution of sequences and system tracts, global sea level curve of the Goru Formation at Chutair Section, Sulaiman Range, Pakistan.

5. Conclusions

The Cretaceous Goru Formation in the study area is comprised of thin to medium bedded, greenish grey, whitish and dark grey color pelagic carbonates. Based on planktonic foraminiferal biostratigraphic studies, eight biozones are established: *Muricohedbergella planispira* Interval Zone; *Ticinella primula* Interval Zone; *Biticinella breggiensis* Interval Zone; *Rotalipora appenninica* Interval Zone; *Rotalipora cushmani* Total Range Zone; *Whiteinella archeocretacea* Partial Range Zone; *Helvetoglobotruncana helvetica* Total Range Zone; and *Marginotruncana sigali* Partial Range Zone. Based on these biozones, an Albian–Turonian (113–89.8 Ma) age is assigned to the Goru Formation in the study area. The detailed petrographic studies revealed five microfacies types, namely: radiolarians rich wack-packestone (deep basinal settings); radiolarians rich wackestone (deep basinal settings); planktonic foraminiferal wack-packestone; planktonic foraminiferal wackestone; and planktonic foraminiferal packestone suggesting deposition of Goru Formation in outer ramp to deep basinal settings. The biostratigraphic and microfacies types are used to establish sequence stratigraphic studies that show transgressive and regressive sea level episodes in the Goru Formation. Six depositional sequences are established that include eleven system tracts including six transgressive and five regressive system tracts ranging from Albian to Turonian age. Two orders of cycles are identified including 2nd and 3rd order. The overall deposition of the Goru Formation takes place at 2nd order of cycle which includes the small-scale 3rd order cycles. The 2nd order local sea level curve of the Goru Formation is matching with the Global Sea level curve whilst the 3rd order curve is not that much matching showing a local tectonic control in the difference in the 3rd order cyclicity.

Author Contributions: S.U., I.U.J. and M.H. proposed the main concept and involved in write up. K.L. and M.M. helped in collection of field data. M.S. and A.A. helped in lab work, preparation of figures and establishing biostratigraphy. U.A. and H.V.T. did provision of relevant literature, review and proofread of the manuscript. All authors have read and agreed to the published version of the manuscript.

Funding: This research work was funded by National Centre of Excellence in Geology, University of Peshawar and PSF-TUBITAK project having project no. PSF-TUBITAK/Earth/KP-UoP (08) by the Pakistan Science Foundation.

Data Availability Statement: Not applicable.

Acknowledgments: The research provided in this paper was conducted as part of the first author's doctoral dissertation. Authors are thankful to the National Centre of Excellence in Geology, University of Peshawar for providing funds for field work. Thanks are extended to Mukhtiar Ghani, Geological Survey Pakistan, for his help during geological field work. The authors acknowledge the PSF-TUBITAK project having project no. PSF-TUBITAK/Earth/KP-UoP (08), which was given to Muhammad Hanif at the National Centre of Excellence in Geology, University of Peshawar by the Pakistan Science Foundation for providing funds for the laboratory analysis.

Conflicts of Interest: The authors declare no conflict of interest.

References

1. Afzal, J.; Williams, M.; Aldridge, R.J. Revised stratigraphy of the lower Cenozoic succession of the Greater Indus Basin in Pakistan. *J. Micropalaeontol.* **2009**, *28*, 7–23. [CrossRef]
2. Kassi, A.M.; Kelling, G.; Kasi, A.K.; Umar, M.; Khan, A.S. Contrasting Late Cretaceous–Palaeocene lithostratigraphic successions across the Bibai Thrust, western Sulaiman Fold–Thrust Belt, Pakistan: Their significance in deciphering the early-collisional history of the NW Indian Plate margin. *J. Asian Earth Sci.* **2009**, *35*, 435–444. [CrossRef]
3. Shah, S. Stratigraphy of Pakistan (memoirs of the geological survey of Pakistan). *Geol. Surv. Pak.* **2009**, *22*, 381.
4. Jenkyns, H. Cretaceous anoxic events: From continents to oceans. *J. Geol. Soc.* **1980**, *137*, 171–188. [CrossRef]
5. Weissert, H.; Erba, E. Volcanism, CO_2 and palaeoclimate: A Late Jurassic–Early Cretaceous carbon and oxygen isotope record. *J. Geol. Soc.* **2004**, *161*, 695–702. [CrossRef]
6. Jones, C.E.; Jenkyns, H.C. Seawater strontium isotopes, oceanic anoxic events, and seafloor hydrothermal activity in the Jurassic and Cretaceous. *Am. J. Sci.* **2001**, *301*, 112–149. [CrossRef]
7. Leckie, R.M.; Bralower, T.J.; Cashman, R. Oceanic anoxic events and plankton evolution: Biotic response to tectonic forcing during the mid-Cretaceous. *Paleoceanography* **2002**, *17*, 13-1–13-29. [CrossRef]
8. Wang, T.; Li, G.; Aitchison, J.C.; Ding, L.; Sheng, J. Evolution of mid-Cretaceous radiolarians in response to oceanic anoxic events in the eastern Tethys (southern Tibet, China). *Palaeogeogr. Palaeoclimatol. Palaeoecol.* **2019**, *536*, 109369. [CrossRef]
9. Premoli Silva, I.; Verga, D. *Practical Manual of Cretaceous Planktonic Foraminifera*; Tipografia Pontefelcino: Perugia, Italy, 2004; p. 283.
10. Leckie, R.M. A paleoceanographic model for the early evolutionary history of planktonic foraminifera. *Palaeogeogr. Palaeoclimatol. Palaeoecol.* **1989**, *73*, 107–138. [CrossRef]
11. Poulsen, C.J.; Barron, E.J.; Arthur, M.A.; Peterson, W.H. Response of the mid-Cretaceous global oceanic circulation to tectonic and CO_2 forcings. *Paleoceanography* **2001**, *16*, 576–592. [CrossRef]
12. Erbacher, J.; Thurow, J. Influence of oceanic anoxic events on the evolution of mid-Cretaceous radiolaria in the North Atlantic and western Tethys. *Mar. Micropaleontol.* **1997**, *30*, 139–158. [CrossRef]
13. Hart, M.B. The evolution and biodiversity of Cretaceous planktonic Foraminiferida. *Geobios* **1999**, *32*, 247–255. [CrossRef]
14. Premoli Silva, I.; Sliter, W. Cretaceous planktonic foraminiferal biostratigraphy and evolutionary trends from the Bottaccione section, Gubbio, Italy. *Palaeontogr. Ital.* **1994**, *82*, 1–89.
15. Shafique, N.A.; Daniels, C. Foraminiferal zonation of Upper Goru Formation–Bawani area, Kirthar Range. *Pak. J. Hydrocarb. Res.* **1990**, *2*, 67–84.
16. Ashraf, U.; Zhang, H.; Anees, A.; Nasir Mangi, H.; Ali, M.; Ullah, Z.; Zhang, X. Application of unconventional seismic attributes and unsupervised machine learning for the identification of fault and fracture network. *Appl. Sci.* **2020**, *10*, 3864. [CrossRef]
17. Ashraf, U.; Zhu, P.; Yasin, Q.; Anees, A.; Imraz, M.; Mangi, H.N.; Shakeel, S. Classification of reservoir facies using well log and 3D seismic attributes for prospect evaluation and field development: A case study of Sawan gas field, Pakistan. *J. Pet. Sci. Eng.* **2019**, *175*, 338–351. [CrossRef]

18. Ashraf, U.; Zhang, H.; Anees, A.; Ali, M.; Zhang, X.; Shakeel Abbasi, S.; Nasir Mangi, H. Controls on reservoir heterogeneity of a shallow-marine reservoir in Sawan Gas Field, SE Pakistan: Implications for reservoir quality prediction using acoustic impedance inversion. *Water* **2020**, *12*, 2972. [CrossRef]
19. Ashraf, U.; Zhang, H.; Anees, A.; Mangi, H.N.; Ali, M.; Zhang, X.; Imraz, M.; Abbasi, S.S.; Abbas, A.; Ullah, Z. A core logging, machine learning and geostatistical modeling interactive approach for subsurface imaging of lenticular geobodies in a clastic depositional system, SE Pakistan. *Nat. Resour. Res.* **2021**, *30*, 2807–2830. [CrossRef]
20. Dar, Q.U.; Pu, R.; Baiyegunhi, C.; Shabeer, G.; Ali, R.I.; Ashraf, U.; Sajid, Z.; Mehmood, M. The impact of diagenesis on the reservoir quality of the early Cretaceous Lower Goru sandstones in the Lower Indus Basin, Pakistan. *J. Pet. Explor. Prod. Technol.* **2022**, *12*, 1437–1452. [CrossRef]
21. Kazmi, A.H.; Abbasi, I.A. *Stratigraphy & Historical Geology of Pakistan*; Department & National Centre of Excellence in Geology: Peshawar, Pakistan, 2008.
22. Williams, M.D. Stratigraphy of the Lower Indus Basin, West Pakistan. In Proceedings of the 5th World Petroleum Congress, New York, NY, USA, 1–5 June 1959.
23. Shafique, N.A. *Spatial Biostratigraphy of NW Pakistan*; Miami University: Oxford, OH, USA, 2001.
24. Smewing, J.D.; Warburton, J.; Daley, T.; Copestake, P.; Ul-Haq, N. Sequence stratigraphy of the southern Kirthar fold belt and middle Indus basin, Pakistan. *Geol. Soc. Lond. Spec. Publ.* **2002**, *195*, 273–299. [CrossRef]
25. Haq, S.S.; Davis, D.M. Oblique convergence and the lobate mountain belts of western Pakistan. *Geology* **1997**, *25*, 23–26. [CrossRef]
26. Jadoon, I.A.; Lawrence, R.D.; Lillie, R.J. Seismic data, geometry, evolution, and shortening in the active Sulaiman fold-and-thrust belt of Pakistan, southwest of the Himalayas. *AAPG Bull.* **1994**, *78*, 758–774.
27. Lawrence, R.; Yeats, R.; Khan, S.; Farah, A.; DeJong, K. Thrust and strike slip fault interaction along the Chaman transform zone, Pakistan. *Geol. Soc. Lond. Spec. Publ.* **1981**, *9*, 363–370. [CrossRef]
28. Riaz, M.S.; Bin, S.; Naeem, S.; Kai, W.; Xie, Z.; Gilani, S.M.M.; Ashraf, U. Over 100 years of faults interaction, stress accumulation, and creeping implications, on Chaman Fault System, Pakistan. *Int. J. Earth Sci.* **2019**, *108*, 1351–1359. [CrossRef]
29. Raza, S.M.; Cheema, I.U.; Downs, W.R.; Rajpar, A.R.; Ward, S.C. Miocene stratigraphy and mammal fauna from the Sulaiman Range, Southwestern Himalayas, Pakistan. *Palaeogeogr. Palaeoclimatol. Palaeoecol.* **2002**, *186*, 185–197. [CrossRef]
30. Flügel, E. *Microfacies of Carbonate Rocks: Analysis, Interpretation and Application*; Springer: Berlin/Heidelberg, Germany, 2013.
31. Postuma, J.A. *Manual of Planktonic Foraminifera*; Elsevier: Amsterdam, The Netherlands, 1971.
32. Sliter, W.V. Biostratigraphic zonation for Cretaceous planktonic foraminifers examined in thin section. *J. Foraminifer. Res.* **1989**, *19*, 1–19. [CrossRef]
33. Embry, A.; Johannessen, E. T–R sequence stratigraphy, facies analysis and reservoir distribution in the uppermost Triassic–Lower Jurassic succession, western Sverdrup Basin, Arctic Canada. In *Norwegian Petroleum Society Special Publications*; Elsevier: Amsterdam, The Netherlands, 1993; Volume 2, pp. 121–146.
34. Haq, B.U.; Hardenbol, J.; Vail, P.R. Chronology of fluctuating sea levels since the Triassic. *Science* **1987**, *235*, 1156–1167. [CrossRef] [PubMed]
35. Amédro, F.; Matrion, B.; Magniez-Jannin, F.; Touch, R. La limite Albien inférieur-Albien moyen dans l'Albien type de l'Aube (France): Ammonites, foraminifères, séquences. *Rev. Paléobiol.* **2014**, *33*, 159–279.
36. Caron, M.; Bolli, H.; Saunders, J.; Perch-Nielsen, K. Cretaceous planktic foraminifera. *Plankton Stratigr.* **1985**, *1*, 17–86.
37. Petrizzo, M.R.; Gilardoni, S.E. Planktonic foraminiferal biostratigraphy of late Albian-Cenomanian pelagic sequences from the Umbria-Marche basin (central Italy) and the Mazagan Plateau (northeast Atlantic Ocean). *Riv. Ital. Paleontol. E Stratigr.* **2020**, *126*, 865–904.
38. Moullade, M. *Étude Stratigraphique et Micropaléontologique du Crétacé Inférieur de La "Fosse Vocontienne"*; Université de Lyon: Lyon, France, 1966; Volume 15.
39. Kennedy, J.W.; Gale, A.S.; Huber, B.T.; Petrizzo, M.R.; Bown, P.; Jenkyns, H.C. The Global Boundary Stratotype Section and Point (GSSP) for the base of the Albian Stage of the Cretaceous, the Col de Pré-Guittard section, Arnayon, Drôme, France. *Epis. J. Int. Geosci.* **2017**, *40*, 177–188. [CrossRef]
40. Khan, S.; Kroon, D.; Wadood, B.; Ahmad, S.; Zhou, X. Marine depositional signatures of the Aptian Oceanic Anoxic Events in the Eastern Tethys, Lower Indus Basin, Pakistan. *Aust. J. Earth Sci.* **2022**, *69*, 251–267. [CrossRef]
41. Brönnimann, P. *Globigerinidae from the Upper Cretaceous (Cenomanian-Maestrichtian of Trinidad, BWI*; Paleontological Research Institution: Ithaca, NY, USA, 1952.
42. Borsetti, A. Foraminiferi planctonici di una serie cretacea dei Dintorni di Piobbico (Prov. di Pesaro). *G. Geol. Ser.* **1962**, *2*, 19–75.
43. Khan, S.; Kroon, D.; Ahmad, S.; Ali, A.; Wadood, B.; Rahman, A. Planktonic foraminiferal biostratigraphy of the Cretaceous strata, Indus Basin, Pakistan, Eastern Tethys: Implications for oceanic anoxic events. *Aust. J. Earth Sci.* **2021**, *68*, 1162–1178. [CrossRef]
44. Bolli, H.M. *Zonation of Cretaceous to Pliocene Marine Sediments Based on Planktonic Foraminifera*; Geologisches Institut der Eidg. Technischen Hochschule und der Universität Zürich: Zürich, Switzerland, 1966.
45. Dalbiez, F. The genus Globotruncana in Tunisia. *Micropaleontology* **1955**, *1*, 161–171. [CrossRef]
46. Barr, F. Cretaceous biostratigraphy and planktonic foraminifera of Libya. *Micropaleontology* **1972**, *18*, 1–46. [CrossRef]
47. Friedrich, O.; Norris, R.D.; Erbacher, J. Evolution of middle to Late Cretaceous oceans—A 55 my record of Earth's temperature and carbon cycle. *Geology* **2012**, *40*, 107–110. [CrossRef]

48. Khosla, A.; Verma, O. Paleobiota from the Deccan volcano-sedimentary sequences of India: Paleoenvironments, age and paleobiogeographic implications. *Hist. Biol.* **2015**, *27*, 898–914. [CrossRef]
49. Verma, O. Cretaceous vertebrate fauna of the Cauvery Basin, southern India: Palaeodiversity and palaeobiogeographic implications. *Palaeogeogr. Palaeoclimatol. Palaeoecol.* **2015**, *431*, 53–67. [CrossRef]
50. Flügel, E.; Munnecke, A. *Microfacies of Carbonate Rocks: Analysis, Interpretation and Application*; Springer: Berlin/Heidelberg, Germany, 2010; Volume 976.
51. Ghabeishavi, A.; Vaziri-Moghaddam, H.; Taheri, A.; Taati, F. Microfacies and depositional environment of the Cenomanian of the Bangestan anticline, SW Iran. *J. Asian Earth Sci.* **2010**, *37*, 275–285. [CrossRef]
52. Gawlick, H.-J.; Schlagintweit, F. Berriasian drowning of the Plassen carbonate platform at the type-locality and its bearing on the early Eoalpine orogenic dynamics in the Northern Calcareous Alps (Austria). *Int. J. Earth Sci.* **2006**, *95*, 451–462. [CrossRef]
53. Haas, J.; Götz, A.E.; Pálfy, J. Late Triassic to Early Jurassic palaeogeography and eustatic history in the NW Tethyan realm: New insights from sedimentary and organic facies of the Csővár Basin (Hungary). *Palaeogeogr. Palaeoclimatol. Palaeoecol.* **2010**, *291*, 456–468. [CrossRef]
54. Heldt, M.; Bachmann, M.; Lehmann, J. Microfacies, biostratigraphy, and geochemistry of the hemipelagic Barremian–Aptian in north-central Tunisia: Influence of the OAE 1a on the southern Tethys margin. *Palaeogeogr. Palaeoclimatol. Palaeoecol.* **2008**, *261*, 246–260. [CrossRef]
55. Jenkyns, H.; Gale, A.; Corfield, R. Carbon-and oxygen-isotope stratigraphy of the English Chalk and Italian Scaglia and its palaeoclimatic significance. *Geol. Mag.* **1994**, *131*, 1–34. [CrossRef]
56. Takashima, R.; Nishi, H.; Huber, B.T.; Leckie, R.M. Greenhouse world and the Mesozoic ocean. *Oceanography* **2006**, *19*, 82–92. [CrossRef]
57. Wilson, J. *Carbonate Facies in Geologic History*; Springer: New York, NY, USA, 1975; p. 471.
58. Anees, A.; Zhang, H.; Ashraf, U.; Wang, R.; Liu, K.; Abbas, A.; Ullah, Z.; Zhang, X.; Duan, L.; Liu, F. Sedimentary facies controls for reservoir quality prediction of lower shihezi member-1 of the Hangjinqi area, Ordos Basin. *Minerals* **2022**, *12*, 126. [CrossRef]
59. Anees, A.; Zhang, H.; Ashraf, U.; Wang, R.; Liu, K.; Mangi, H.; Jiang, R.; Zhang, X.; Liu, Q.; Tan, S. Identification of Favorable Zones of Gas Accumulation Via Fault Distribution and Sedimentary Facies: Insights from Hangjinqi Area, Northern Ordos Basin. Front. *Earth Sci.* **2022**, *9*, 822670. [CrossRef]
60. Thanh, H.V.; Sugai, Y. Integrated modelling framework for enhancement history matching in fluvial channel sandstone reservoirs. *Upstream Oil Gas Technol.* **2021**, *6*, 100027. [CrossRef]
61. Vo Thanh, H.; Lee, K.-K. 3D geo-cellular modeling for Oligocene reservoirs: A marginal field in offshore Vietnam. *J. Pet. Explor. Prod. Technol.* **2022**, *12*, 1–19. [CrossRef]
62. Tucker, M.E.; Wright, V.P. *Carbonate Sedimentology*; Blackwells: Oxford, UK, 1990; p. 481.
63. Omidvar, M.; Safari, A.; Vaziri-Moghaddam, H.; Ghalavand, H. Facies analysis and paleoenvironmental reconstruction of Upper Cretaceous sequences in the eastern Para-Tethys Basin, NW Iran. *Geol. Acta* **2014**, *14*, 363–384.
64. Adabi, M.H.; Zohdi, A.; Ghabeishavi, A.; Amiri-Bakhtiyar, H. Applications of nummulitids and other larger benthic foraminifera in depositional environment and sequence stratigraphy: An example from the Eocene deposits in Zagros Basin, SW Iran. *Facies* **2008**, *54*, 499–512. [CrossRef]
65. Emery, D.; Myers, K. *Sequence Stratigraphy*; John Wiley & Sons: Hoboken, NJ, USA, 2009; p. 264.
66. Duval, B.; Cramez, C.; Vail, P. Types and hierarchy of stratigraphic cycles. *Seq. Stratigr. Eur. Basins* **1992**, 44.
67. Vail, P.R.; Audemard, F.; Bowman, S.A.; Eisner, P.N.; Perez-Cruz, C. The stratigraphic signatures of tectonics, eustasy and sedimentology—An overview. In *Cycles and Events in Stratigraphy*; Einsele, G., Ricken, W., Seilacher, A., Eds.; Springer: Berlin/Heidelberg, Germany, 1991; pp. 617–659.
68. Catuneanu, O. *Principles of Sequence Stratigraphy*; Elsevier: Amsterdam, The Netherlands, 2006.
69. Posamentier, H.; Jervey, M.T.; Vail, P. Eustatic controls on clastic deposition I—Conceptual framework. *SEPM Spec. Publ.* **1988**, *42*, 125–154.
70. Miall, A.D. *The Geology of Stratigraphic Sequences*; Springer Science & Business Media: New York, NY, USA, 2010.
71. Haq, B.U. Cretaceous eustasy revisited. *Glob. Planet. Change* **2014**, *113*, 44–58. [CrossRef]
72. Hallam, A. *Phanerozoic Sea-Level Changes*; Columbia University Press: New York, NY, USA, 1992.
73. Arthur, M.; Brumsack, H.-J.; Jenkyns, H.; Schlanger, S. Stratigraphy, geochemistry, and paleoceanography of organic carbon-rich Cretaceous sequences. In *Cretaceous Resources, Events and Rhythms*; Springer: Berlin/Heidelberg, Germany, 1990; pp. 75–119.

minerals

Article

Geochemical Study of the Iron Age Settlement Occupational Layer and the Early Roman Time Agricultural Layer at Voorthuizen, The Netherlands

Olga Druzhinina [1,2,3,*], Laura Gedminienė [4] and Kasper van den Berghe [3]

[1] Faculty of Geography, Herzen State Pedagogical University of Russia, 191186 Saint-Petersburg, Russia
[2] Shirshov Institute of Oceanology, Russian Academy of Sciences, 117997 Moscow, Russia
[3] FindX Research Center, 8031 VK Zwolle, The Netherlands; berghekj@hotmail.com
[4] Nature Research Centre, Institute of Geology and Geography, 08412 Vilnius, Lithuania; laura.gedminiene@gamtc.lt
* Correspondence: olga.alex.druzhinina@gmail.com

Abstract: The application of geochemical analysis in archaeology provides a better understanding of ancient human activities. This paper presents the results of geochemical multi-element, LOI, MS, and geochronological analyses of a cultural layer at the Voorthuizen (The Netherlands) archaeological site. The study has revealed a difference in the geochemical composition of an occupational surface in the Iron Age settlement and an early Roman Time agricultural horizon. The former is enriched in Ba, Sr, Rb, Th, Cl, and Mg, while the latter displays elevated P, Pb, Ni, Mn, and V values. The study has provided a deeper insight into the agricultural techniques applied at the Voorthuizen site and on the so-called Celtic Fields, known as ancient field systems dating from the same period and widespread throughout north-west Europe. It seems that household waste was not used as a fertilizer at Voorthuizen, while the application of manure is characteristic of Celtic Fields. However, phosphorous values in the Voorthuizen agricultural horizon are comparable to those in the Celtic Fields, suggesting similar sources of P in both cases. Elevated Si and "mobile" Fe, Mn, V, Pb, As, and Sn, along with higher MS measurements, are indicative of the use of extra mineral matter for the fertilization of the Voorthuizen ancient arable field.

Keywords: geochemistry; archaeology; ancient human activities; pre- and protohistoric agriculture; settlement occupational layer; Celtic Fields; Iron Age; Roman Time; phosphorous value; Veluwe

Citation: Druzhinina, O.; Gedminienė, L.; van den Berghe, K. Geochemical Study of the Iron Age Settlement Occupational Layer and the Early Roman Time Agricultural Layer at Voorthuizen, The Netherlands. *Minerals* **2022**, *12*, 373. https://doi.org/10.3390/min12030373

Academic Editor: Massimo D'Antonio

Received: 6 February 2022
Accepted: 16 March 2022
Published: 18 March 2022

1. Introduction

The application of geochemical analysis to archaeology provides a deeper understanding of the functional meaning of the different zones of settlements and within structures, the identification of economic activities, including the types of an anthropogenic impact on the environment, the recognition of phases in the existence of settlements, as well as some other practical problems approached by archaeologists [1–6].

In recent decades, various geochemical methods have been developed to study traces of human presence, and a number of reliable anthropogenic geochemical indicators have been identified, allowing geochemistry to take its rightful place in scientific research [3,6–10]. However, there is a gap between the use of geochemical analysis in scientific projects and its use during rescue archaeological excavations, at least when it comes to rescue excavations in The Netherlands. Financial and time pressures certainly limit the choice of research methods, but the main reason why geochemistry is seldom used in Dutch rescue archaeology is that field archaeologists are often unaware of the advantages and potential of the method [5]. This, in turn, is due to the current lack of publications presenting regional data and the results of the use of geochemistry in rescue archaeological investigations.

This paper is part of a scientific study undertaken by a field team during rescue excavations of an Iron Age settlement at Voorthuizen in the central Netherlands (Figure 1).

In addition to geochemistry, the entire study comprised geochronological, phytolith, pollen, and non-pollen palynomorph (NPP) analyses with the aim to obtain more information on ancient economic activities at the site and their impact on the surrounding landscape. While this study is still in progress [11], the present contribution discusses the selective results of the geochemical investigation of Voorthuizen anthropogenic layers, such as the occupational surface of the settlement and the ancient agricultural horizon.

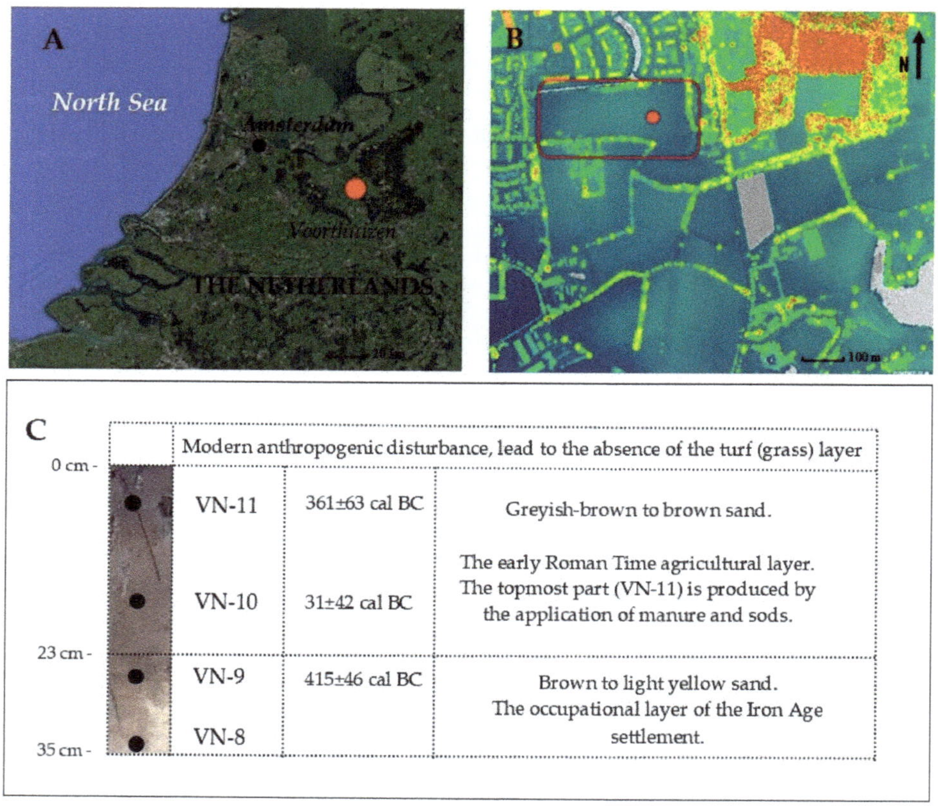

Figure 1. (**A**) Location of Voorthuizen in The Netherlands. (**B**) Northern part of the excavated area in Voorthuizen and the location of the profile, where samples VN-8–VN-11 were taken. (**C**) Profile of the Voorthuizen cultural layer and sampling sites.

2. Study Area and Archaeological Setting

Voorthuizen is situated on one of the coversand ridges in the transition zone of the ice-pushed Veluwe massif to the Geldese Vallei formed by ice tongue erosion. The modern vegetation around the Voorthuizen site is dominated by pine and mixed forest, heather land, and meadows on the various types of podzol soils. The study area has a temperate maritime climate. The annual average temperatures range between 5 °C and 14.4 °C and the average annual precipitation, evenly distributed over the year, varies from 750–850 mm.

2018–2019 excavations in Voorthuizen have revealed the remains of several settlements (or several phases in habitation) over an area of 3 ha [12]. The earliest date obtained for the area goes back to the (late) Bronze Age (1100–750 BC), though the majority of archaeological finds represent the Iron Age–Roman Time (750 BC–50 AD) periods of habitation.

The geochemical results reported in this paper were obtained for samples taken from soil layers at various depths in the northern part of the excavation terrain. The remains of the settlement found here are dated on the basis of archaeological finds and AMS ^{14}C dates.

The archaeological material is composed of ceramics. The ceramic finds consist mainly of hand-shaped pottery, which, based on technical aspects, such as fabric, leaning, and the diversity of shapes, can be dated at the Middle–Late Iron Age. Single fragments of wheel-thrown pottery from Roman Time are present, too [12].

The location, where samples are taken, is situated at the northern edge zone of the ancient settlement discovered. The edge zone as such contains few finds and features. Close to the sampling point, some structures used for storing the crops harvest were present. The structures can be split in so-called spiekers, built as four-posted square structures, and a large multi-posted, almost hangar-like, storage for crops or grain, a so-called horreum. Such types of structures are common on the margins of the settlements. The soil profile studied during the excavation visually consists of a brown, nearly homogeneous ~35 cm thick sand horizon overlying light yellow coversand (Figure 1).

The number of samples taken and the choice of research methods were dependent on the project budget. Samples were taken from the brown layer and the transitional layer as deep as the coversand mother material, where remains of archaeological features (mostly postholes) were found. The study has revealed three interlayers in the brown horizon dated at Iron Age–early Roman Time.

According to geoarchaeological interpretation of the soil built-up at the Voorthuizen site, the layers in the lower part of the profile were the living surface (sample VN-9) of the Iron Age settlement and the underlying horizon (sample VN-8) with traces of former structures in it. During the next habitation phase, in Roman Time, the settlement shifted several hundred metres to the south, while the study territory was involved in agricultural activities (samples VN-10, VN-11). The Roman Time agricultural layer is an early historical example of man-made arable horizon produced by the application of manure and sods, which were cut in the surrounding and brought to the field.

3. Methods

3.1. Radiocarbon Analysis

Terrestrial soil samples from anthropogenic layers were analyzed by AMS ^{14}C survey after acid-base-acid pre-treatment. The analysis was performed at the Centre for Physical Sciences and Technology, Mass Spectrometry Laboratory, Vilnius, Lithuania. The calibrated data (calAD) were obtained with OxCal v. 4.4.2. [13].

3.2. Geochemical Analysis

One sample from each sublayer was taken for geochemical analysis. The samples were prepared at the Open Access Centre of the Nature Research Centre in Vilnius using the standard procedures described previously in [14]. Dried samples were powdered and 20 mm tablets pressed previously were added to each specimen Licowax binder (Fluxana). The tablets were analyzed on EDXRF equipment Spectro Xepos using the Turboquant for a pressed pellet calibration procedure [6] elaborated by the manufacturers at Klaipeda University, Marine Research Institute. The equipment was used to identify 31 elements (Si, Al, Fe, K, Na, Ti, Ba, Ca, Mg, Mn, P, Zr, Br, Cr, Cu, Cl, Y, S, Rb, Sr, Zn, As, Co, Ga, Nb, Ni, Pb, Sn, Th, V, and Tl) in each subsample.

Loss-on-ignition analysis was carried out on four samples. Organic content (LOI 550 °C) was measured using the basic loss-on-ignition methodology [15] at the Nature Research Centre (Vilnius) and heating the sample at 550 °C and carbonates (LOI 900 °C) at 900 °C, respectively.

Since 2007, quality control has been performed by participation in the "International Soil-Analytical Exchange" (ISE) programme organised by Wageningen University [16]. More than 50 ISE reference samples and other certified reference materials were used for the recalibration of geochemical results.

Energy-dispersive X-ray fluorescence (EDXRF) equipment Spectro Xepos HE (Kleve, Germany) uses the TurboQuant (TQ) II calibration method elaborated by the manufacturers for samples with various matrices. The TQ method combines different procedures: calculation of the mass attenuation coefficient, using the extended Compton model, and final calibration based on fundamental parameters method. Calibration for major, minor and trace elements performed by measuring a series of international geological and mineral reference materials (e.g., GSR-01–GSR-10, AC-E, AL-I, AN-G, BE-N, BX-N, DR-N, DT-N, FK-N, GA, GH, GS.N, IF-G, MA-N, MICA-Fe, MICA-Mg, PM-S, UB-N, WS-E, ZW-C, SARM-18–SARM-20, NIST-2709, NIST-2711, NIST-2780–NIST-282, BCR-142–BCR-146, TB, GnA, SY-3, NOD-P1, GXR-03, G303-06, NCSDC-86306, NCSDC-86315, NIST-120C, GSS-5, AMIS-84, AMIS-0118, AMIS-0122, AMIS-0185, and AMIS-0191).

The Spearman correlation coefficients between the content of each element and other variables (the amount of organic matter, carbonates, mineral matter, MS, and geochemical elements) were calculated for each soil sample using StatSoft Statistica 8 MR-3. Significance levels (P values) for the correlations were calculated as well.

3.3. Magnetic Susceptibility (MS)

Four samples were analyzed at the Palaeomagnetic Laboratory of the Nature Research Centre in Vilnius, Lithuania, using the MFK1-B kappa bridge (AGICO) static specimen method with a manual holder facility. The data obtained were analyzed using the SAFYR software, and MS values were expressed in SI units (10^{-8} m^3 kg^{-1}).

4. Results

4.1. Radiocarbon Analysis

All the samples contained a sufficient amount of carbon for accurate measurement and produced a sufficient ion beam during AMS ^{14}C measurement. The δ ^{13}C values are within the normal range for organic samples, confirming the significant reliability of the results.

AMS ^{14}C measurements yielded the following results (Table 1):

Table 1. Results of the geochronological study of the Voorthuizen archaeological site.

Sample Designation	Lab. Code	Radiocarbon Age, BP	Calibrated Date (cal BC)	Median, cal BC	Av, cal BC
VN-11	FTMC-BJ24-1	2278 ± 26	68.3% probability 395–359 calBC (48.1%) 276–261 calBC (11.9%) 244–234 calBC (8.3%) 95.4% probability 400–351 calBC (54.2%) 291–209 calBC (41.2%)	361 ± 63	323
VN-10	FTMC-BJ24-2	2041 ± 26	68.3% probability 89–81 calBC (5.4%) 53 calBC-10 calAD (62.8%) 95.4% probability 150–135 calBC (1.9%) 114 calBC- 29 calAD (91.3%) 43–59 calAD (2.2%)	31 ± 42	34
VN-9	FTMC-BJ24-3	2360 ± 27	68.3% probability 465–437 calBC (18.4%) 421–391 calBC (49.8%) 95.4% probability 516–387 calBC (95.4%)	415 ± 46	434
V-14-sp363	FTMC-LY85-1	2549 ± 27	68.3% probability 794–754 calBC (47.4%) 682–669 calBC (9.6%) 610–593 calBC (11.3%) 95.4% probability 800–747calBC (49.8%) 689–664calBC (12.9%) 644–551calBC (32.7%)	750 ± 80	701

Samples VN-9–VN-11 represent soil profile horizons, while sample V-14-sp363 shows the date obtained for the remains of the structure found in the northern part of excavation.

4.2. Geochemical Analysis

As the differences in the values obtained were considerable, the elements were divided into macro (>100 mg/kg) and micro (<100 mg/kg) types for a better visual comparison (Figure 2). The results of geochemical multi-element analysis are shown below:

	Si	Al	K	Na	Fe	Ti	Ca	Mg	Zr	Ba	P	Mn
VN-11	442,003	13,649	6277	5496	4252	880	719	399	413	238	465	191
VN-10	423,816	17,491	6585	5617	3952	868	759	463	374	253	389	193
VN-9	418,897	19,524	6655	5932	4104	1022	754	503	396	255	299	188
VN-8	420,899	21,457	6888	5611	3779	925	904	571	350	268	149	103

(a)

	Cl	S	Sr	Rb	Cr	Zn	Y	Br	Cu	V	Pb	Ga	Ni	Nb	Th	As	Co	Sn	Tl
VN-11	43.5	66.7	30.7	25.5	12.3	7.88	13.0	8.27	10.0	8.95	7.56	4.91	2.89	3.95	2.51	2.00	0.98	1.75	0.20
VN-10	45.3	95.1	32.2	26.3	18.7	10.5	12.1	10.6	11.4	8.47	5.78	5.06	2.68	3.93	2.51	1.62	0.95	1.54	0.22
VN-9	51.6	89.7	33.0	26.8	16.8	11.2	13.1	11.4	11.9	7.39	5.55	5.25	2.63	4.02	2.58	1.78	1.49	1.75	0.23
VN-8	52.0	45.7	36.3	27.7	16.3	11.9	11.4	9.19	8.95	7.90	5.48	5.00	4.53	3.55	2.51	1.93	1.36	1.30	0.21

(b)

Figure 2. Comparison of element values (mg kg^{-1}) in the soils samples (VN-8–VN-11) of the Voorthuizen archaeological site: (**a**) Macro-elements (Si, Al, K, Na, Fe, Ti, Ca, Mg, Zr, Ba, P, and Mn) values; (**b**) Micro-elements (Cl, S, Sr, Rb, Cr, Zn, Y, Br, Cu, V, Pb, Ga, Ni, Nb, Th, As, Co, Sn, and Tl). Values are expressed in logarithmic scales, for visual expression.

Sample VN-11, the upper part of the soil profile studied, is enriched in Si and Zr, as well as in Fe, P, V, Pb, As, and Sn. Al, K, Na, Ca, Mg, Ba, Cl, Sr, Rb, Cr, Zn, Br, Ga, Th, and Tl display the lowest values in this part of the profile. Sample VN-8, on the contrary, exhibits the highest Al, K, Ca, Mg, Ba, Sr, and Rb values accompanied by low to the lowest Si, Zr, Ti, and Th values. In comparison to this, sample VN-10 shows the highest Mn, S, and Cr values.

The dynamics of individual chemical elements, commonly regarded as indicators of anthropogenic activity, does not show a uniform pattern. For example, P value displays a well-defined trend decreasing towards the lower part of the profile, with the amount of P (149 mg kg^{-1}) in sample VN-8 two to three times as low as that in other samples. The K value increases slightly from sample VN-11 (6277 mg kg^{-1}) to a maximum (6888 mg kg^{-1})

in sample VN-8. A similar trend is observed for N, but the difference is that maximum Na value (5932 mg kg^{-1}) is in sample VN-9.

Based on increasing Zn and Mg dynamics with a maximum of 11.9 mg kg^{-1} and 571 mg kg^{-1}, respectively, in sample VN-8, the Cu and Mn distribution displays a more complicated pattern: elevated values are characteristic of samples VN-10 and VN-9, while the lowest part of the profile exhibits minimum Cu and Mn values (8.95 mg kg^{-1} and 103 mg kg^{-1}, respectively).

Sr, Ba, and Rb share the same profile distribution trend. The values of these elements grow towards the lower part of the profile, reaching a maximum in sample VN-8: Sr (36.3 mg kg^{-1}), Ba (268 mg kg^{-1}), and Rb (27.7 mg kg^{-1}). Th displays a more stable dynamics throughout the profile, except for sample VN-9, where its amount shows a maximum value of 2.58 mg kg^{-1}. A significant increase in the Ca value (904 mg kg^{-1}) in sample VN-8 is remarkable.

The soil-sediments consist of terrigenous material with a small admixture of organic matter in the bottom layer, which increases towards the topmost layer from 3.03% to 6.61%, respectively. LOI 900 °C is very insignificant and was revealed only in VN-8 and VN-10 layers. VN-8 consists 0.32% and VN-10 0.25% of LOI 900 °C (Table 2).

Table 2. Results of the LOI and microcharcoal analyses at the Voorthuizen archaeological site.

	VN-8	VN-9	VN-10	VN-11
Microcharcoal	803	1855	1800	1300
LOI 550 °C, %	3.03	4.82	5.46	6.61
LOI 900 °C, %	0.32	0	0.25	0
LOI residue, %	96.65	95.36	94.29	93.39

The amount of microcharcoal is significant in all samples; the highest concentration is in sample VN-9 (Table 2).

4.3. Magnetic Susceptibility (MS)

The MS values are comparably high, following LOI 550 °C trends. They increase upwards from 1.9 to 5.3 × 10^{-8} m^3 kg^{-1}. The biggest change is fixed between VN-8 and VN-9 samples (Table 3).

Table 3. Results of the MS measurements at the Voorthuizen archaeological site.

	VN-8	VN-9	VN-10	VN-11
MS (×10^{-8} m^3 kg^{-1})	1.9	3.8809	4.43993	5.2832

5. Discussion

The case study of the Voorthuizen archaeological site provides the opportunity to compare the geochemical composition of two important archaeological phenomena: the settlement occupational surface and the ancient agricultural horizon. We can thus understand whether there are chemical elements that better indicate particular anthropogenic activities related to the settlement itself, as well as agricultural activities per se.

Indicators of human activities. The range of the elements that have been reported to be indicative of human occupation consists of two main groups: elements with generally elevated values as compared to their concentrations in natural soils, and those which can also decrease during the transformation of the soil horizon into a cultural layer. The frequently used indicators of anthropogenic activities are phosphorus (P), calcium (Ca), potassium (K), sodium (Na), and magnesium (Mg), as well as trace metals such as manganese (Mn), copper (Cu), lead (Pb), and zinc (Zn) [8,10,17,18]. P is associated with general organic (both human and animal) refuse, including kitchen and residential waste, animal pens, and agricultural activity. Animal bones, shells, and building or agricultural activities with lime application enrich cultural layers in Ca. High Ca concentrations at metal working sites are due to the use of ashes of calcined bones in this process [7]. Ca, K, and Mg may indicate the presence of ash in intense burning zones, and K is also prevalent in plant tissues. Therefore, higher K value indicate the existence of settlements and agricultural activities. While Cu and Pb enrichment have been reported as an indicator of mining, smelting and fuel combustion at a wide range of archaeological sites [10,19], according to [20], the considerable elevation of Pb and Zn was observed in ancient arable soils in Scotland. Pb isotope signatures and analysis of waste streams have shown that the pollution was caused by peat and turf ash (Pb and Zn), and by bird carcasses (Zn). Strontium (Sr), thorium (Th), barium (Ba), and rubidium (Rb) are present in household waste material [10,17,21].

However, the transformation of natural soil into a cultural layer can be accompanied by a decrease in Cu, Zn, Co, Fe, and Mn concentrations [22]. In addition, site effects appear to influence elements such as Ti, Ni, and Fe [10], though these elements are successfully used for functional area interpretation on archaeological sites by [23].

It has been noted that similar associations related to anthropogenic activities, such as P–Mg–Ca–Sr–Ba–Cl–Mn–Zn–Cu [24], have been revealed even in other areas and sedimentary environments. This association is well-defined (maybe, except Mn) in the lower part of the Voorthuizen archaeological site profile (samples VN-8 and VN-9), representing the cultural occupational surface of the Iron Age settlement. Simultaneously, the values of all the above elements besides Mn and P, decrease in the upper—ancient agricultural—part of the profile, while manganese and phosphorus concentrations grow.

Fe and Mn depletion revealed in [22] during the study of Bronze Age and Roman Time cultural layers at three archaeological sites in The Netherlands has also been reported for the Iron Age settlement layer in Voorthuizen. Simultaneously, this layer is enriched in a group of household waste indicators, such as Sr, Ba, Rb, and Th [21]. The latter shows slightly lowered values in the agricultural layer of Roman Time.

Charcoal and bone are essential for geochemical elements, such as Ca, Ba, Cu, Sr, Zn, P, and Pb, in the loading and post depositional retention of soils [10,18] and, thus, may be significant for the formation of soil element concentration patterns. Both the occupational surface and agricultural horizon in Voorthuizen are enriched in microcharcoal, and the relationship between microcharcoal finds and the geochemical composition of the soil layers in this case is expressed through a positive correlation with Cu and S (Table 4).

Table 4. Correlation matrix of chosen (Cu, P, S, MS, and microcharcoal) elements. Marked correlations are significant at $p < 0.05000$, $n = 4$.

	Cu	P	S	MS	Charcoal
Cu	1.0000 p = -				
P	0.4121 p = 0.588	1.0000 p = -			
S	0.9664 p = 0.034	0.5304 p = 0.470	1.0000 p = -		
MS	0.4975 p = 0.502	0.9915 p = 0.009	0.5877 p = 0.412	1.0000 p = -	
Charcoal	0.9895 p = 0.011	0.5239 p = 0.476	0.9892 p = 0.011	0.5964 p = 0.404	1.0000 p = -

-: absence of data.

No significant enhancement of Pb in arable fields, as compared to that in farmyards, was observed by [10]. This pattern is also true for the Voorthuizen archaeological site, though the soil profile still displays a slight increase of Pb upwards with the highest concentration in the Roman Time agricultural layer (7.56 mg/kg). This value is comparable with the values obtained for the Nistelrode Roman Time settlement, which is located on sandy soils in the south-eastern Netherlands (7.66 and 6.22 mg/kg on-site and off-site, respectively [22]). A broader comparison of the geochemical composition of Voorthuizen and Nistelrode shows a relative similarity in the values of major anthropogenic indicators with the exception of Cu and Zn, whose concentrations are remarkably higher at Nistelrode, and Ba and Sr with significantly higher values at Voorthuizen (Table 5). Other factors, in addition to possible differences in lithology, could be to a different range of activities carried out on these two locations and imprinted to the soil: geochemical measurements at Nistelrode are indicative of a dwelling house with its functional zones, while the situation at Voorthuizen is related to the study of the occupational surface close to but outside the structure, which was located at the edge of the settlement and seems to have been used as a crop storage.

Table 5. Major anthropogenic geochemical indicators at Voorthuizen (samples VN 8-11) and Nistelrode (samples NS46) archaeological sites, The Netherlands. Unmarked element values expressed in mg/kg, * marked element oxides values expressed in %.

Element/Site mg/kg, * oxide %	VN-8, XRF	VN-9, XRF	VN-10, XRF	VN-11, XRF	Element	NS46 On-Site, XRF	NS46 Off-Site, XRF	Element	NS46 On-Site, ICP-OES	NS46 Off-Site, ICP-OES
Ca	904	754	759	719	CaO	* 0.06%	* 0.08%	Ca	119.89	82.28
Cu	8.95	11.92	11.40	10.01	Cu	17.75	13.51	Cu	2.77	2.40
Mn	103	188	193	191	MnO	* 0.02%	* 0.01%	Mn	185.71	150.31
P	149	299	389	465	P_2O_5	* 0.06%	* 0.03%	P	285.96	240.86
Pb	5.48	5.55	5.78	7.56	Pb	7.66	6.22	Pb	-	-
Ti	925	1022	868	880	TiO_2	* 0.17%	* 0.14%	Ti	21.04	15.84
Sr	36.34	33.07	32.29	30.74	Sr	27.82	31.55	Sr	-	-
Ba	268	255	253	238	Ba	171.97	179.97	Ba	4.94	3.18
Th	2.51	2.58	2.51	2.51	Th	2.70	1.85	Th	-	-
Rb	27.71	26.87	26.32	25.57	Rb	26.54	27.00	Rb	-	-
Ni	4.53	2.63	2.68	2.89	Ni	3.93	2.20	Ni	-	-
Sn	1.30	1.75	1.54	1.75	Sn	2.34	1.04	Sn	-	-
Zn	11.94	11.24	10.57	7.88	Zn	17.90	12.03	Zn	8.12	7.07
MS ($\times 10^{-8}$ m^3 kg^{-1})	1.9	3.8809	4.43993	5.2832						
Microcharcoal	803	1855	1800	1300						

Phosphorus. It has been noted in [8] that P is unique as a sensitive and persistent indicator of human activity. This conclusion is based on (1) a long list of the anthropogenic sources of phosphorus: human waste refuse; ash; manure—guano, dung; animal husbandry, etc.; (2) accumulation of anthropogenic P at the site of deposition directly connected to the duration of occupation; and (3) relative P stability in the soil in terms of biochemical weathering. Though some natural and cultural factors, such as "certain parent materials and organic amendments, redoxymorphic conditions ... geomorphic forces, soil formation, and disturbance ... , may redistribute or remove particles that host P compounds" [8], (p. 325), the majority of cases of soil enrichment in phosphorus remains a useful geochemical indicator of human presence and activities. Numerous studies have clearly shown that the highest phosphorus concentration within a settlement is found in byres, as well as in middens and hearths, while in arable fields manure (dung and waste) provokes elevated P values [4,8,9,25–28]. In The Netherlands, studies of P as an indicator of human activity have been carried out at several ancient settlements and on the Celtic Fields, which represent an ancient system of agricultural activity dating back to the late Bronze Age up to the Roman Time (1100 BC–200 AD) and widespread throughout north-west Europe [14,22].

The geochemical study of Voorthuizen has revealed a remarkable difference in P values between the Iron Age settlement occupational surface and the Roman Time agricultural horizon. The P value of the latter is 2–3 times as high, reaching 465 mg/kg. Comparison of Voorthuizen has yielded P values for natural sandy heathland soils (Noordseveld, Zeijen, The Netherlands), prehistoric Celtic Fields soils (Zeijen, The Netherlands) and the Nistelrode Roman Time settlement (Table 5), showing that P values of the agricultural layer of Voorthuizen are the highest and are comparable to values obtained for the banks (high ridges) of the Celtic Fields [14,22]. This seems to reflect the similarity of the agricultural (manure) technique applied in both cases. Simultaneously, the lower values of the household waste indicators in the Voorthuizen agricultural layer seem to indicate that litter was not used as a fertilizer in this case, though this type of manure is known to occur on Celtic Fields in Zeijen [14].

Comparative analysis of living layers in two settlements, Voorthuizen and Nistelrode, has revealed that P values are relatively similar as well (299 mg/kg and 285.96 mg/kg, respectively; Table 6). These values can serve as a first P reference number for the occupational layers of the "sandy" prehistoric settlements in The Netherlands Further research is expected to provide more data to make this reference more accurate.

Table 6. Comparison of P values: (i) Voorthuizen archaeological site cultural layer (VN-8–VN-11) (mg P kg^{-1} dry soil, XRF, and real total); (ii) Natural heathland soils and Celtic Field soils at Noordseveld (Zeijen, The Netherlands) (mg P kg^{-1} of dry soil, detected spectrophotometrically using molybdene blue, [14]); and (iii) Nistelrode (NS46) (mg P kg^{-1}, detected using ICP-OES [22]).

Object	Mean	Standard Deviation	Spread
	Voorthuizen, The Netherlands		
VN-8	149	79	-
VN-9	299	5	-
VN-10	389	20	-
VN-11	465	29	-
	Natural heathland soils at Noordseveld, Zeijen, The Netherlands		
Ah-horizon podzol	131	62	48–231
	Celtic Field soils at Zeijen, The Netherlands		
Field plots	235	64	153–371
Basic layer ridges	211	42	135–275
Top layer ridges	323	94	170–498
	Nistelrode (NS46), The Netherlands		
On site	285.96	-	-
Of site	240.86	-	-

Indicators of extra mineral matter input. The Voorthuizen samples have Si values most similar to other sandy soil locations in The Netherlands, for which Si values are available in ISE reports [16]. However, soil sample VN-11 has the highest Si value (442,003 mg/kg), as compared to the data reported and other samples from Voorthuizen. A similar phenomenon, consisting of unusually high concentrations of Si in the layers with anthropogenic contexts, was observed by [29] at an Iron Age *Viereckschanze* type of site in Bohemia. The authors concluded that this may have been due to a particular type or intensity of human activity that required the use of additional amounts of sand. Sample VN-11 also displays "mobile" Fe, Mn, V, Pb, As, and Sn values that are higher than those in the lower part of the profile and the highest organic matter concentration (up to 6.61%). In addition, carbonate matter (LOI 900 °C and, i.e., Ca) in soil samples VN-8 and VN-10 increases slightly. This, along with high Si value, seems to be due to the manure technique, when "a new portion" of elements with sods and dung was brought in, distorting natural podzolization processes, which lead to the replacement of mobile elements from the upper parts of the soil profile and their deposition in deeper horizons.

The results of the MS analysis also seem to indicate an input of an additional mineral component into the ancient agricultural layer. This assumption is based on the fact that magnetic susceptibility can be used as a tracer of mineral matter in sediment: higher MS values show higher enrichment in a mineral component. The Voorthuizen MS results show an increasing trend of mineral content in soil samples VN-8-11, respectfully, with the highest value 5283 × 10^{-8} m^3 kg^{-1} in the topmost layer.

6. Conclusions

The geochemical survey of anthropogenic horizons at the Voorthuizen archaeological site has led us to conclude that:

Comparison of the geochemical composition of the occupational surface of the Iron Age settlement and the early Roman Time agricultural horizon shows that the former is enriched in geochemical indicators such as Ba, Sr, Rb, Th, Cl, and Mg, while the agricultural horizon has elevated P, Pb, Ni, Mn, and V values.

The occupational surface of the settlement displays the higher values of household waste geochemical indicators than those of the agricultural horizon studied. It is assumed that waste as a fertilizer was not used in this case, while household garbage was used as a component of agricultural technology in the Celtic Fields. At the same time, P value in the Voorthuizen agricultural layer is comparable to that of the Celtic Fields, suggesting similar P sources in both manure techniques.

Elevated Si and "mobile" Fe, Mn, V, Pb, As, and Sn values, along with higher MS results, seem to be indicative of extra mineral matter input during fertilization of the ancient arable field at Voorthuizen.

Author Contributions: Conceptualization, O.D., K.v.d.B. and L.G.; methodology, L.G.; formal analysis, L.G.; data curation, L.G.; writing—original draft preparation, O.D.; writing—review and editing, O.D., K.v.d.B. and L.G.; visualization, L.G.; supervision, O.D.; project administration, K.v.d.B. All authors have read and agreed to the published version of the manuscript.

Funding: The geochemical research was carried out with the support of State Assignment no. FSZN-2020-0016.

Data Availability Statement: Raw data are available upon request, please contact the corresponding author.

Acknowledgments: The authors would like to thank the reviewers and editors for their valuable suggestions.

Conflicts of Interest: The authors declare no conflict of interest.

References

1. Cuenca-García, C. Inorganic Geochemical Methods in Archaeological Prospection. In *Best Practices of GeoInformatic Technologies for the Mapping of Archaeolandscapes*; Sarris, A., Ed.; Archaeopress: Oxford, UK, 2015; pp. 219–231.
2. Entwistle, J.A.; Abrahams, P.W. Multi-element analysis of soils and sediments from Scottish historical sites. The potential of inductively coupled plasma-mass spectrometry for rapid site investigation. *J. Archaeol. Sci.* **1997**, *24*, 407–416. [CrossRef]
3. Entwistle, J.A.; McCaffrey, K.J.W.; Dodgshon, R.A. Geostatistical and multielemental analysis of soils to interpret land-use history in the Hebrides, Scotland. *Geoarchaeology* **2007**, *22*, 391–415. [CrossRef]
4. Middleton, W.D. Identifying chemical activity residues on prehistoric house floors: A methodology and rationale for multi-elemental characterization of a mild extract of anthropogenic sediments. *Archaeometry* **2004**, *46*, 47–65. [CrossRef]
5. Save, S.; Issenmann, R.; Kovacik, J.; Poirier, S.; Demarly-cresp, F.; Sedlbauer, S.; Teyssonneyre, Y. Large-scale geochemical survey by pXRF spectrometry of archaeological settlements and features: New perspectives on the method. *Archaeol. Prospect.* **2020**, *27*, 1–16. [CrossRef]
6. Schramm, R.; Heckel, J. Fast analysis of traces and major elements with ED (P) XRF using polarized X-rays: TURBOQUANT. *J. Phys.* **1998**, *8*, 335–342. [CrossRef]
7. Cook, S.F.; Heizer, R.F. Studies on the Chemical Analysis of Archaeological Sites. In *University of California Publications in Anthropology 2*; University of California Press: Berkeley, CA, USA, 1965.
8. Holliday, V.T.; Gartner, W.G. Methods of soil P analysis in archaeology. *J. Archaeol. Sci.* **2007**, *34*, 301–333. [CrossRef]
9. Schlezinger, D.R.; Howes, B.L. Organic Phosphorus and Elemental Ratios as Indicators of Prehistoric Human Occupation. *J. Archaeol. Sci.* **2000**, *27*, 479–492. [CrossRef]
10. Wilson, C.A.; Davidson, D.A.; Cresser, M.S. Multi-element soil analysis: An assessment of its potential as an aid to archaeological interpretation. *J. Archaeol. Sci.* **2008**, *35*, 412–424. [CrossRef]
11. Druzhinina, O.A.; van den Berghe, K.J.; Golyeva, A.A.; Gedminienė, L.; Karpińska-Kołaczek, M. The results of the geoarchaeological study of the Voorthuizen archaeological site (The Netherlands) (in prep.). In Proceedings of the vEGU21: The 23rd EGU General Assembly, Online, 19–23 April 2021.
12. van den Berghe, K.J. *Evaluatierapport Proefsleuven en Opgraving Kromme Akker Zuid te Voorthuizen, Gemeente Barneveld*; Econsultancy Rapport 7241.003; Econsultancy: Zwolle, The Netherlands, 2020.

13. Bronk Ramsey, C. *OxCal Project*; Research Lab for Archaeology: Oxford, UK, 2020. Available online: https://c14.arch.ox.ac.uk (accessed on 1 February 2021).
14. Spek, T.; Groenman-van Waateringe, W.; Kooistra, M.; Bakker, L. Formation and land-use history of Celtic fields in north-west Europe–an interdisciplinary case study at Zeijen, The Netherlands. *Eur. J. Archaeol.* **2003**, *6*, 141–173. [CrossRef]
15. Dean, W.E. Determination of carbonate and organic matter in calcareous sediments and sedimentary rocks by loss on ignition: Comparison with other methods. *J. Sediment. Res.* **1974**, *44*, 242–248.
16. Dijk, D.V. Wageningen evaluating programmes for analytical laboratories (WEPAL): A world of experience. *Commun. Soil Sci. Plant Anal.* **2002**, *33*, 2457–2465. [CrossRef]
17. Wilson, C.A.; Davidson, D.A.; Cresser, M.S. An evaluation of multielement analysis of historic soil contamination to differentiate space use and former function in and around abandoned farms. *Holocene* **2005**, *15*, 1094–1099. [CrossRef]
18. Wilson, C.A.; Davidson, D.A.; Cresser, M.S. An evaluation of the site specificity of soil elemental signatures for identifying and interpreting former functional areas. *J. Archaeol. Sci.* **2009**, *36*, 2327–2334. [CrossRef]
19. Monna, F.; Hamer, K.; Lévêque, J.; Sauer, M. Pb isotopes as a reliable marker of early mining and smelting in the northern Harz province (Lower Saxony, Germany). *J. Geochem. Explor.* **2000**, *68*, 201–210. [CrossRef]
20. Meharg, A.A.; Deacon, C.; Edwards, K.J.; Donaldson, M.; Davidson, D.A.; Spring, C.; Scrimgeour, C.M.; Feldmann, J.; Rabb, A. Ancient manuring practices pollute arable soils at the St Kilda World Heritage site, Scottish North Atlantic. *Chemosphere* **2006**, *64*, 1818–1828. [CrossRef] [PubMed]
21. Horák, J.; Janovský, M.; Hejcman, M.; Šmejda, L.; Klír, T. Soil geochemistry of medieval arable fields in Lovětín near Třešť, Czech Republic. *Catena* **2018**, *162*, 14–22. [CrossRef]
22. Oonk, S.; Slomp, C.P.; Huisman, D.J.; Vriend, S.P. Effects of site lithology on geochemical signatures of human occupation in archaeological house plans in the Netherlands. *J. Archaeol. Sci.* **2009**, *36*, 1215–1228. [CrossRef]
23. Janovský, M.P.; Karlík, P.; Horák, J.; Šmejda, L.; Asare Opare, M.; Beneš, J.; Hejcman, M. Historical land-use in an abandoned mountain village in the Czech Republic is reflected by the Mg, P, K, Ca, V, Cr, Mn, Fe, Ni, Cu, Zn, Rb, Zr, and Sr content in contemporary soils. *Catena* **2020**, *187*, 104347. [CrossRef]
24. Taraškevičius, R.; Stančikaitė, M.; Bliujienė, A.; Stakėnienė, R.; Zinkutė, R.; Kusiak, J. Search for geochemical indicators of pre-urban habitation sites: Case study from the Skomantai hill-fort and settlement, western Lithuania. *Geochem. Explor. Environ. Anal.* **2012**, *12*, 265–275. [CrossRef]
25. Roos, C.I.; Nolan, K.C. Phosphates, plowzones, and plazas: A minimally invasive approach to settlement structure of plowed village sites. *J. Archaeol. Sci.* **2012**, *39*, 23–32. [CrossRef]
26. Salisbury, R.B. Interpolating geochemical patterning of activity zones at Late Neolithic and Early Copper Age settlements in eastern Hungary. *J. Archaeol. Sci.* **2013**, *40*, 926–934. [CrossRef]
27. Šmejda, L.; Hejcman, M.; Horák, J.; Shai, I. Ancient settlement activities as important sources of nutrients (P, K, S, Zn and Cu) in Eastern Mediterranean ecosystems—The case of biblical Tel Burna, Israel. *Catena* **2017**, *156*, 63–72. [CrossRef]
28. Wells, E.C.; Terry, R.E.; Parnell, J.J.; Hardin, P.J.; Jackson, M.W.; Houston, S.D. Chemical analyses of ancient anthrosols in residential areas at Piedras Negras, Guatemala. *J. Archaeol. Sci.* **2000**, *27*, 449–462. [CrossRef]
29. Danielisová, A.; Horák, J.; Janovský, M.; Strouhalová, B.; Bursák, D. Geochemical approach to determine the anthropogenic signal at non-intensively settled archaeological sites—The case of an Iron Age enclosure in Bohemia. *Catena* **2022**, *210*, 105895. [CrossRef]

minerals

MDPI

Article

Elemental Enrichment in Shallow Subsurface Red Sea Coastal Sediments, Al-Shuaiba, Saudi Arabia: Natural vs. Anthropogenic Controls

Ibrahim M. Ghandour [1,2,*] and Mohammed H. Aljahdali [1]

1 Marine Geology Department, Faculty of Marine Sciences, King Abdulaziz University, Jeddah 80200, Saudi Arabia; maljahdli@kau.edu.sa
2 Geology Department, Faculty of Science, Tanta University, Tanta 31527, Egypt
* Correspondence: ighandour@kau.edu.sa

Abstract: Geochemical analysis of the 23 sediment samples collected from a short (0.6 m long) core retrieved from the coastal creek that was previously connecting the northern and southern Al-Shuaiba Lagoons, Red Sea, Saudi Arabia, was accomplished to assess the elemental enrichment levels and the natural and anthropogenic driving forces for this enrichment. Statistical analysis and upcore variation in elemental concentrations enabled subdivision of the core formally into three units, lower, middle, and upper. The enriched elements in the lower and middle units display poor to negative correlations with the enriched elements in the upper unit. The lower unit is enriched in elements (Mo, As, U, and Re) suggesting deposition under anoxic conditions, possibly related to the Medieval Climate Anomaly. The middle unit is enriched in the carbonate-related constituents ($CaCO_3$, Ca, and Sr). The upper unit is enriched in elements that co-vary significantly with Al suggesting increased terrigenous supply associated with the construction of the road between the two lagoons. The enrichment of elements in the lower and middle units is naturally driven, whereas the enrichment of lithogenic elements in the upper unit, though of geogenic origin, is induced after the road construction.

Keywords: Al-Shuaiba Lagoon; anthropogenic vs. natural; Red Sea; sediment geochemistry; anoxic bottom conditions; redox-sensitive elements; terrigenous elements

Citation: Ghandour, I.M.; Aljahdali, M.H. Elemental Enrichment in Shallow Subsurface Red Sea Coastal Sediments, Al-Shuaiba, Saudi Arabia: Natural vs. Anthropogenic Controls. *Minerals* **2021**, *11*, 898. https://doi.org/10.3390/min11080898

Academic Editors: Marianna Kulkova and Dmitry Subetto

Received: 27 July 2021
Accepted: 17 August 2021
Published: 19 August 2021

1. Introduction

Coastal lagoons are the most used, fragile, vulnerable, and threatened coastal ecosystems [1–5]. They hold varieties of important living resources of commercial and ecological interests. They are affected by several natural and anthropogenic pressures and threats. Climate changes, rock weathering, soil erosion, and volcanic eruptions are the most common natural threats. On the other hand, anthropogenic perturbations vary from various industrial and agricultural activities, burning of fossil fuels, untreated waste discharges, increasing nutrient supply, unplanned coastal urbanization, road runoff, dumping, and leachate from landfills [5–7]. These threats have a massive direct and indirect negative effect and increase the pressure on the marine coastal ecosystem. Water stagnation and poor flushing rates that are caused either by natural and/or anthropogenic processes are among the most serious environmental problems in coastal lagoons. Water stagnation puts pressure on the coastal water and sediment quality encourages the appearance of harmful algae and bacteria and eliminates oxygen from the ecosystem by anaerobic bacteria with the subsequent release of H_2S that can kill a variety of invertebrates and fishes and causes anomalous enrichment of the redox-sensitive metals [8]. Consequently, these coasts become unsuitable for aquaculture, fisheries, and coastal touristic and recreation activities.

The natural and anthropogenic threats and their environmental consequences are imprinted in the chemical composition of the lagoonal bottom sediments. These threats lead to the enrichment of bottom sediments with potentially toxic elements. Metal enrichment can endanger aquatic life and destroy ecosystems [9,10]. The concentrations of trace elements in the sediments provide key information about potential anthropogenic pressures on the ecosystem, as well as potential internal sources in these ecosystems. Sediments are a possible source and/or a carrier for metals [11,12] and they may limit their bioavailability or allow their remobilization and re-suspension in the water column depending on bottom redox conditions [13,14]. Under certain conditions, the enriched toxic metals can be released into the water column and become a threat to aquatic ecosystems as well as a health risk to humans due to their toxicity, persistence, and ability to accumulate in living organisms [5,15,16]. Therefore, the study of sediment geochemistry is crucial to monitor the aquatic ecosystem as it provides information about the redox bottom conditions, contamination level, and ecotoxicological risk [5,17]. The chemical composition of shallow subsurface coastal sediments provides a historic record of natural and anthropogenic influences on coastal systems [18–20].

Along the Saudi Red Sea coast, there are over 20 hypersaline coastal lagoons with unique environmental characteristics [21]. Among them, Al-Shuaiba and Al-Kharrar Lagoons are the most studied [22–27]. At Al-Shuaiba area some 80 km south of Jeddah City, there are two lagoons, the northern and southern Al-Shuaiba Lagoons (Figure 1a), that are sites for rare traditional fishing. In the past, water was freely circulated between the two lagoons through a shallow and small channel/creek and the oxygen concentration in the water was normally replenished. In the early 1980s, the two lagoons were separated by a paved road, which ended water circulation and the area has undergone dramatic environmental changes (Figure 2). Water circulation in the channel to the north of the southern Al-Shuaiba Lagoon is sluggish and essentially stagnant, resulting in the accumulation of organic matter from dead organisms and plants from the nearby mangroves.

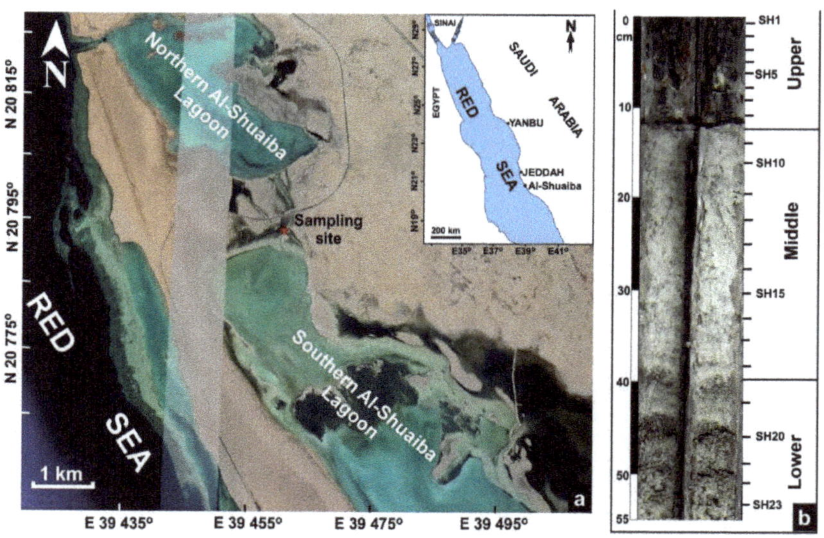

Figure 1. Location map of the area of study (a) and the core sample (b).

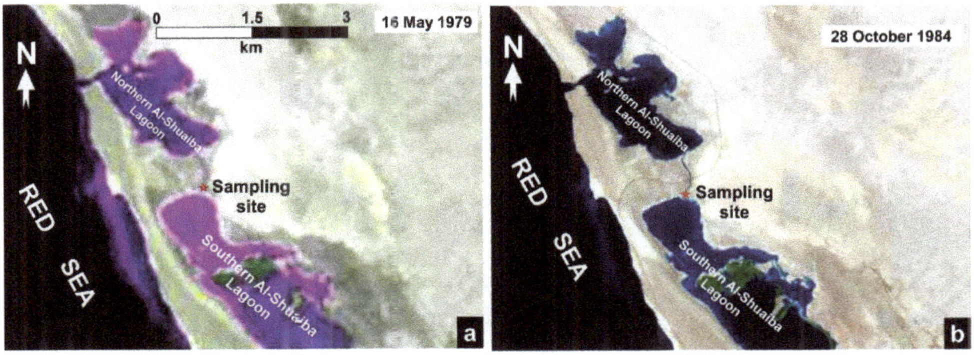

Figure 2. Historic land-sat images of Al-Shuaiba area showing the evolution of the area before (**a**) and after the road construction (**b**) between the two lagoons.

Most of the previous studies in the Al-Shuaiba area were carried out on the southern Al-Shuaiba Lagoon bottom sediments and the shallow subsurface sediments of the tidal flat to the east. The effect of meteorological forcing and tide on the flushing of Al-Shuaiba Lagoon was studied by Ahmad and Sultan [28] and Al-Barakati [29]. The mineralogical and chemical composition of the southern Al-Shuaiba Lagoon bottom sediment was studied by Al-Washmi and Gheith [22] and Basaham et al. [26], whereas Abu-Zied and Bantan [24] studied the benthic foraminiferal distribution and their environmental controls and usefulness in sea-level reconstruction. Abu-Zied and Bantan [30] studied the vertical variations of benthic foraminifera and major and trace elements and their environmental and paleoclimatic significances in shallow subsurface sediments of the tidal flat east of the southern Al-Shuaiba Lagoon. Al-Farawati et al. [31] studied the speciation and distribution pattern of phosphorus in the sediments of the northern Al-Shuaiba Lagoon.

None of the previous studies has committed an investigative effort to document the potential impacts of road construction on the water and sediment quality of the area. The environmental deterioration and adverse effect of road construction on sediment chemical composition are therefore a target of this work. It is crucial to determine sediment geochemistry and assess if any metal enrichment has been caused or amplified by the construction of the road in the early 1980s and the resultant stagnation. The present work, therefore, aims to (1) reveal the vertical profiles and assess the degree of enrichment of the environmentally sensitive major and trace elements prior to and after the construction of the road; (2) determine the reason(s) and sources of elemental enrichment; and (3) evaluate the most suitable redox-sensitive elements. To investigate the history of the channel and the beginning of the anthropo-pressure, coring of the bottom sediments of the channel/creek connecting the two lagoons was conducted.

2. Area of Study

The northern and southern Al-Shuaiba Red Sea hyper-saline coastal lagoons (Figure 1a) were formed by the sub-pluvial to pluvial Pleistocene erosion and subsequently submerged by a sea-level rise during the early-middle Holocene [21,32,33]. They are bounded to the west by raised coral terraces of Pleistocene age and the east by a vast tidal flat. They show unique environmental characteristics as they are connected to the Red Sea through narrow inlets with water depths of 7 and 15 m, respectively. The climate is tropical with warm and dry conditions throughout the year, scarce rainfall (63 m/year), and no riverine inflow. The annual average evaporation rate over the Red Sea is 2.06 ± 0.22 m/year [34,35].

The lagoons are separated by a paved road that ended the water circulation. They lack industrial, freshwater, and terrigenous sediments influx. The lagoonal bottom sediments consist of calcareous biogenous sediments admixed with relatively rare siliciclastic and traces of evaporite deposits [21]. The texture of the bottom sediments ranges between sandy mud and gravelly sand. The sand to gravel size fraction is dominated by skeletal remains of coral debris, coralline algae, molluscan shells, foraminifera, sponges, and bryozoans. The sediment distribution is controlled by the shape and size of the inlet, water depths, and tidal currents. The colour of sediments varies in the two lagoons, with dark sediments mainly found at the periphery of the lagoons. The bottom sediments of the northern lagoon are brown to grey, whereas they range from yellow to grey in the southern lagoon. Fine sediments are dark with shades of grey, and the coarse sediments are light with shades of yellowish to greyish brown. The coarse sediments including shells are stained grey to black due to the reducing environment and formation of authigenic pyrite. Stagnant conditions prevail inside the lagoons due to insufficient water exchange with the open sea, and the lack of rainfall causes hyper-saline conditions [21]. Dark grey sediments occur in quiescent reducing areas where currents are almost negligible and incapable of re-suspending and transporting sediments. The reducing environment is further enhanced by restricted water exchange between the Red Sea and the lagoons. Organic-rich sediments with a pungent smell are dark-coloured and enriched in mangrove and dark stained shells and coral debris remains reflecting the development of reducing conditions [21].

Basaham et al. [26] distinguished two groups of elements in the southern Al-Shuaiba lagoonal bottom sediments. The first group defines elements that are associated with detrital minerals such as silicates (Al-Fe-Si), feldspars (K-Rb-Ba), and heavy minerals (V-Cr-Zr, Ti-Y-Nb). They are highly enriched in the shallow nearshore sediments. The second group defines elements Ca and Sr that are associated with carbonate minerals and are concentrated in the deeper parts of the lagoon. Mangrove (*Avicennia marina*) stands are scattered around the small islands and along the eastern margins of the lagoons. Seagrass and macro-algae dominate the shallow parts of the lagoons [36]. Abu-Zied and Bantan [30] recorded benthic foraminiferal assemblages, *Monalysidium acicularis*, *Quinqueloculina* cf. *Q. limbata*, and *Coscinospira hemprichii* that are sensitive indicators of sea-level changes.

3. Materials and Methods

A short core (0.6 m long) was obtained by simple manual push and rotation from the channel that connects the two Al-Shuaiba Lagoons in a water depth of 0.6 m. The core was collected by a simple manual push with rotation using a PVC tube with a 6.35 cm internal diameter. Twenty-three samples were collected from the upper 0.55 m at 1.5 to 2-cm sampling intervals (Figure 1b). Dissolved oxygen (DO), water salinity, and pH were measured for the surface water, sediment-water interface, and interstitial waters using a handheld multi-parameters instrument YSI 556 MPS (YSI Environmental, Yellow Springs, OH, USA). The samples were dried at room temperature then at 105 °C for 24 h to remove water. The concentrations of major (>1%), minor (between 1% and 100 µg/g), and trace (equal to or below 100 µg/g) were performed using the multi-acid digestion package ICP-ES/MS (MA200) at the Bureau Veritas Mineral Laboratories (BVML, Vancouver, Canada). A 0.25 g of the sample was heated in HNO_3, $HClO_4$, and HF to fuming and taken to dryness. The residue was then dissolved in HCl. The concentrations of elements were determined using the ICP-MS with a Perkin Elmer Elan 6000 ICP (Perkin Elmer, Waltham, MA, United States) mass spectrometer and the concentrations were expressed as µg/g. Procedural blanks and certified reference materials OREAS 25A-4A and OREAS 45E were prepared using the same analytical procedure and reagents and analyzed for the BVML quality assurance/control (QA/QC) protocol. Organic matter content was determined by loss of weight by ignition (LOI_{550}) in a muffle furnace at 550 °C for 2 h [37]. To assess the possible elemental enrichments, the enrichment factor (*EF*; [38]) is determined as follows:

$$EF = ([X]/[R])sample/([X]/[R])reference$$

where X is the concentration of an element and R is the concentration of the normalizer. The reference material used in this study is local reference corresponding to uncontaminated sediments in the same core. It represents the average chemical composition of three samples from the core assumed to correspond to sediments deposited prior to the complete closure of the canal. Since it is conservative and of natural origin, titanium (Ti) was chosen as a normalizer. Ti is contributed mainly by and is located entirely in heavy and clay minerals. Five categories of enrichments are defined [39]: $EF < 1.5$ no enrichment; $1.5 < EF < 5$ moderate; $5 < EF < 20$ significant; $20 < EF < 40$ very high; and $EF > 40$ extreme. In addition, the data of some elements were also normalized against the most abundant lithogenic element—aluminum (Al). Pearson correlation and multivariate analysis were performed using IBM SPSS Statistics for Windows, Version 20.0. (IBM Corp., Armonk, NY, USA). The principal component analysis (PCA) was used to determine relationships among variables and simplify the analytical dataset. The sample matrix consists of 23 sediment samples and 30 variables.

4. Results

The water salinity, pH, and dissolved oxygen (DO) vary in the channel with depth. The surface water salinity, pH, and DO are 51.56‰, 8.4 and 6.08 mg/L, respectively, and they are 39.1‰, 8.08 and 1.92 mg/L, respectively, at the sediment-water interface. A remarkable decrease in salinity (28.88‰), pH (8.01), and DO (0.51 mg/L) is recorded for the interstitial waters.

Based on the colour, sediment composition, texture and organic matter, $CaCO_3$, and elemental concentrations, the core sediments can be subdivided into three units, upper, middle, and lower (Figure 1b). The upper unit spans the top 12 cm, marked by dark grey to black colour with stench smell. It consists of argillaceous sands with abundant algal mats. The middle unit extends from 12 to 39 cm and consists of light grey to white calcareous sands enriched in molluscan shell fragments. The lower unit extends from depths 39 to 55 cm and it consists of colour banded, massively bedded, light grey sands (Figure 1b).

4.1. Chemical Composition

The elemental concentrations and their enrichment factors, LOI_{550}, and $CaCO_3$ contents of 23 samples are summarized in Table 1 and the down-core variations of their concentrations are shown in Figure 3. The geochemical composition of the sediments shows a downcore co-variation of elements Al, Fe, Mg, K, Na, P, LOI_{550}, Mn, V, Cr, Co, Ni, Zn, Zr, Rb, La, Li, Nb, Sc, and Ce. Their concentrations are higher in the upper than the middle and lower parts of the core (Figure 3). The downcore distribution pattern of these elements shows an opposite trend to Ca, Sr, $CaCO_3$, Mo, As, U, and Re (Figure 3). The value of the LOI_{550} increases monotonically upcore with the highest content in the upper unit which contains the highest proportion of terrigenous mud. It varies from 3% to 15% (mean 9%). The $CaCO_3$ decreases upcore ranging from 75% to 36% (mean 59%). On average, the most abundant major elements are Ca (22.95%), Mg (2.9%), Al (2.48%), Na (1.51%), and Fe (1.34%). Ca varies in the lower and middle parts of the core between 23.65% and 27.09 %, whereas it varies in the upper part from 14.55% to 20.51%. The Al content varies in the lower and middle parts of the core from 1.62% to 2.66%, whereas it ranges between 3.11% and 4.19% in the upper part. Similarly, the highest concentrations of Mg (4.45%), Na (2.33%), and Fe (2.09%) were recorded from the upper part of the core (Figure 3).

Table 1. Results of chemical analysis of the shallow subsurface coastal sediments of the creek that was connecting the two Al-Shuaiba Lagoons. Ranges and average (in brackets) of concentrations, enrichment factor (*EF*), Al-normalized elements, and the calculated background values.

Element/ Constituent	Concentrations	Enrichment Factor	Al-Normalized	Background Values
Ti%	0.14–0.33 (0.21)	-	0.059–0.173 (0.090)	0.215
Al%	1.62–4.19 (2.48)	-	-	2.15
Fe%	0.81–2.09 (1.34)	0.62–1.72 (1.19)	0.497–0.590 (0.543)	1.17
Ca%	14.55–27.09 (22.95)	0.47–1.55 (0.98)	3.473–16.722 (10.292)	24.95
Mg%	1.95–4.45 (2.9)	0.61–1.71 (1.20)	0.942–1.484 (1.199)	2.53
Na%	0.83–2.33 (1.51)	0.66–1.97 (1.21)	0.441–0.778 (0.620)	1.3
K%	0.43–0.96 (0.59)	0.63–1.55 (1.18)	0.213–0.340 (0.245)	0.53
P%	0.03–0.05 (0.04)	0.68–1.51 (1.18)	0.013–0.019 (0.016)	0.033
S%	0.3–0.6 (0.44)	0.54–1.80 (1.28)	0.095–0.227 (0.185)	0.37
$CaCO_3$%	36–75 (59)	-	-	-
LOI_{550}%	3–15 (9)	-	-	-
Mn µg/g	153–405 (264)	0.79–1.39 (1.11)	93.67–156.91 (108.99)	245
V µg/g	25–70 (42)	0.68–1.60 (1.18)	15.43–20.22 (17.11)	36.7
Cr µg/g	15–44 (26.96)	0.67–1.83 (1.29)	9.26–14.07 (10.97)	21.7
Co µg/g	1.9–11.2 (5.45)	0.51–2.23 (1.30)	1.17–2.69 (2.10)	4.3
Ni µg/g	4.2–22.9 (12.54)	0.54–2.37 (1.37)	2.59–6.59 (4.84)	9.4
Cu µg/g	5.30–24 (11.07)	0.53–3.15 (1.47)	3.14–8.47 (4.36)	7.8
Zn µg/g	9–49 (24.6)	0.58–2.95 (1.49)	5.56–14.85 (9.41)	17
Pb µg/g	2.7–24.6 (10.13)	0.26–2.82 (1.24)	1.47–12.36 (4.08)	8.5
Zr µg/g	8.5–30.7 (19.72)	0.63–1.59 (1.10)	5.25–9.79 (7.98)	18.5
Ce µg/g	10–22 (15.3)	0.60–1.29 (1.04)	5.06–7.91 (6.34)	15.3
Nb µg/g	1.9–4.6 (3.12)	0.83–1.45 (1.12)	1.04–2.02 (1.30)	2.87
Rb µg/g	11.9–26.5 (16.87)	0.59–1.60 (1.19)	6.32–7.96 (6.87)	14.8
La µg/g	5.2–11 (7.7)	0.59–1.25 (1.00)	2.53–4.05 (3.19)	8
Y µg/g	4.5–11.7 (7.43)	0.74–1.42 (1.12)	2.69–4.20 (3.06)	6.87
Sc µg/g	2–9 (5.09)	0.66–1.93 (1.32)	1.23–2.31 (2.04)	4
Li µg/g	4–13.6 (8.21)	0.56–1.89 (1.25)	2.47–3.92 (3.28)	6.8
Sr µg/g	1618–3957 (3201)	0.36–1.32 (0.93)	386.16–2043.83 (1416.84)	3631
Sb µg/g	0.3–2.7 (0.63)	0.72–5.97 (1.75)	0.12–0.64 (0.24)	0.37
Sn µg/g	0.2–2.6 (0.64)	0.46–6.25 (1.54)	0.12–0.83 (0.24)	0.43
Th µg/g	0.8–3.2 (1.49)	0.69–1.51 (1.08)	0.44–1.70 (0.62)	1.4
Mo µg/g	2.1–7.5 (4.83)	0.45–3.10 (1.64)	0.67–4.44 (2.09)	3.2
As µg/g	5–17 (9.35)	0.81–3.34 (1.61)	1.91–9.04 (4.25)	6.3
U µg/g	3.2–8 (5.34)	0.57–2.55 (1.26)	0.76–4.55 (2.40)	4.6
Re µg/g	0.003–0.035 (0.013)	0.19–4.45 (1.38)	0.0007–0.0188 (0.0060)	0.01

The most abundant trace elements in the sediments are Sr (mean 3201 µg/g) and Mn (mean 264 µg/g), whereas other trace elements occur in concentrations less than 100 µg/g (Table 1). Compared with the samples in the middle and lower units, the samples of the upper unit contain relatively high trace element content except for Mo, As, U, and Re (Figure 3). The concentration ranges of V, Cr, Co, Ni, Cu, and Zn are 25–70, 15–44, 1.9–11.2, 4.2–22.9, 5.30–24, and 9–49 µg/g, respectively (Table 1). Trends for Mo, As, U, and Re are all characterized by pronounced positive peaks marking the lower part of the core (Figure 3), corresponding to the low concentrations of other trace elements. Although the overall concentration of Cu and Zn increases upcore, the highest concentrations of the two elements appear at 7–9 cm depth (Figure 3). Lead (Pb) showed irregular variations compared with the other heavy metals (Figure 3). Sb and Sn were relatively constant in the lower part of the core and increased markedly at the depths of 3 and 6 cm, respectively (Figure 3). A clear upcore change in Al-normalized trace elements is apparent. A significant feature of the upper part is the remarkable enrichment in Al-normalized concentrations of Co, Ni, and Zn, whereas the lower part notably shows enrichment in Al-normalized Mo, As, U, and Re (Figure 4). The profiles of Al-normalized Mn and V are relatively similar showing peaks in the middle unit (Figure 4).

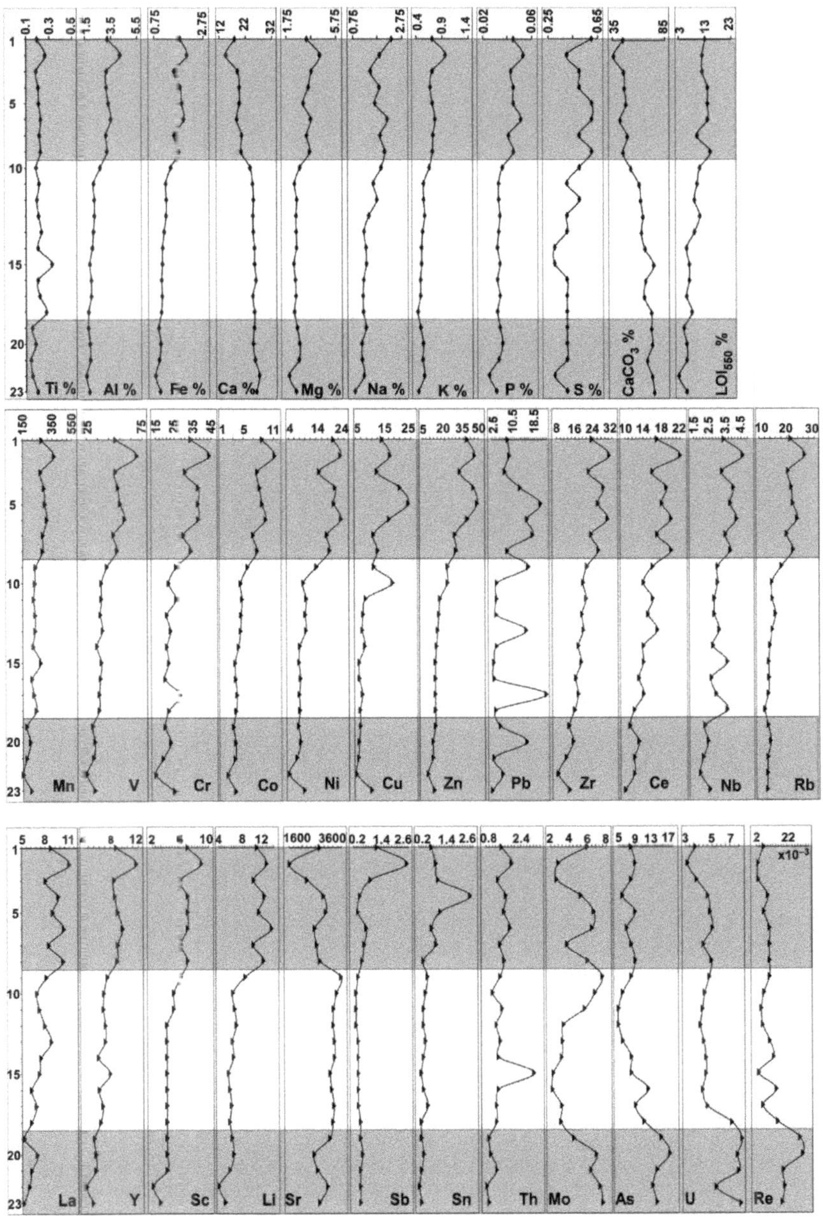

Figure 3. The upcore variation in the concentrations of major (%) and trace elements (μg/g). Elements Al, Fe, Mg, K, Na, OM, Mn, V, Cr, Co, Ni, Zn, Zr, Ce, Nb, Rb La, Y, Li, and Sc display the same profile enriched in the upper unit, Ca, CaCO3, and Sr behave similarly and Mo, As, U, and Re show similar vertical distribution pattern.

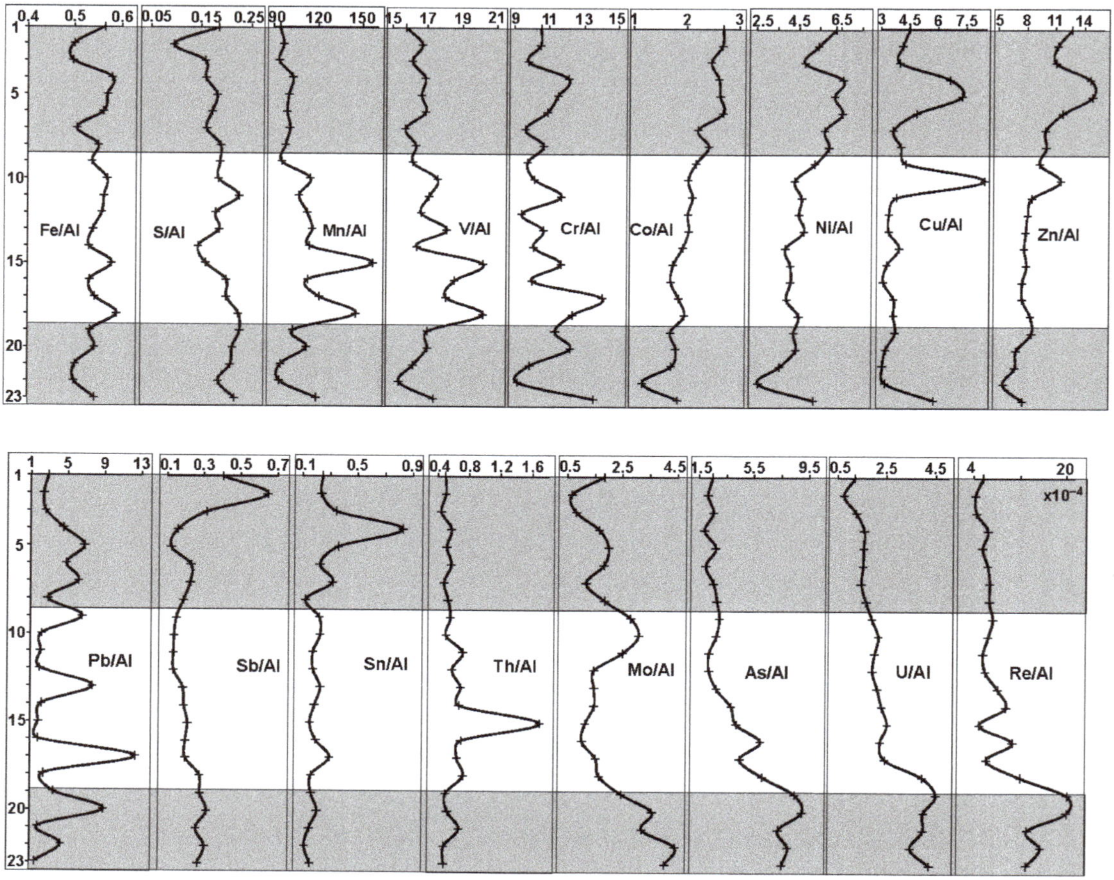

Figure 4. The vertical profiles of Al-normalized elements. Note the upcore increase in the Al-normalized Co, Ni, and Zn values, and the upcore decrease in the Mo/Al, As/Al, U/Al, and Re/Al values. Values of minor and trace elements are multiplied by 10^{-4}.

Aluminum (Al) shows a significant positive correlation ($r \geq 0.9$) with Fe, Mg, K, P, Mn, V, Cr, Co, Ni, Zn, Zr, Ce, Rb, La, Y, Sc, and Li (Table 2). It correlates positively with OM ($r = 0.86$), Na ($r = 0.8$), Nb ($r = 0.82$), and Cu ($r = 0.78$). The significant positive correlations between Al and trace elements Mn, V, Cr, Co, Ni, Cu, and Zn (Figure 5) implies that these elements were derived from a terrigenous source and were not measurably enhanced by reduction from the water column. On the other hand, Ca shows a significant positive interrelationship with $CaCO_3$ ($r = 0.94$) and Sr ($r = 0.7$). These three components negatively correlated with elements characteristic of the detrital fraction (Table 2), suggesting that they are present in the carbonate fraction. Trace elements Mo, As, U, and Re correlate negatively ($r \leq 0$) with Al (Table 2). The significant positive correlation between Ti and Th ($r = 0.75$) indicates detrital fluxes and most Th is held in clay and heavy minerals [40,41].

Table 2. Pearson correlation matrix between Ti, Al, Fe, Ca, $CaCO_3$, and LOI_{550} and major, minor, and trace elements in the shallow subsurface coastal sediments of the creek that was connected with the two Al-Shuaiba Lagoons.

Element/ Constituent	Ti	Al	Fe	Ca	$CaCO_3$	LOI_{550}
Ti	1.00	-	-	-	-	-
Al	0.19	1.00	-	-	-	-
Fe	0.24	0.98 **	1.00	-	-	-
Ca	−0.20	−0.98 **	−0.96 **	1.00	-	-
Mg	0.13	0.94 **	0.89 **	−0.96 **	-	-
Na	0.10	0.81 **	0.85 **	−0.77 **	-	-
K	0.12	0.96 **	0.91 **	−0.94 **	-	-
P	0.24	0.95 **	0.96 **	−0.97 **	-	-
S	−0.10	0.7 **	0.76 **	−0.67 **	-	-
$CaCO_3$	−0.11	−0.98 **	−0.96 **	0.94 **	1.00	-
LOI_{550}	0.24	0.86 **	0.92 **	−0.81 **	−0.88 **	1.00
Mn	0.58 **	0.9 **	0.92 **	−0.89 **	−0.84 **	0.83 **
V	0.35	0.98 **	0.98 **	−0.96 **	−0.94 **	0.86 **
Cr	0.23	0.93 **	0.95 **	−0.94 **	−0.89 **	0.82 **
Co	0.21	0.99 **	0.99 **	−0.98 **	−0.97 **	0.88 **
Ni	0.20	0.97 **	0.99 **	−0.94 **	−0.94 **	0.9 **
Cu	0.11	0.78 **	0.84 **	−0.75 **	−0.77 **	0.8 **
Zn	0.17	0.94 **	0.97 **	−0.94 **	−0.93 **	0.88 **
Zr	0.39	0.92 **	0.96 **	−0.88 **	−0.91 **	0.91 **
Ce	0.32	0.91 **	0.92 **	−0.85 **	−0.91 **	0.85 **
Nb	0.67 **	0.82 **	0.86 **	−0.82 **	−0.77 **	0.80 **
Rb	0.15	0.99 **	0.98 **	−0.96 **	−0.97 **	0.87 **
La	0.34	0. 9 **	0.91 **	−0.85 **	−0.89 **	0.86 **
Y	0.47 *	0.94 **	0.94 **	−0.93 **	−0.9 **	0.83 **
Sc	0.23	0.96 **	0.97 **	−0.95 **	−0.95 **	0.87 **
Li	0.15	0.96 **	0.97 **	−0.94 **	−0.95 **	0.9 **
Sr	−0.04	−0.6 **	−0.52 *	0.7 **	0.50 *	−0.29
Sb	0.21	0.68 **	0.59 **	−0.75 **	−0.59 **	0.32
Sn	0.04	0.63 **	0.67 **	−0.64 **	−0.64 **	0.61 **
Pb	0.01	0.4	0.41	−0.37	−0.42 *	0.33
Th	0.75 **	0.43 *	0.47 *	−0.41	−0.36	0.42 *
Mo	−0.5 *	0.00	0.05	0.04	−0.04	0.08
As	−0.38	−0.43 *	−0.49 *	0.34	0.51 *	−0.61 **
U	−0.35	−0.57 **	−0.55 **	0.51 *	0.61 **	−0.51 *
Re	−0.5 *	−0.5 **	−0.55 **	0.47 *	0.56 **	−0.59 **

Note: ** Correlation is significant at the 0.01 level; *correlation is significant at the 0.05 level.

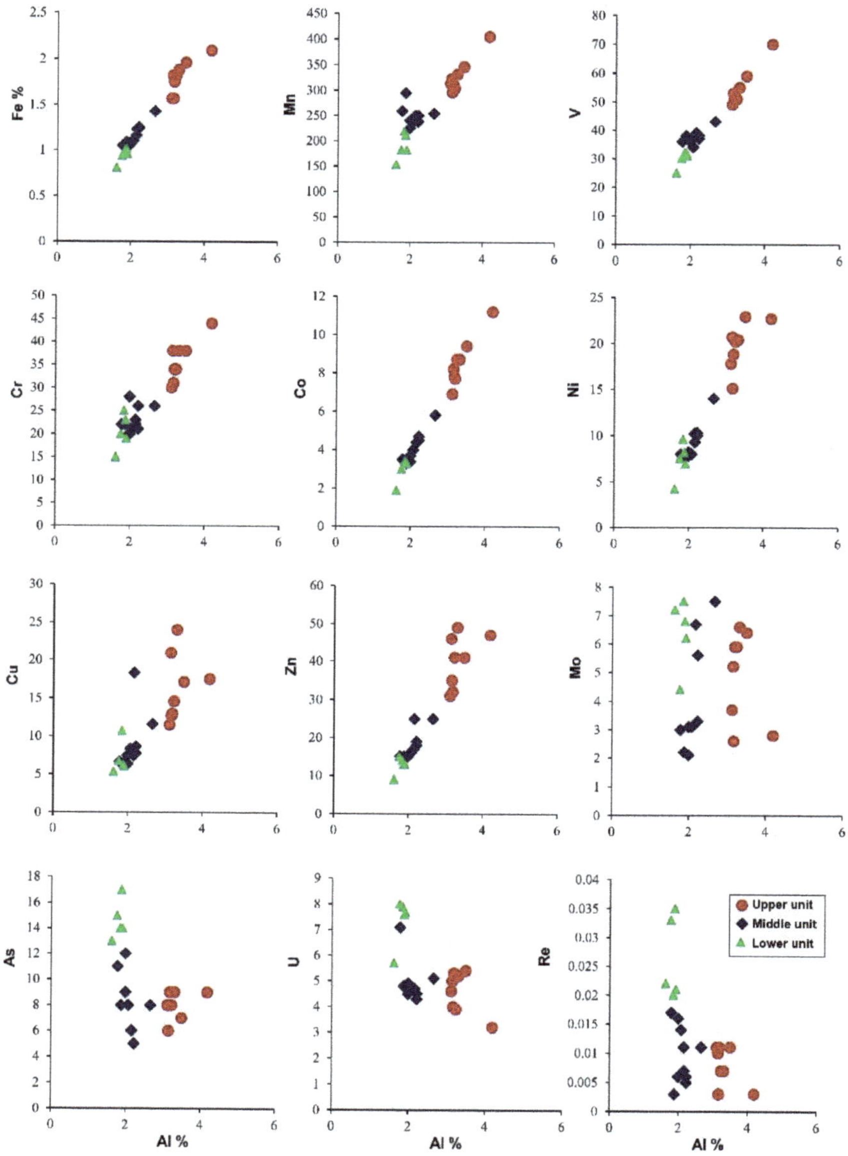

Figure 5. Bivariate plots between Al (%) and major (Fe %), minor (Mn) and trace elements V, Cr, Co, Ni, Cu, Zn, Mo, As, U, and Re (μg/g). Note the positive correlation with the first eight elements and the poor and negative correlation with the last four elements.

4.2. Enrichment Factor

The enrichment factors (*EF*) for Fe, Mn, V, Cr, Co, Ni, Cu, Zn, Mo, As, U, and Re were calculated using the average concentration of local background and Ti as a normalizer. The *EF* values are shown in Table 1 and the vertical profiles for some trace elements are displayed in Figure 6. Most trace elements are depleted relative to the average local background, except for Mo, U, As, and Re in the lower unit. Elements Fe, Mn, V, Cr,

Co, Ni, Cu, and Zn are generally depleted in the lower and middle units of the core, whereas Fe, V, Cr, Co, Ni, Cu, and Zn are slightly enriched (1.5 < *EF* < 5) in the upper unit (Figure 6). Manganese (Mn), Sb, and Sn are below the local background level (*EF* < 1.5). However, some samples in the upper unit show slight to moderate enrichment for Sb and Sn (1.5 < *EF* < 10). Some samples in the lower and upper parts show slight enrichment of Pb (1.5 < *EF* < 5). The lower part shows slight enrichment (1.5 < *EF* < 5) of Mo, As, U, and Re, whereas As, U, and Re are depleted in the middle and upper parts of the core (Figure 6).

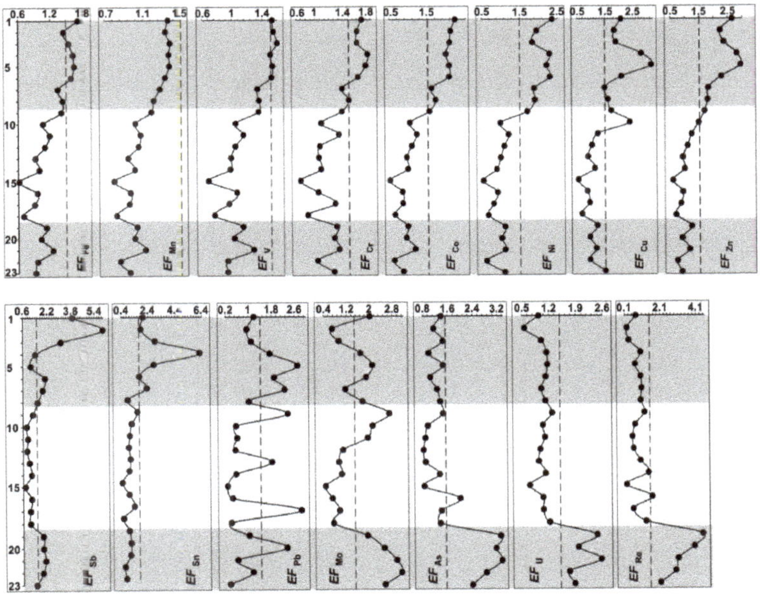

Figure 6. The vertical profile of *EF* values of environmentally significant major (Fe), minor (Mn), and trace elements. The *EF* values for Cr, Co, Ni, Cu, and Zn are slightly enriched in the upper unit, whereas the lower unit shows higher *EF* values for Mo, As, U, and Re than the other two units.

4.3. Statistical Analysis

To determine the similarities and differences between sampling sites and to identify a potential source(s) of elements and the controlling factors on their concentration and distribution, the geochemical dataset was treated statistically using principal component analysis (PCA). The matrix includes 23 sediment samples and 30 variables. Components with eigenvalues greater than 1 were retained. The positive higher loadings of elements were used for factor interpretations.

Four components, with eigenvalues greater than 1, account for 89.28% of the total variance (Table 3). The first two components explain 68% and 10.67% of the total variance, respectively, and thus, account for most of the variance. Components 3 and 4 explaining 7.03% and 3.58% are less important. According to the loadings of components 1 and 2, elements can be roughly clustered into three primary groups. Component 1 is characterized by positive scores (>0.9) for lithogenic constituents (Fe, V, Co, Al, Zr, Y, Ni, Rb, Sc, Zn, Cr, P, Mn, Ce, Nb, and LOI_{550}). These elements are enriched in the sediments of the upper unit. Component 2 characterizes hydrogenous elements (Mo, As, U, and Re as the main positive descriptors) and corresponds to the sediments of the lower unit (Figure 7a). The elements Mo, As, U, and Re are projected in the quadrant of negative values of component 1 and positive values of component 2, while calcareous biogenic components (Ca, $CaCO_3$, and Sr) are in an adjacent quadrant with negative values for components 1 and 2. Elements Ti and Th are clustered into one group indicating their possible similar source and behaviour. They are projected in the quadrant with positive values of component 1 and negative values of component 2. Elements Nb, Mn, Y, Zr, Ce, and V are projected in the quadrant of positive values of component 1 and negative values of component 2 (Figure 7a). The rest of the metals (Cu, Pb, Sb, and Sn) are gathered together and may indicate another possible anthropogenic source.

Table 3. Results of principal component (R-mode) analysis, four components (PC1-PC4) extracted with the PC1 and PC2 are the most important. Significant values are in bold.

Variable	PC1	PC2	PC3	PC4
Fe	**0.99**	0.10	0.04	−0.01
V	**0.99**	−0.01	−0.11	0.00
Co	**0.99**	0.13	−0.04	−0.07
Al	**0.98**	0.12	−0.07	−0.11
Zr	**0.97**	−0.09	0.10	0.07
Y	**0.97**	−0.16	−0.12	0.03
Ni	**0.97**	0.19	0.04	0.05
Rb	**0.97**	0.18	−0.04	−0.08
Sc	**0.97**	0.10	−0.03	−0.02
Ca	−0.96	−0.15	0.19	0.09
Zn	**0.96**	0.17	0.08	−0.01
$CaCO_3$	−0.96	−0.12	−0.08	0.19
Cr	**0.95**	0.14	−0.06	0.12
P	**0.95**	0.17	−0.16	0.09
Mn	**0.95**	−0.21	−0.11	0.15
Ce	**0.93**	−0.09	0.07	−0.04
Nb	**0.90**	−0.26	−0.08	0.27
LOI_{550}	**0.90**	0.03	0.29	0.02
Cu	0.81	0.22	0.27	0.07
S	0.70	0.38	0.38	0.08
Sn	0.68	0.13	0.21	0.05
U	−0.60	**0.55**	−0.15	0.48
Pb	0.42	0.23	0.30	0.31
Ti	0.33	−0.75	−0.23	0.47
Mo	−0.05	**0.72**	0.36	0.12
Re	−0.61	**0.62**	−0.25	0.24
Th	0.53	−0.60	−0.13	0.35
As	−0.53	**0.58**	−0.52	0.24
Sr	−0.54	−0.32	0.71	0.10
Sb	0.61	0.08	−0.69	−0.25
Eigen value	20.402	3.201	2.108	1.075
% Variance	68.01	10.67	7.03	3.58
Cumulative variance	68.01	78.67	85.70	89.28

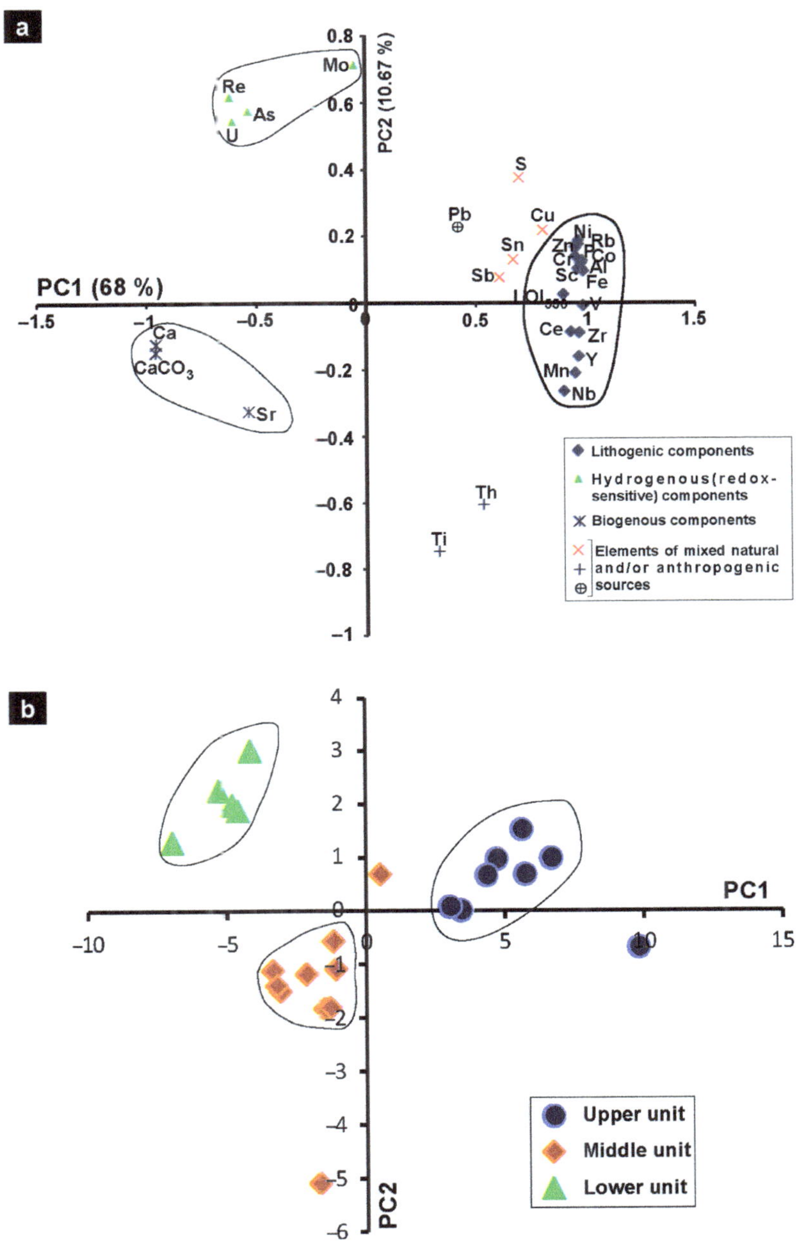

Figure 7. (**a**) Biplot of PC1 and PC2 (R-mode) loading and clustering of elements (variables) into lithogenous having higher PC1 and PC2 loading, redox-sensitive elements having negative PC1 and positive PC2 values, and carbonate related elements having negative PC1 and PC2 loadings. (**b**) Biplot of PC1 and PC2 (Q-mode) loading and subdivision of the core into three units. Samples (cases) in the upper unit having high component 1 scores are enriched in lithogenous elements, and other sediments (middle and lower units) are enriched in biogenous and hydrogenous elements.

The plot (Figure 7b) of component scores (Table 4) enabled clustering samples into three major groups, corresponding to the upper, middle, and lower units. Component 1 cluster samples from the upper unit that show positive scores (>1), whereas the samples of the middle and lower units display negative scores. These samples occupy the right upper quadrant and are enriched in the elements of lithogenic origin Al, Fe, Mg, Na, K, P, LOI_{550}, Mn, V, Cr, Co, Ni, Zn, Zr, Rb, and Sc. Component 2 characterizes samples from the lower unit that show positive scores (>1). These samples occupy the upper left quadrant and are enriched in Mo, U, As, and Re. The third group of samples occupies the lower left quadrant of the figure and are dominated by carbonate-and biogenous-related elements Ca, $CaCO_3$, and Sr (Figure 7b). The second and third groups are water derived constituents.

Table 4. Results of principal component (Q-mode) analysis, four components (PC1-PC4) are extracted with the PC1 and PC2 are the most important. Significant values are in bold.

Sample	PC1	PC2	PC3	PC4
SH-1	4.72	0.98	0.09	−1.11
SH-2	9.83	−0.67	−4.35	−0.95
SH-3	3.43	0.02	−0.77	−1.76
SH-4	5.75	0.69	1.65	0.66
SH-5	5.64	1.53	1.91	1.27
SH-6	6.69	1.00	0.32	1.16
SH-7	3.03	0.08	−0.06	0.02
SH-8	4.35	0.68	0.32	0.46
SH-9	0.55	0.70	2.13	0.08
SH-10	−1.16	−0.56	1.83	−0.69
SH-11	−1.06	−1.08	1.58	−0.95
SH-12	−1.48	−1.83	0.99	−1.29
SH-13	−1.28	−1.80	0.87	0.04
SH-14	−3.06	−1.49	−0.02	−0.84
SH-15	−1.67	−5.08	−1.29	2.01
SH-16	−3.33	−1.10	−0.50	−0.69
SH-17	−2.12	−1.17	0.71	0.36
SH-18	−3.21	−1.39	−0.91	1.54
SH-19	−5.25	2.27	−1.00	0.27
SH-20	−4.12	3.02	−1.51	1.34
SH-21	−4.76	2.00	−0.98	−0.37
SH-22	−6.92	1.30	−0.23	−1.24
SH-23	−4.56	1.90	−0.78	0.68

5. Discussion

The components of the shallow subsurface sediments of the coastal creek north of the southern Al-Shuaiba Lagoon consist mainly of lithogenous, organic matter, and water column-derived (hydrogenous and biogenous) constituents. Based on the geochemical data and statistical analysis, three levels of elemental enrichments corresponding to the lower, middle, and upper units of the core can be distinguished. The enriched elements/constituents are different in these levels reflecting the different driving forces and sources. Variations in element concentrations reflect their behaviour in the water column, redox conditions, and the rate of terrigenous and organic matter influx. The significant positive correlation between Al and trace elements (Mn, V, Cr, Co, Ni, Cu, Zn, Rb, Sc,

Zr, Y, and Nb) suggests that these elements are hosted essentially by the lithogenous constituents [42]. Elements of lithogenic origin are enriched in the upper unit. The high organic carbon content in the upper unit is attributed to the high organic matter influx rather than to the depletion of oxygen in the bottom water. A clear downcore depletion in lithogenous elements is apparent below the dark grey upper unit. This depletion is attributed to the general low detrital influx and the significant dilution of the detrital component with biogenic component ($CaCO_3$). The middle unit is enriched in biogenous constituents such as Sr, Ca, and $CaCO_3$ [43,44]. The covariation of Sr with Ca indicates that Sr is fixed primarily in the calcareous skeletal material. The most important hydrogenous constituents are Mo, As, U, and Re, which are relatively enriched in the sediments of the lower unit.

The elemental enrichments in the upper and lower units are the most distinctive. Both natural and anthropogenic drivers are responsible for elemental enrichments. To interpret the driving forces for these enrichments it is important to take the climatic setting of the area into account. The area, in general, is extremely hot and arid with no source of terrigenous influx except for aeolian transportation. Along the Red Sea coast, lithogenic elements are supplied from land to the coastal area mainly through wadis and by aeolian transportation [21,45]. However, the area of study is extremely arid and lacks terrigenous influx since the nearby wadis are generally inactive. Noteworthy, the increased enrichment levels of lithogenous elements in the upper part probably coincide with the road construction between the southern and northern Al-Shuaiba Lagoons. Therefore, this enrichment is induced by human activities particularly by landfills during road construction and the subsequent road runoff. Ghandour et al. [7] recorded a level of enrichment of geogenic elements in the southern nearshore zone area of Sharm Obhur, Jeddah, and they attributed this enrichment to dumping and landfills. The slight enrichment of Cu, Zn, and Pb in the upper part may be related to anthropogenic activities such as atmospheric fallout from the fuel combustion or rare boat activity.

The second most distinct enrichment level is recorded for elements Mo, As, U, and Re from the lower unit. They covary in the core suggesting that they have had a similar mechanism of precipitation. They are easily reduced and precipitated from anoxic bottom water [42,46] and, therefore, they are enriched in the much more restricted O_2-deficient setting [47]. Both natural and anthropogenic drivers can be responsible for oxygen depletion. However, this event precedes the beginning of anthropo-pressure. Oxygen decline is related to a lack of oxygen supply due to density stratification. High-temperature anomalies lead to water column stratification and the development of bottom anoxia in this shallow water. In such a setting, a solar heating-induced density gradient within the water column and water stratification prevents efficient mixing. In addition, warming reduces the solubility of oxygen in the water and enhances microbial activity. The warming effect tends to be severe in semi-enclosed basins. Abundant *Sorites orbiculus*, high palaeo-tidal elevation, and enrichments in both $\delta^{18}O$ and $\delta^{13}C$ values in the shallow subsurface coastal sediments at Al-Shuaiba during the AD 1000–1550 interval were attributed to the Medieval climate anomaly (MCA) warming event [30,48]. The MCA was a period of climate perturbation during which the Earth's temperature was similar to or slightly higher than the temperature of the 20th century [49]. The key drivers for this warming event include solar activity, ocean cycles, aerosols from volcanic eruptions, and greenhouse gases [48–52].

Redox-Sensitive Elements

Trace metal enrichment and trace metal ratios in the lagoonal and ocean bottom sediments have been used to address paleoredox conditions [53–58]. Molybdenum (Mo), V, Cr, Co, Ni, Cu, Zn, As, U, and Re are currently of high interest in this regard since they are reduced to a lower valency and become reactive or insoluble under anoxic conditions [54]. The trace elements in the core sediments include both hydrogenous and lithogenous components. To use trace elements as paleoredox proxy, it is important first to determine whether the elements are hydrogenous or lithogenous. A significant positive correlation between Al and trace elements indicates that the element is contained in the terrigenous component. In the present study, Al displays moderate to low correlation with Mo ($r < 0.3$) and negative correlation with U, As, and Re suggesting that these elements are of authigenic origin, and they are reliable paleoredox proxies. Hydrogenous elements are the ones that can be used for paleoenvironmental interpretations. Multi-elements proxies are more suitable for the analysis of paleo-redox conditions rather than a single elemental proxy [58]. Mo, As, Re, and U concentrations and their Al-normalized values show positive shifts in the lower unit, whereas no such positive shifts were observed for other elements. This suggests that the creek has experienced suboxic/anoxic bottom water conditions during the deposition of the lower unit. Enrichments in Re, Mo, As, and U are well known for numerous Phanerozoic organic-rich shales and their enrichment is usually associated with anoxic to euxinic bottom-water conditions [59–61]. The lithogenic origin of elements V, Cr, Co, Ni, Cu, and Zn obliterates their significances as reliable paleoredox indicators. The low concentrations and low Al-normalized, as well as EF values for these elements in the lower unit, may be explained by the "basin reservoir effect" [62]. This effect suggests that the concentrations of these elements decreased under an increasingly stagnant bottom water body due to their removal to the sediment without adequate resupply by water renewal. The depletion of Mn in the lower unit suggests that under anoxic conditions Mn occurs as Mn (II) in the dissolved phase [54]. The high solubility of Mn^{2+} in sediments deposited under reducing conditions may result in Mn being depleted if fixation in carbonate minerals is not possible [63].

The results presented in this study so far show that it is possible to distinguish clearly between anoxic and oxic depositional conditions for the shallow subsurface coastal sediments of the northern Al-Shuaiba Lagoon. The study helps in understanding the process of shallow water hypoxia and the most relevant redox sensitive elements.

6. Conclusions

The present study investigates the elemental enrichments and interprets them in the context of redox bottom conditions and increased terrigenous influx. The chemical data from the shallow subsurface sediments of the coastal creek that was connecting the southern and northern Al-Shuaiba Lagoons reflect environmental changes through time either due to natural and/or anthropogenic influences.

1. Based on the colour of sediments, vertical variation in elemental concentrations, and statistical analysis, the core was subdivided into three units, upper, middle, and lower.
2. The upper organic-rich unit shows enrichment of lithogenous elements and values of LOI_{550} (organic matter) content. Trace elements Mn, V, Cr, Co, Ni, and Zn display significant positive correlations with Al suggesting their lithogenic source. Though these elements are derived from a lithogenic source, their enrichment is related to human activity. This level is attributed to the road construction between the two lagoons. The relatively high concentrations of Pb and Cu in the upper unit are possibly related to atmospheric road dust and from the increasing movement of cars. Though the road construction limited the water circulation between the two lagoons, the sediments were deposited under oxic conditions as shown by the depletion of redox-sensitive elements.
3. The lower and middle units of the core contain relatively higher carbonate content than the upper part. Strontium (Sr) distribution appears to be controlled by the presence of biogenic carbonate minerals. Calcium (Ca) distribution follows largely the spatial distribution of the carbonate content since Ca, Sr, and $CaCO_3$ are biogenic components.
4. The chemical composition of the lower unit suggests deposition in poorly circulated bottom water, with prevailing suboxic or even anoxic conditions probably related to the solar forcing (Medieval climate anomaly). This is confirmed by depletions in Mn and Co and relative enrichment of Mo, As, U, and Re. The distribution pattern of the Al-normalized redox sensitive elements is characterized by markedly high values

in the lower unit. They display a negative to poor correlation with Al excluding lithogenic sources. Molybdenum (Mo), As, U, and Re are reliable and most promise proxies for redox conditions since they behave conservatively in oxygenated waters and are enriched in anoxic sediments.

Author Contributions: I.M.G., proposal of the idea, conceptualization, sample collection and coordination for analysis, data analysis, and writing the manuscript; M.H.A., data analysis, reviewing the manuscript and participation in the discussion. All authors have read and agreed to the published version of the manuscript.

Funding: This project was funded by the Deanship of Scientific Research (DSR) at King Abdulaziz University, Jeddah, under grant No. G: 617-150-1441.

Data Availability Statement: The data are available upon request at KAU and can be requested by contacting the corresponding author.

Acknowledgments: The authors thank the DSR for this technical and financial support. The authors are grateful to Aaid Al-Zubairi (KAU) for his assistance in the fieldwork. The authors would like to thank Brian Jones (Wollongong Univ.) for his constructive comments and suggestions that improved the manuscript. We are very grateful to the editor and the reviewers for their constructive comments and editorial handling.

Conflicts of Interest: The authors declare no conflict of interest.

References

1. Danovaro, R. Pollution threats in the Mediterranean Sea: An overview. *Chem. Ecol.* **2003**, *19*, 15–32. [CrossRef]
2. Barbier, E.B.; Hacker, S.D.; Kennedy, C.; Koch, E.W.; Stier, A.C.; Silliman, B.R. The value of estuarine and coastal ecosystem services. *Ecol. Monogr.* **2011**, *81*, 169–193. [CrossRef]
3. Robb, C.K. Assessing the impact of human activities on British Columbia's estuaries. *PLoS ONE* **2014**, *9*, e99578. [CrossRef]
4. Coynel, A.; Gorse, L.; Curti, C.; Schafer, J.; Grosbois, C.; Morelli, G.; Ducassou, E.; Blanc, G.; Maillet, G.M.; Mojtahid, M. Spatial distribution of trace elements in the surface sediments of a major European estuary (Loire Estuary, France): Source identification and evaluation of anthropogenic contribution. *J. Sea Res.* **2016**, *18*, 77–91. [CrossRef]
5. Badassan, T.E.; Avumadi, A.M.; Ouro-Sama, K.; Gnandi, K.; Jean-Dupuy, S.; Probst, J.L. Geochemical Composition of the Lomé Lagoon Sediments, Togo: Seasonal and Spatial Variations of Major, Trace and Rare Earth Element Concentrations. *Water* **2020**, *11*, 3026. [CrossRef]
6. McAlister, J.J.; Smith, B.J.; Neto, J.B.; Simpson, J.K. Geochemical distribution and bioavailability of heavy metals and oxalate in street sediments from Rio de Janeiro, Brazil: A preliminary investigation. *Environ. Geochem. Health* **2005**, *5*, 429–441. [CrossRef]
7. Ghandour, I.M.; Basaham, S.; Al-Washmi, A.; Masuda, H. Natural and anthropogenic controls on sediment composition of an arid coastal environment: Sharm Obhur, Red Sea, Saudi Arabia. *Environ. Monit. Assess* **2014**, *3*, 1465–1484. [CrossRef] [PubMed]
8. Weeks, S.J.; Currie, B.; Bakun, A. Massive emissions of toxic gas in the Atlantic. *Nature* **2002**, *415*, 493–494. [CrossRef] [PubMed]
9. Morina, A.; Morina, F.; Djikanović, V.; Spasić, S.; Krpo-Ćetković, J.; Kostić, B.; Lenhardt, M. Common barbel (Barbus barbus) as a bioindicator of surface river sediment pollution with Cu and Zn in three rivers of the Danube River Basin in Serbia. *Environ. Sci. Pollut. Res.* **2016**, *23*, 6723–6734. [CrossRef] [PubMed]
10. Schintu, M.; Marrucci, A.; Marras, B.; Galgani, F.; Buosi, C.; Ibba, A.; Cherchi, A. Heavy metal accumulation in surface sediments at the port of Cagliari (Sardinia, western Mediterranean): Environmental assessment using sequential extractions and benthic foraminifera. *Mar. Pollut. Bull.* **2016**, *111*, 45–56. [CrossRef] [PubMed]
11. Shi, Z.; Wang, X.; Ni, S. Metal contamination in sediment of one of the upper reaches of the Yangtze River: Mianyuan River in Longmenshan region, Southwest of China. *Soil Sediment Contam.* **2015**, *24*, 368–385. [CrossRef]
12. Wang, X.; Shi, Z.; Shi, Y.; Ni, S.; Wang, R.; Xu, W.; Xu, J. Distribution of potentially toxic elements in sediment of the Anning River near the REE and V-Ti magnetite mines in the Panxi Rift, SW China. *J. Geochem. Explor.* **2018**, *184*, 110–118. [CrossRef]
13. Al-Rousan, S.; Al-Taani, A.A.; Rashdan, M. Effects of pollution on the geochemical properties of marine sediments across the fringing reef of Aqaba, Red Sea. *Mar. Pollut. Bull.* **2016**, *110*, 546–554. [CrossRef]
14. Krishnakumar, S.; Ramasamy, S.; Chandrasekar, N.; Peter, T.S.; Gopal, V.; Godson, P.S.; Magesh, N.S. Trace element concentrations in reef associated sediments of Koswari Island, Gulf of Mannar biosphere reserve, southeast coast of India. *Mar. Pollut. Bull.* **2017**, *117*, 515–522. [CrossRef] [PubMed]
15. Bonanno, G.; Di Martino, V. Trace element compartmentation in the seagrass Posidonia oceanica and biomonitoring applications. *Mar. Pollut. Bull.* **2017**, *116*, 196–203. [CrossRef] [PubMed]
16. Ternengo, S.; Marengo, M.; El Idrissi, O.; Yepka, J.; Pasqualini, V.; Gobert, S. Spatial variations in trace element concentrations of the sea urchin, Paracentrotus lividus, a first reference study in the Mediterranean Sea. *Mar. Pollut. Bull.* **2018**, *129*, 293–298. [CrossRef]

17. Zhang, H.; Walker, T.R.; Davis, E.; Ma, G. Spatiotemporal characterization of metals in small craft harbour sediments in Nova Scotia, Canada. *Mar. Pollut. Bull.* **2019**, *140*, 493–502. [CrossRef]
18. Vinodhini, R.; Narayanan, M. Bioaccumulation of heavy metals in organs of fresh water fish *Cyprinus carpio* (Common carp). *Int. J. Environ. Sci. Technol.* **2008**, *5*, 179–182. [CrossRef]
19. Badr, N.B.; El-Fiky, A.A.; Mostafa, A.R.; Al-Mur, B.A. Metal pollution records in core sediments of some Red Sea coastal areas, Kingdom of Saudi Arabia. *Environ. Monit. Assess* **2009**, *155*, 509–526. [CrossRef]
20. Ma, L.; Wu, J.; Abuduwaili, J.; Liu, W. Geochemical responses to anthropogenic and natural influences in Ebinur Lake sediments of arid Northwest China. *PLoS ONE* **2016**, *11*, e0155819. [CrossRef]
21. Rasul, N.M. Lagoon sediments of the Eastern Red Sea: Distribution processes, pathways and patterns. In *The Red Sea*; Springer: Berlin/Heidelberg, Germany, 2015; pp. 281–316.
22. Al-Washmi, H.A.; Gheith, A.M. Recognition of diagenetic dolomite and chemical surface features of the quartz grains in coastal sabkha sediments of the hypersaline Shuaiba Lagoon, Eastern Red Sea Coast, Saudi Arabia. *J. King Abdulaziz Univ Mar. Sci.* **2003**, *14*, 101–112. [CrossRef]
23. Basaham, A.S. Mineralogical and chemical composition of the mud fraction from the surface sediments of Al-Kharrar, a Red Sea coastal lagoon. *Oceanologia* **2008**, *50*, 557–575.
24. Abu-Zied, R.H.; Bantan, R.A. Hypersaline benthic foraminifera from the Shuaiba Lagoon, eastern Red Sea, Saudi Arabia: Their environmental controls and usefulness in sea-level reconstruction. *Mar. Micropaleontolgy* **2013**, *103*, 51–67. [CrossRef]
25. Basaham, A.S.; El Sayed, M.A.; Ghandour, I.M.; Masuda, H. Geochemical background for the Saudi Red Sea coastal systems and its implication for future environmental monitoring and assessment. *Environ. Earth Sci.* **2015**, *74*, 4561–4570. [CrossRef]
26. Basaham, A.S.; Ghandour, I.M.; Haredy, R. Controlling factors on the geochemistry of Al-Shuaiba and Al-Mejarma coastal lagoons, Red Sea, Saudi Arabia. *Open Geosci.* **2019**, *11*, 426–439. [CrossRef]
27. Youssef, M.; El-Sorogy, A. Environmental assessment of heavy metal contamination in bottom sediments of Al-Kharrar lagoon, Rabigh, Red Sea, Saudi Arabia. *Arab. J. Geosci.* **2016**, *9*, 474. [CrossRef]
28. Ahmed, F.C.; Sultan, S.A.R. The effect of meteorological forcing on the flushing of Shuaiba Lagoon on the eastern coast of the Red Sea. *J. King Abdulaziz Univ. Mar. Sci.* **1992**, *3*, 3–9. [CrossRef]
29. Al-Barakati, A.M.A. Application of 2-D tidal model, Shuaiba Lagoon, eastern Red Sea coast. *Can. J. Comput. Math. Nat. Sci. Med.* **2010**, *1*, 9–20.
30. Abu-Zied, R.H.; Bantan, R.A. Palaeoenvironment, palaeoclimate and sea-level changes in the Shuaiba Lagoon during the late Holocene (last 3.6 ka), eastern Red Sea coast, Saudi Arabia. *Holocene* **2015**, *25*, 1301–1312. [CrossRef]
31. Al-Farawati, R.; El Sayed, M.; Shaban, Y.; El-Maradney, A.; Orif, M. Phosphorus Speciation in the Coastal Sediments of Khawr Ash Shaibah Al-Masdudah: Coastal Lagoon in the Eastern Red Sea, Kingdom of Saudi Arabia. *Arab Gulf J. Sci. Res.* **2014**, *32*, 93–101.
32. Braithwaite, C.J.R. Geology and paleography of the Red Sea region. In *Key Environments: Red Sea*; Edwards, A.J., Mead, S.M., Eds.; Pergamon Press: New York, NZ, USA, 1987; pp. 22–24.
33. Brown, G.F.; Schimdt, D.L.; Huffman, A.C. *Shield Area of Western Saudi Arabia, Geology of the Arabian Peninsula*; Professional Paper 560-A; US Geological Survey: Reston, VA, USA, 1989.
34. Hötzl, H.; Zötl, J.G. Climatic changes during the Quaternary Period. In *Quaternary Period in Saudi Arabia*; Al-Sayari, S.S., Zötl, J.G., Eds.; Springer: New York, NY, USA, 1978; pp. 301–311.
35. Sofianos, S.S.; Johns, W.E.; Murray, S.P. Heat and freshwater budgets in the Red Sea from direct observations at Bab el Mandeb. *Deep Sea Research Part II: Top. Stud. Oceanogr.* **2002**, *49*, 1323–1340. [CrossRef]
36. Abu-Zied, R.H.; Bantan, R.A.; El Mamoney, M.H. Present environmental status of the Shuaiba Lagoon, Red Sea Coast, Saudi Arabia. *J. King Abdulaziz Univ. Mar. Sci.* **2011**, *22*, 159–179. [CrossRef]
37. Heiri, O.; Lotter, A.F.; Lemcke, G. Loss-on-ignition as a method for estimating organic and carbonate content in sediments: Reproducibility and comparability of results. *J. Paleolimnol.* **2001**, *25*, 101–110. [CrossRef]
38. Chester, R.; Stoner, J.H. Pb in particulates from the lower atmosphere of the eastern Atlantic. *Nature* **1973**, *245*, 27–28. [CrossRef]
39. Zhang, J.; Liu, C.L. Riverine composition and estuarine geochemistry of particulate metals in China—Weathering features, anthropogenic impact and chemical fluxes estuarine. *Coast. Shelf Sci.* **2002**, *54*, 1051–1070. [CrossRef]
40. Tribovillard, N.P.; Desprairies, A.; Lallier-Vergès, E.; Bertrand, P.; Moureau, N.; Ramdani, A.; Ramanampisoa, L. Geochemical study of organic-matter rich cycles from the Kimmeridge Clay Formation of Yorkshire (UK): Productivity versus anoxia. *Palaeogeogr. Palaeoclimatol. Palaeoecol.* **1994**, *108*, 165–181. [CrossRef]
41. Murphy, A.E.; Sageman, B.B.; Hollander, D.J.; Lyons, T.W.; Brett, C.E. Black shale deposition and faunal overturn in the Devonian Appalachian basin: Clastic starvation, seasonal water-column mixing, and efficient biolimiting nutrient recycling. *Paleoceanography* **2000**, *15*, 280–291. [CrossRef]
42. Hild, E.; Brumsack, H.-J. Major and minor element geochemistry of Lower Aptian sediments from the NW German Basin (core Hoheneggelsen KB 40). *Cretac. Res.* **1998**, *19*, 615–633. [CrossRef]
43. Whitfield, M. Interactions between Phytoplankton and Trace Metals in the Ocean-Introduction. *Adv. Mar. Biol.* **2001**, *41*, 3–130.
44. Ho, T.Y.; Quigg, A.; Finkel, Z.V.; Milligan, A.J.; Wyman, K.; Falkowski, P.G.; Morel, F.M. The elemental composition of some marine phytoplankton 1. *J. Phycol.* **2003**, *39*, 1145–1159. [CrossRef]

45. Haredy, R.; Ghandour, I.M. Geochemistry and mineralogy of the shallow subsurface Red Sea coastal sediments, Rabigh, Saudi Arabia: Provenance and paleoenvironmental implications. *Turk. J. Earth Sci.* **2020**, *29*, 257–279. [CrossRef]
46. Landing, W.M.; Lewis, B.L. Collection, processing, and analysis of marine particulate and colloidal material for transition metals. In *Marine Particles: Analysis and Charterization*; Hurd, D.C., Spencer, D.W., Eds.; American Geophysical Union Geophysical Monograph Series; American Geophysical Union: Washington, DC, USA, 1991; Volume 63, pp. 263–272.
47. Lézin, C.; Andreu, B.; Pellenard, P.; Bouchez, J.L.; Emmanuel, L.; Fauré, P.; Landrein, P. Geochemical disturbance and paleoenvironmental changes during the Early Toarcian in NW Europe. *Chem. Geol.* **2013**, *341*, 1–5. [CrossRef]
48. Lüning, S.; Gałka, M.; Vahrenholt, F. Warming and cooling: The Medieval Climate Anomaly in Africa and Arabia. *Paleoceanography* **2017**, *32*, 1219–1235. [CrossRef]
49. Diaz, H.; Trigo, R.; Hughes, M.; Mann, M.; Xoplaki, E.; Barriopedro, D. Spatial and Temporal Characteristics of Climate in Medieval Times Revisited. *Bull. Am. Meteorol. Soc.* **2011**, *92*, 1487–1500. [CrossRef]
50. Moberg, A.; Sonechkin, D.M.; Holmgren, K.; Datsenko, N.M.; Karlén, W.; Lauritzen, S.-E. Highly variable Northern Hemisphere temperatures reconstructed from low- and high-resolution proxy data. *Nature* **2005**, *433*, 613–617. [CrossRef]
51. Zhou, L.; Tinsley, B.; Huang, J. Effects on winter circulation of short- and long-term solar wind changes. *Adv. Space Res.* **2014**, *54*, 2478–2490. [CrossRef]
52. Andrews, M.B.; Knight, J.R.; Gray, L.J. A simulated lagged response of the North Atlantic Oscillation to the solar cycle over the period 1960–2009. *Environ. Res. Lett.* **2015**, *10*, 054022. [CrossRef]
53. Wignall, P.B.; Meyers, K.J. Interpreting benthic oxygen levels in mudrocks: A new approach. *Geology* **1988**, *16*, 452–455. [CrossRef]
54. Calvert, S.E.; Pedersen, T.F. Geochemistry of Recent oxic and anoxic marine sediments: Implications for the geological record. *Mar. Geol.* **1993**, *113*, 67–88. [CrossRef]
55. Morford, J.L.; Emerson, S. The geochemistry of redox sensitive trace metals in sediments. *Geochim. Cosmochim. Acta* **1999**, *63*, 1735–1750. [CrossRef]
56. Algeo, T.J.; Maynard, J.B. Trace-element behavior and redox facies in core shales of Upper Pennsylvanian Kansas-type cyclothems. *Chem. Geol.* **2004**, *206*, 289–318. [CrossRef]
57. Rimmer, S.M.; Thompson, J.; Goodnight, S.; Robl, T.L. Multiple controls on the preservation of organic matter in Devonian–Mississippian marine black shales: Geochemical and petrographic evidence. *Palaeogeogr. Palaeoclimatol. Palaeoecol.* **2004**, *215*, 125–154. [CrossRef]
58. Tribovillard, N.; Algeo, T.J.; Lyons, T.; Riboulleau, A. Trace metals as paleoredox and paleoproductivity proxies: An update. *Chem. Geol.* **2006**, *232*, 12–32. [CrossRef]
59. Brumsack, H.-J. The trace metal content of recent organic carbon-rich sediments: Implications for Cretaceous black shale formation. *Palaeogeogr. Palaeoclimatol. Palaeoecol.* **2006**, *232*, 344–361. [CrossRef]
60. Warning, B.; Brumsack, H.-J. Trace metal signatures of eastern Mediterranean sapropels. *Palaeogeogr. Palaeoclimatol. Palaeoecol.* **2000**, *158*, 293–309. [CrossRef]
61. Wilde, P.; Lyons, T.W.; Quinby-Hunt, M.S. Organic carbon proxies in black shales: Molybdenum. *Chem. Geol.* **2004**, *206*, 167–176. [CrossRef]
62. Algeo, T.J.; Lyons, T.W. Mo-total organic carbon covariation in modern anoxic marine environments: Implications for analysis of paleoredox and paleohydrographic conditions. *Paleoceanography* **2006**, *21*, PA1016. [CrossRef]
63. Scholz, F.; McManus, J.; Sommer, S. The manganese and iron shuttle in a modern euxinic basin and implications for molybdenum cycling at euxinic ocean margins. *Chem. Geol.* **2013**, *355*, 56–68. [CrossRef]

MDPI
St. Alban-Anlage 66
4052 Basel
Switzerland
Tel. +41 61 683 77 34
Fax +41 61 302 89 18
www.mdpi.com

Minerals Editorial Office
E-mail: minerals@mdpi.com
www.mdpi.com/journal/minerals

www.ingramcontent.com/pod-product-compliance
Lightning Source LLC
Chambersburg PA
CBHW061610210926
51850CB00009B/2272